Sustainable Civil Infrastructures

Sustainable Infrastructure impacts our well-being and day-to-day lives. The infrastructures we are building today will shape our lives tomorrow. The complex and diverse nature of the impacts due to weather extremes on transportation and civil infrastructures can be seen in our roadways, bridges, and buildings. Extreme summer temperatures, droughts, flash floods, and rising numbers of freeze-thaw cycles pose challenges for civil infrastructure and can endanger public safety. We constantly hear how civil infrastructures need constant attention, preservation, and upgrading. Such improvements and developments would obviously benefit from our desired book series that provide sustainable engineering materials and designs. The economic impact is huge and much research has been conducted worldwide. The future holds many opportunities, not only for researchers in a given country, but also for the worldwide field engineers who apply and implement these technologies. We believe that no approach can succeed if it does not unite the efforts of various engineering disciplines from all over the world under one umbrella to offer a beacon of modern solutions to the global infrastructure. Experts from the various engineering disciplines around the globe will participate in this series, including: Geotechnical, Geological, Geoscience, Petroleum, Structural, Transportation, Bridge, Infrastructure, Energy, Architectural, Chemical and Materials, and other related Engineering disciplines.

More information about this series at http://www.springer.com/series/15140

Hany Shehata · Braja Das
Editors

Advanced Research on Shallow Foundations

Proceedings of the 2nd GeoMEast International Congress and Exhibition on Sustainable Civil Infrastructures, Egypt 2018 – The Official International Congress of the Soil-Structure Interaction Group in Egypt (SSIGE)

Springer

Editors
Hany Shehata
Soil-Structure Interaction Group in Egypt
Cairo, Egypt

Braja Das
California State University
Sacramento, CA, USA

ISSN 2366-3405 ISSN 2366-3413 (electronic)
Sustainable Civil Infrastructures
ISBN 978-3-030-01922-8 ISBN 978-3-030-01923-5 (eBook)
https://doi.org/10.1007/978-3-030-01923-5

Library of Congress Control Number: 2018957571

This Springer imprint is published by the registered company Springer Nature Switzerland AG
The registered company address is: Gewerbestrasse 11, 6330 Cham, Switzerland

Contents

About the Editors

Braja Das Professor Braja Das is the Dean Emeritus of the College of Engineering and Computer Science, California State University, USA. He is a geotechnical engineer by profession and received his Ph.D. degree in 1972 from the University of Wisconsin-Madison, USA. For more than three decades, he has worked as a faculty member and in academic administration in several universities. He is a Fellow and Life Member of the American Society of Civil Engineers; Life Member of the American Society for Engineering Education; and Emeritus Member of TRB's AFS-80 Committee on Stabilization of Geomaterials and Stabilized Materials.

As a teacher, he has received many teaching awards including the Distinguished Achievement Award for Teaching Excellence, University of Texas at El Paso (1983); AMOCO Foundation Award for Outstanding Teaching, University of Texas at El Paso (1983); Western Electric Fund Award, ASEE (1984); Ralph R. Teetor Educational Award, Society of Automotive Engineers Inc. (1985).

He has published more than 300 papers in various journals and peer-reviewed conference proceedings. They are mostly in the areas of shallow foundations, earth anchors, and geosynthetics. He published some of the early founding studies on the bearing capacity of shallow foundations on geogrid-reinforced sand. His past editorial activities in geotechnical engineering journals include Editorial Board Member of Journal of Geotechnical Engineering, ASCE (1988–1992); Lowland Technology International, Japan (1999–2007); Geotextiles and Geomembranes, Elsevier, UK

(2008 – present). He was the Associate Editor of the International Journal of Offshore and Polar Engineering (1995–1998) and Co-Editor of Geotechnical and Geological Engineering, Springer (2001–2006). He is the Founder and Editor-in-Chief of the International Journal of Geotechnical Engineering (Taylor and Francis, UK) since 2007 up to the present.

He has been a very popular keynote or invited speaker in many conferences and special meetings in various countries. These countries include Mexico, Dominican Republic, Costa Rica, El Salvador, Peru, Colombia, Ecuador, India, University of Birmingham, UK, Korea, Bolivia, Venezuela, Turkey, Turkish Republic of North Cyprus, and Tunisia. He has been named as the first Eulalio Juárez Badillo Lecturer by the Mexican Society of Geotechnical Engineers (November 2016).

He is the author/co-author of several text and reference books in the area of geotechnical engineering, probably having written more than any other individual in geotechnical engineering. His textbooks have been translated into several languages and are used throughout the world. His clear, concise, and singularly unique style of presentation is very much admired, thereby attracting many students over the last 25 years who might not have considered geotechnical engineering as a profession. This is truly his most outstanding achievement which has endeared him to many young geotechnical engineers around the world. This is his legacy—now and for years to come.

Hany Farouk Shehata Shehata is the Founder and CEO of the Soil-Structure Interaction Group in Egypt "SSIGE." He is a partner and Vice-President of EHE-Consulting Group in the Middle East and Managing Editor of the "Innovative Infrastructure Solutions" journal, published by Springer. He worked in the field of civil engineering early, while studying, with Bechtel Egypt Contracting & PM Company, LLC. His professional experience includes working in culverts, small tunnels, pipe installation, earth reinforcement, soil stabilization, and small bridges. He also has been involved in teaching, research, and consulting. His areas of specialization include static and dynamic soil–structure interactions involving buildings, roads, water structures, retaining walls, earth reinforcement, and bridges, as well as different disciplines of project management and contract administration. He is the author of an Arabic practical book titled "Practical Solutions for Different Geotechnical Works: The Practical Engineers' Guidelines." He is currently working on a new book titled "Soil-Foundation-Superstructure Interaction: Structural Integration." He is the contributor of more than 50 publications in national and international conferences and journals. He served as a Co-Chair of the GeoChina 2016 International Conference in Shandong, China. He serves also as a Co-Chair and Secretary General of the GeoMEast 2017 International Conference in Sharm El-Sheikh, Egypt. 2016 Outstanding reviewer of the ASCE as selected by the Editorial Board of International Journal of Geomechanics.

Honors Lecture: Biological Cementation of Unstable Soils and Grounds for Civil Infrastructure Developments

Mohamed A. Shahin[1(✉)] and Liang Cheng[2]

[1] School of Civil and Mechanical Engineering, Curtin University, Perth, Australia
m.shahin@curtin.edu.au
[2] Biofuels Institute of Jiangsu University, Jiangsu University, Zhenjiang, People's Republic of China

Abstract. Problematic soils (e.g., loose, erodible, and collapsible) are unstable and pose significant challenges to the geotechnical engineering communities due to their low bearing capacity and high compressibility. Such unstable soils are widespread in the world and greatly hinder civil infrastructure developments such as building foundations, roads, retaining walls, etc. Engineered ground improvement is often necessary to increase soil load-carrying capacity and prevent excessive post-construction deformations. However, existing techniques of ground improvement can be highly expensive (e.g., pile foundations) and some have significant environmental, sometimes toxic effects (e.g., chemical stabilization by Portland cement). In this presentation, an innovative technology for ground improvement, through biologically induced calcite cementation (bio-cementation), will be presented and discussed. This technology should replace most conventional soil stabilization methods, for superior urban and coastal infrastructure developments.

1 Introduction

Unstable soils encountered in civil and resource constructions possess poor geotechnical engineering properties, and thus pose significant construction, maintenance and safety issues. Among the conventional methods of ground improvement, chemical stabilization by grouts (e.g., lime or cement) is widely employed, where the chemical additives alter the mineralogical structure of soil to improve its physical and mechanical properties. However, most chemical grouts increase the pH of groundwater to highly alkaline levels and reduce soil permeability, thereby causing serious environmental problems and contributing to disturbance of local ecosystems. For instance, acrylamide grout was associated with five cases of water poisoning in Japan resulting in the ban of nearly all chemical grouts (Karol 2003). Recently, American agencies have proposed to ban most synthetic grouting materials (DeJong et al. 2010), both because of their toxicity and contribution to global carbon dioxide emissions (Li et al. 2013). As there are more than 40,000 worldwide projects/year employing chemically stabilized soils at a total cost exceeding US$6 billion (DeJong et al. 2010), there is an immense

H. Shehata and B. Das (Eds.): GeoMEast 2018, SUCI, pp. 1–9, 2019.
https://doi.org/10.1007/978-3-030-01923-5_1

need for alternative environmentally-friendly and sustainable technologies to fulfil the increasing demand on ground improvement.

Bio-cementation via microbial-induced calcite precipitation (MICP) consists of injecting or mixing soil with cultivated ureolytic earth bacteria, typically *Bacillus pasteurii* (also known as *Bacillus sphaericus*) or *Bacillus megaterium*, in urea and calcium-rich solution. The microbial hydrolysis of urea generates carbonate and precipitates calcite (calcium carbonate) on the surfaces of the constituent soil grains. This process forms cementing bridges that bind soil particles together, leading to increased soil strength and stiffness. The chemical reactions of bio-cementation involves two main steps (Cheng et al. 2013):

First, microbial urease hydrolyzes the urea to form ammonium and carbonate ions:

$$CO(NH_2)_2 + 2H_2O \rightarrow 2NH_4^+ + CO_3^{2-} \tag{1}$$

Second, in the presence of a calcium source (e.g., CaCl2) the produced carbonate ions react with calcium ions and precipitate as calcium carbonate crystals:

$$Ca^{2+} + CO_3^{2-} \rightarrow CaCO^3 \tag{2}$$

Because calcite occurs on the soil grains, bio-cemented soils remain relatively permeable, which is desirable in most geotechnical applications. Advanced imagery revealed that bio-cemented soils have unique fabric and structure that differ from other artificially cemented soils, because the mineralogy and size of the calcium carbonate crystals attached to the native soil particles are dependent on the bio-chemical conditions (Al Qabany and Soga 2013; Cheng et al. 2017; Mujah et al. 2017). It has also been observed that calcite is as effective as other cements in increasing strength and stiffness, and that bio-cementation has proved its sustainability and capability to alter and improve most soil engineering properties (Cheng et al. 2013). This technology is eco-friendly and sustainable as it uses a bacteria-driven process that results in calcite precipitation, which mimics the natural digenesis from sand to sandstone or lime to limestone (see Fig. 1), only within a short time instead of million years. This paper presents some features of the research undertaken at Curtin University (Australia) on bio-cementation technology.

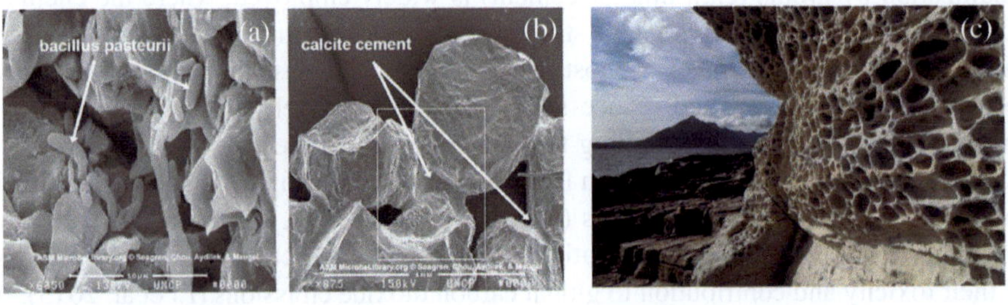

Fig. 1. Natural bacteria induced bio-cementation: (a–b) 3dp.bio/research-labs/mica; and (c) rock formation (protocolarchitecture.wordpress.com)

2 Experimental Program

This section briefly presents and discusses some experimental results obtained from the work done by the authors, including the practicality of this technique towards its usage for ground improvement. The effects of some salient factors affecting bio-cementation are examined and presented, including the initial soil pH, degree of saturation of treated soil, degree of temperature at which bio-cementation occurs, presence of fines in the treated soil matrix, and attainment of treatment uniformity. In addition, the possibility of using the seawater as a calcium source for bio-cementation is also investigated, which can significantly reduce the cost of bio-cementation treatment in marine environment. Moreover, the paper presents the use of bio-cementation to produce sand bricks "bio-bricks" as a construction material.

2.1 Preparation of Materials

Bio-cementation via MICP is usually more successful for sand soils, which have pore sizes within the range of injected bacterial solutions (Mitchell and Santamarina 2005; Mujah et al. 2017). However, the feasibility of treating sand with fines and the corresponding maximum cementation strength have also been assessed by the authors (see Cheng and Shahin 2015). The authors also investigated the use of bio-cementation for treating oil-contaminated soils, utilizing a modified approach of MICP technique (Cheng and Shahin 2017), and enhancing fiber/matrix bonding in polypropylene fiber reinforced cementitious composites (Hao et al. 2018).

In this paper, sand specimens of 50 mm in dimeter and 150 mm in height were prepared and used for the experimental program. Before treatment, highly ureolytic bacteria were cultivated aerobically in the laboratory as follows. The urease active strain used was *Bacillus sphaericus*. The liquid medium was prepared using 20 gm per liter of yeast extract added to deionized water. The following substances were added to the media: 0.17 M of Ammonium Sulphate and 0.1 mM of Nickel Chloride. Before inoculation, the growth medium was sterilized by supplying an inoculum of approximately 2–5% of pure bacteria culture into the medium to initiate microbial growth, then the medium was incubated for an extended period of time. After 24 h of cultivation at 28 °C, the bacteria culture was collected and stored at 4 °C prior to use. The optical density (OD600) of the culture varied between 0.6–1.0, and the urease activity was approximately 5 U/mL (1 U = 1 μmol urea hydrolyzed per min). Reagent solutions containing a mixture of 1 M calcium chloride (111 g/L) and 1 M urea (60 g/L) were flushed through the sand specimens at an injection flow rate of 1 L/h. Throughout the flushing applications, fully saturated condition (i.e., 100% degree of saturation) was maintained (except for studying the impact of degree of saturation) using a pressurized vacuum to remove the previously supplied solution, leaving the next solution application as residual. For the seawater experiments, the cementation solution was prepared by adding 10 mM urea (0.6 g/L) into seawater. Figure 2 shows a typical example of the prepared sand specimens treated with MICP bio-cementation and the corresponding SEM image displaying the formation of calcite crystals at the soil contact points.

Fig. 2. Typical example of bio-treated sand (left) and its SEM image (right)

2.1.1 Initial pH of Soil

The results of the effect of initial pH of soil indicated that the acidity and alkalinity conditions of soil have negative impact on the bio-treated soil, resulting in decreased soil strength even in the presence of high content of calcite crystals. As mentioned by many researchers (e.g., Harkes et al. 2010), the soil pH value can influence the bacteria transport and adhesion, which is an essential factor for achieving homogeneously improved strength. The initial pH of soil can also affect the formation of crystals as the solubility of calcite varies according to the pH value.

2.1.2 Degree of Saturation

To provide the desired degree of saturation (i.e., 20%, 40%, 80%, and 100%) within the soil matrix, the reagent solution (i.e., calcium chloride and urea) was flushed through the sand specimens at various amounts of water. Throughout the flushing applications, a specified degree of saturation was maintained using a pressurized vacuum to remove the previously supplied solution, leaving the next solution application as residual. The results indicated that at any degree of saturation, there is an increase in the soil compressive strength with the increase in the produced calcite content, at any degree of saturation (Cheng et al. 2013). The results also proved that at the same calcite content, the soil treated at lower degree of saturation exhibited significantly higher values of shear strength. This can be attributed to the effectiveness of the calcite crystals formation within the soil matrix, which is clearly demonstrated by the images taken from the scanning electron microscopy (SEM) shown in Fig. 3, for soil treated at fully and partially saturated conditions. For fully saturated treatment condition (Fig. 3a), the produced calcite is not fully formed at the inter-particle contact points of the soil grains but rather floccules either on the grain surface or suspends in the pore space between the soil grains. These nucleation sites are ineffective; hence, the calcite formation provides no significant soil strength improvement. In the case of partially saturated treatment condition (Fig. 3b), the calcite formation effectively coats over the soil particles and predominantly occurs at the effective areas of the granular contact points. This calcite formation provides rationale to the significant reduction in the calcite content, obtaining similar strength to that witnessed for the saturated condition. The above results indicate that the bio-treated soils gain its mechanical strength mainly through the effectiveness of the calcite crystals formation rather than the total amount of produced calcite.

Fig. 3. SEM images of bio-cemented soil treated at: (a) fully saturated condition; and (b) partially saturated condition (Cheng et al. 2013)

2.1.3 Degree of Temperature

The values of degree of temperature used in the present study simulate the subsurface soil temperature in the cold (4 °C), tropical (25 °C), and arid (50 °C) regions. During the process of MICP treatment, the sand samples used were placed: (1) inside 4 °C refrigerator; (2) at the room temperature ±25 °C; and (3) inside 50 °C oven. The results indicated that at any similar produced amount of calcite, the strength improvement was higher at 25 °C compared to that at either lower temperature of 4 °C or higher temperature of 50 °C (Cheng et al. 2017). The calcite crystals formed at 50 °C were the least efficient to gain strength improvement. The microstructure examination indicated that treatment at 50 °C leads to typical individual small crystals of 2–5 μm, covering the entire sand grain surface as a coating layer (see Fig. 4a). Small calcite crystals cannot effectively connect the soil grains, resulting in low strength improvement. The formation of such small crystals is probably due to the high temperature at which the rate of calcite nucleation becomes much faster, leading to production of abundant small crystals. For soil treated at 25 °C, the average crystal size increased by 10 times (individual crystals size between 20–50 μm) compared to those formed at 50 °C (see Fig. 4b). These large calcite crystals precipitate on the grain surface, covering the contact areas of the sand grains. At low temperature of 4 °C, small individual calcite crystals of 5–10 μm were observed, similar to those observed at 50 °C, and this is due to the slow crystal growth rate due to the slow urea hydrolysis process.

Fig. 4. SEM images of bio-cemented soil treated at different temperature: (a) hot weather 50 °C; and (b) room temperature ±25 °C (Cheng et al. 2017)

2.1.4 Percentage of Fines

Sand specimens contained clay fines of 5% and 10% were treated. The results of bio-treated sand containing 5% clay showed a slight increase in the UCS values compared to those obtained from bio-treated pure sand (Cheng and Shahin 2015). This is due to the increased cohesion of the sand with fines, or the increase in the larger contact surface area provided by the clay fine particles, which may facilitate the bridging formation between the sand particles via the calcite crystals. The specimens of sand containing 10% clay fines encountered clogging by the excessive cementation at the injection end and the minor calcite precipitation inside the treated sand specimens (Cheng and Shahin 2015). This observation occurred after three flushes of treatment, and further treatment became difficult to conduct due to the serious clogging. This is due to the high amount of hydrolysed urea and the calcite formed at the injection end. The sand samples containing 10% clay content, which have smaller pores, acted as a filter to the bacteria and resulted in accumulation of the bacterial cells around the injection end. The accumulated urease activity associated with the low infiltration rate of the cementation solution resulted in excessive bio-cementation occurred at the injection end rather than uniformly distributed along the sand columns. This indicates that bio-cementation using the injection method may not be applicable to treat soils that contain more than 5% fines due to the immediate bio-clogging at the injection end (the bacterial cells block the pores) and thus alternative treatment processes are necessary for bio-cementation treatment of coarse grained soils containing fines.

2.1.5 Attainment of Treatment Uniformity

One of the main intrinsic obstacles of MICP bio-cementation technology as a practical ground improvement methodology is the uniformity of calcite formation and the corresponding mechanical strength achieved, which are due to the non-uniform transport and attachment of bacteria to the surface of soil particles. This is because when the bacteria travel through the pore space of soils, they are likely to be filtered through the soil grains with long-linear reduction of microbe concentration along the injection path (Ginn et al. 2002). Furthermore, the attachment of bacteria to porous materials is usually influenced by many physical, chemical and biological factors, and adsorbed bacteria can be remobilized from the soil surface into the liquid phase by flushing of low salinity solutions (Harkes et al. 2010). In this regard, the authors proposed a novel approach that can achieve a good MICP treatment uniformity (see Cheng and Shahin 2016). The approach uses pre-formed urease active crystals, named as "bio-slurry", as a source of urease activity to induce a homogeneous bio-cementation for soil stabilization. In contrast to the other usually adopted MICP treatment methods, this treatment approach involves premixing of bio-slurry with soil, similar to the traditional cement mixing method, followed by flushing of the cementation solution. Compared to the use of bacterial cells, the advantages of using bio-slurry include more uniform and controllable activity distribution, high urease activity retention and firm attachment against water flushing. Figure 5 shows photos of the bio-slurry used in the bio-slurry treatment approach, taken during the reaction period and settling period.

Fig. 5. Photos of bio-slurry taken during the reaction period (left) and settling period (right) (Cheng and Shahin 2016)

2.1.6 Use of Seawater

The cost of MICP process including bacterial cultivation, chemical usage, equipment, and labour may prevent the progress of further commercial development of this emerging ground improvement technique. Consequently, the authors have made an attempt (see Cheng et al. 2014) to exploit the potential benefit of using the seawater as a calcium source in the cementation solution instead of the commercially available calcium chloride, for marine environment. This attempt reduces the cost of bio-cementation and brings it closer to be cost-effective and commercially acceptable ground improvement alternative. The results clearly demonstrated the feasibility of using the seawater as a calcium reagent for bio-cementation. However, the use of seawater as a relatively dilute calcium solution requires several number of subsequent treatments (probably 60 or 80 flushes) to get similar UCS values of those obtained from the traditional cementation solution of urea and calcium chloride.

2.1.7 Production of Bio-bricks

Bricks are usually used as building blocks for structures; however, the manufacturing of commonly used bricks is either energy intensive (e.g., fired clay bricks) or relies on unsustainable or un-environmentally-friendly materials that contribute to the global carbon dioxide emissions (e.g., cement bricks). The authors used MICP to form sustainable bricks called "bio-bricks" that do not utilize energy consumption. The work aimed to determine the viability of using bio-bricks as a construction material based on the associated strength, durability, and masonry properties, by means of comparison to relevant standards. The properties investigated include the uniformity of calcite precipitation, compressive strength, salt attack resistance, water absorption, and fire resistance. Figure 6 shows typical examples of the produced bio-bricks.

Fig. 6. Photo of produced bio-bricks

3 Conclusions

This paper demonstrated the use of bio-cementation via microbial-induced calcite precipitation (MICP) as a promising technique for ground improvement and production of bricks. However, the efficiency of bio-cementation in improving the soil strength varies significantly according to the physical and environmental treatment conditions. Findings presented in this paper confirmed that higher soil strength can be obtained at lower degree of saturation, challenging the widely belief that bio-cemented soils need to be treated under full saturated conditions. The results also indicated that bio-cementation is able to process in different environmental conditions, such as extreme high temperature. However, the compressive strength of bio-treated soils can vary significantly, depending on the treatment environmental condition, with lower strength gained at hot temperature of 50 °C compared to lower temperature of 4 °C or room temperature of 25 °C. The paper also indicated that bio-cementation of sand with fines (e.g., clayey sand) is not successful as it faces a tremendous challenge; thus, requires further investigation in the future. The paper also emphasised the potential use of the bio-slurry in providing a better treatment uniformity for MICP bio-treated soil compared to the usually used injection method of bacteria-cementation solution. The paper also confirmed that it is possible to use the seawater as a chemical reagent for bio-cementation to replace the calcium chloride, to provide a considerable strength improvement after repeated treatments. This finding is interesting as it extends the application of bio-cementation to broader areas, such as ground improvement in marine environment. Marine bio-cementation can potentially contribute to more sustainable human activities and significantly benefit society in areas of offshore and onshore infrastructure protection and maintenance, as well as coastline erosion prevention. Finally, the paper demonstrated that bio-cementation was successful in producing bio-bricks that possess good construction material attributes, complying with available codes and standards.

Acknowledgments. The authors would like thank their former and current PhD and undergraduate students who contributed significantly to the experimental program of this work.

References

Al Qabany, A., Soga, K.: Effect of chemical treatment used in MICP on engineering properties of cemented soils. Geotechnique **63**(4), 331–339 (2013)

Cheng, L., Cord-Ruwisch, R., Shahin, M.A.: Cementation of sand soil by microbially induced calcite precipitation at various degrees of saturation. Can. Geotech. J. **50**(1), 81–90 (2013)

Cheng, L., Shahin, M.A. Assessment of different treatment methods of micobial-induced calcite precipitation for clayey soil improvement. In: 68th Canadian Geotechnical Conference (GeoQuebec), Quebec, Canada (2015)

Cheng, L., Shahin, M.A.: Urease active bio-slurry: a novel soil improvement approach based on microbially induced calcite precipitation. Can. Geotech. J. **53**(9), 1376–1385 (2016)

Cheng, L., Shahin, M.A.: Stabilisation of oil-contaminated soils using microbially induced calcite crystals by bacterial flocs. Geotech. Lett. **7**(2), 1–6 (2017)

Cheng, L., Shahin, M.A., Cord-Ruwisch, R.: Bio-cementation of sandy soil using microbial-induced carbonate precipitation (MICP) for marine environments. Geotech. **64**(12), 1010–1013 (2014)

Cheng, L., Shahin, M.A., Mujah, D.: Influence of key environmental conditions on microbially induced cementation for soil stabilization. J. Geotech. Geoenviron. Eng. **143**(1), Article Number 04016082 (2017)

DeJong, J.T., Mortensen, B.M., Martinez, B.C., Nelson, D.C.: Bio-mediated soil improvement. Ecol. Eng. **36**(1), 197–210 (2010)

Ginn, T.R., Wood, B.D., Nelson, K.E., Scheibe, T.D., Murphy, E.M., Clement, T.P.: Processes in microbial transport in the natural subsurface. Adv. Water Resour. **25**, 1017–1042 (2002)

Hao, Y., Cheng, L., Hao, L.H., Shahin, M.A.: Enhancing fiber/matrix bonding in polypropylene fiber reinforced cementitious composites by microbially induced calcite precipitation pre-treatment. Cement Concr. Compos. **88**, 1–7 (2018)

Harkes, M.P., van Passen, L.A., Booster, J.L., Whiffin, V.S., van Loosdrecht, M.C.: Fixation and distribution of bacterial activity in sand to induce carbonate precipitation for ground reinforcement. Ecol. Eng. **36**(2), 112–117 (2010)

Karol, R.H.: Chemical Grouting and Soil Stabilisation. Marcel Dekker, New York (2003)

Li, J., Tharakan, P., Macdonald, D., Liang, X.: Technological, economic and financial prospects of carbon dioxide capture in the cement industry. Energy Policy **61**, 1377–1387 (2013)

Mitchell, J.K., Santamarina, J.C.: Biological considerations in geotechnical engineering. J. Geotech. Geoenviron. Eng. **131**(19), 1222–1233 (2005)

Mujah, D., Shahin, M.A., Cheng, L.: State-of-the-art review of biocementation by microbially induced calcite precipitation (MICP) for soil stabilization. Geomicrobiol J. **34**(6), 524–537 (2017)

Advanced New Methodology for the Identification of Stiffness and Strength of Weak Rock as Basis for Economic Foundation Design

Rolf Katzenbach[1(✉)], Steffen Leppla[1], and Marwan Alzaylaie[2]

[1] Ingenieursozietaet Professor Dr.-Ing. Katzenbach GmbH, Frankfurt am Main, Germany
katzenbach@katzenbach-ingenieure.de
[2] Dubai Creative Clusters Authority, Dubai, United Arab Emirates

Abstract. The most important aspects for the design of any foundation system are safety, optimisation and the sustainability. An optimised and safe design of foundation systems for high-rise structures in difficult soil and groundwater conditions is based on a reduction of construction material used, construction time spent, energy consumed and the adequate consideration of the soil-structure interaction. This is also important for the high-rise structures like skyscrapers and bridge piers in weak rock like in Dubai, UAE. Due to the large loads most of these structures are founded in the Dubai Sandstone and Dubai Siltstone. Up to now the rock mechanical parameters for these rock layers have been defined on the very conservative side which led to over-dimensioned foundations in many cases. In a large research program the bearing behaviour of Dubai Sandstone and Dubai Siltstone has been investigated by field and laboratory tests, by in-situ pile load tests and the numerical back-analysis using the Finite-Element-Method (FEM). The comprehensive research investigations show, that the stiffness of Dubai Sandstone and Dubai Siltstone is more than 20 times higher as it is assumed up to now. The paper presents the scope of research, the epoch-making results and the significance for the engineering practice.

1 Introduction

Dubai in the United Arab Emirates (UAE) is one of the cities with the biggest density of high-rise buildings in the world (Fig. 1). All these high-rise buildings are founded on conventional deep foundation in the Dubai Sandstone and the Dubai Siltstone. The load of the high-rise structures into the subsoil is transferred by bored piles or barrettes.

In the scope of a large research program the bearing behavior of the Dubai Sandstone and the Dubai Siltstone due to high-rise structures has been analyzed at the Institute and Laboratory of Geotechnics of Technische Universitaet Darmstadt (TU Darmstadt), Germany. This research program included a high amount of field and laboratory tests. The area for the research program was the Business Bay District of Dubai (Fig. 2). This area is about 3.5 km long and about 2 km wide. The subsoil conditions in the Business Bay District are representative for whole Dubai.

H. Shehata and B. Das (Eds.): GeoMEast 2018, SUCI, pp. 10–20, 2019.
https://doi.org/10.1007/978-3-030-01923-5_2

Fig. 1. Skyline of Dubai, UAE

Fig. 2. Business Bay District, Dubai (UAE)

At the moment the highest high-rise building of the world, the Burj Khalifa, is located in this area (Fig. 3). The Burj Khalifa has 163 floors and a height of 828 m. This impressive high-rise building is surrounded by a water basin where every 30 min fountains accompanied with music enthuse the inhabitants and the many guests and tourists (Fig. 4). In a few years the Burj Khalifa will be only the second highest high-rise building in the world. The Jeddah Tower in Saudi Arabia will be 1,007 m high. The Business Bay District is located about 2.5 km landwards of the coast line (Fig. 5).

Fig. 3. Burj Khalifa, Dubai (UAE), h = 828 m

Fig. 4. Fountains at Burj Khalifa in Business Bay District, Dubai (UAE)

Fig. 5. Palm Island (left) and Business Bay District (centre), Dubai (UAE)

2 Subsoil and Groundwater Conditions

The subsoil and groundwater conditions in Business Bay District in Dubai have been investigated by 166 core drillings. The results can be summarized as follows:

Under the surface middle dense sands and partly gravels can be detected unto a depth of about 15 m. Below the sands and gravels follows the Dubai Sandstone with a thickness of 15 m to 40 m. Under the Dubai Sandstone follows the Dubai Siltstone. The two rock layers are also named as "Weak Sandstone" and "Weak Siltstone" (Sharif and Ahmed 2010; Stipho 1984). The groundwater level is similar to the sea water level. That means that the groundwater level is about 3 m to 4 m below the surface. The subsoil and groundwater conditions are shown in Fig. 6.

The subsoil reports are normally developed by geotechnical consulting engineers from UAE, UK, USA and Australia. In these reports the following shear parameters and deformation parameters are given:

- angle of friction φ'
  - Dubai Sandstone: 30–45°
  - Dubai Siltstone: 38–45°
- cohesion c′
  - Dubai Sandstone: 20–105 kN/m^2
  - Dubai Siltstone: 50–155 kN/m^2
- Young's Modulus E
  - Dubai Sandstone: 24–120 MN/m^2
  - Dubai Siltstone: 60–180 MN/m^2

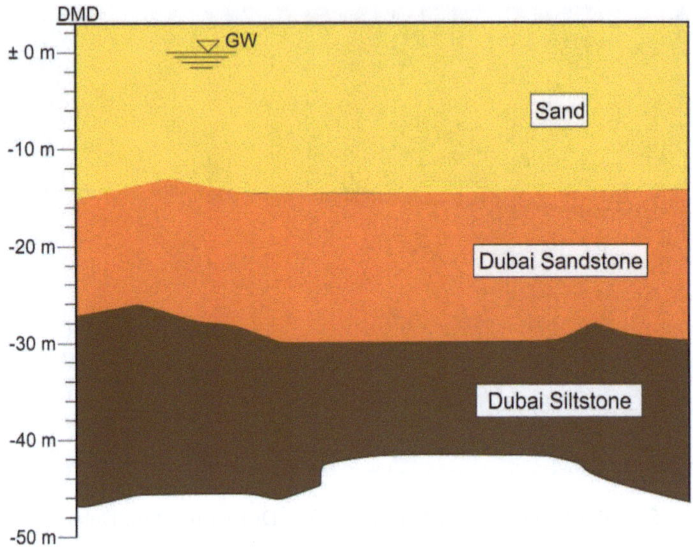

Fig. 6. Subsoil and groundwater conditions in Dubai (UAE)

3 Pile Load Tests

The scientific research of the authors covers 28 large construction projects with more than 40 high-rise buildings. For these large construction projects 116 static pile load tests have been carried out, which have been analyzed scientifically by the authors.

108 of the static pile load tests have been carried out on the final building piles. 8 of the static pile load tests have been carried out on separated piles that have not been used for the load transfer. Only 3 of the test piles have been instrumented with strain gauges, extensometers or load cells. Due to this fact for the most static pile load tests only the load and the pile head settlement have been recorded. A typical example of the pile load tests is shown in Fig. 7.

The test pile No. 73 is 1.3 m thick and 40 m long. The pile goes through the Dubai Sandstone and ends in the Dubai Siltstone. The maximum load was about 32 MN. The measured settlement at the pile head was only 0.7 cm.

The pile load test No. 73 was back-analyzed with numerical simulations using the Finite-Element-Method (FEM). For the first back-analysis the mechanical parameters of the soil reports have used. The comparison of the first back-analysis (grey line in Fig. 8) and the measurement data (red line in Fig. 8) shows no good accordance. Based on this insight additional numerical simulations have been carried out to calibrate the mechanical parameters (green line in Fig. 8). As a result of the calibration the stiffness of the Dubai Sandstone respectively of the Dubai Siltstone is about $E = 5{,}500\ MN/m^2$. Using the information of the existing soil reports, the stiffness of the rocky subsoil is definitely under estimated. Based on this follow a totally wrong estimation of the bearing capacity and the load deformation behavior of the pile.

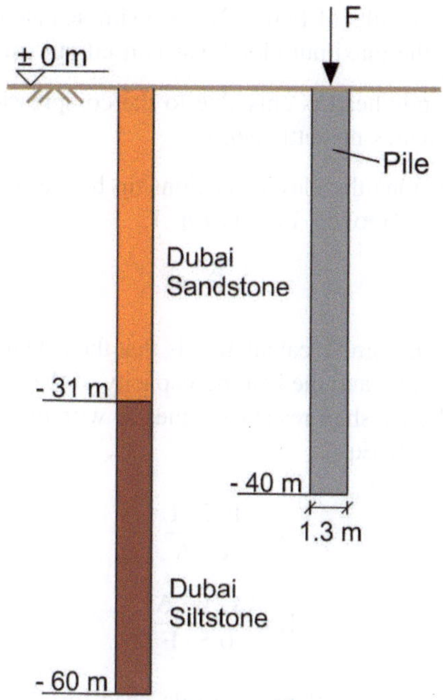

Fig. 7. Pile load test at pile No. 73

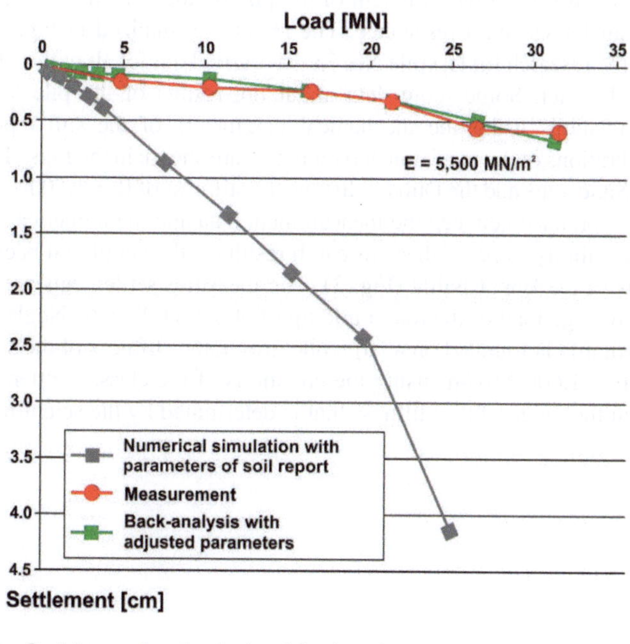

Fig. 8. Measured and calculated load-settlement curves of pile No. 73

To proof the research results of Figure No. 8 a simple calculation of the necessary shaft area of the pile for the maximum load was carried out using two simplifications:

- the settlement of the pile head is only due to the compression of the concrete
- the shaft resistance creates no settlements

With these simplifications the elastic relationship between stress and stiffness can be used which is given by Hooke's Law in Eq. 1.

$$\sigma = E \cdot \varepsilon \tag{1}$$

The requirement for this simple calculation is that the calculated length L is smaller than the length of the test pile and the bearing capacity of the test pile is not reached. If the pile load is reduced by the shaft resistance linearly with the depth Eq. 3 can be used which is a transformation of Eq. 2.

$$s = \frac{F/2 \cdot L}{E \cdot A} \tag{2}$$

$$L = \frac{s \cdot E \cdot A}{0.5 \cdot F} \tag{3}$$

In these Equations are: s = settlement at pile head [m]
E = Young's Modulus of the concrete [MN/m^2]
F = load at pile head [MN]

With a maximum load at the pile head of 32 MN and a maximum settlement at the pile head of 0.7 cm Eq. 3 gives a length of the pile of about 18 m, which is necessary for the load transfer via shaft resistance. The result is visualized in Fig. 9.

The scientific research for test pile No. 73 was carried out for all of the 116 test pile in Business Bay District. Some more data about the results of the pile load tests, the numerical simulations using the mechanical parameters of the soil reports and the numerical calibrations of the mechanical parameters are shown in the Figs. 10, 11 and 12. For the Dubai Sandstone and the Dubai Siltstone the stiffness of E = 1,500–2,700 MN/m^2 gives a good accordance between the measurement data and the numerical results.

Another possibility to proof these research results is the simplified back-analysis of the settlements of the Burj Khalifa (Fig. 3). The measures settlements are about 5 cm, caused by an average load of the total building of about 8,000 MN. Neglecting the fact that the Burj Khalifa is founded on a large pile group, the stiffness of the subsoil can be calculated to E = 1,600 MN/m^2 using the equations of the classic settlement analysis. This result is in the range of the stiffness that is determined by the scientific research of the authors.

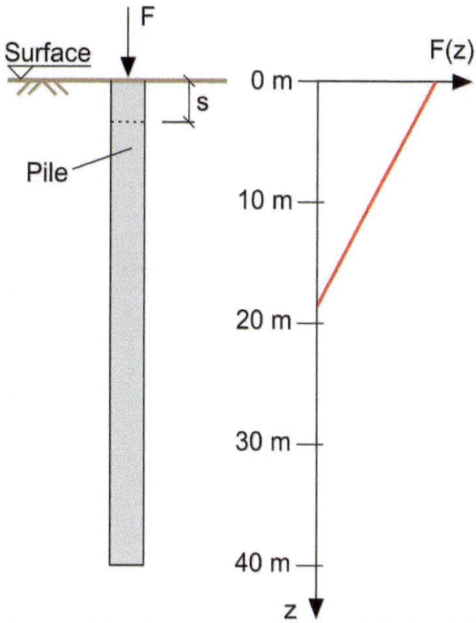

Fig. 9. Calculation of the necessary length L based on the maximum load and the measured settlement at the pile head

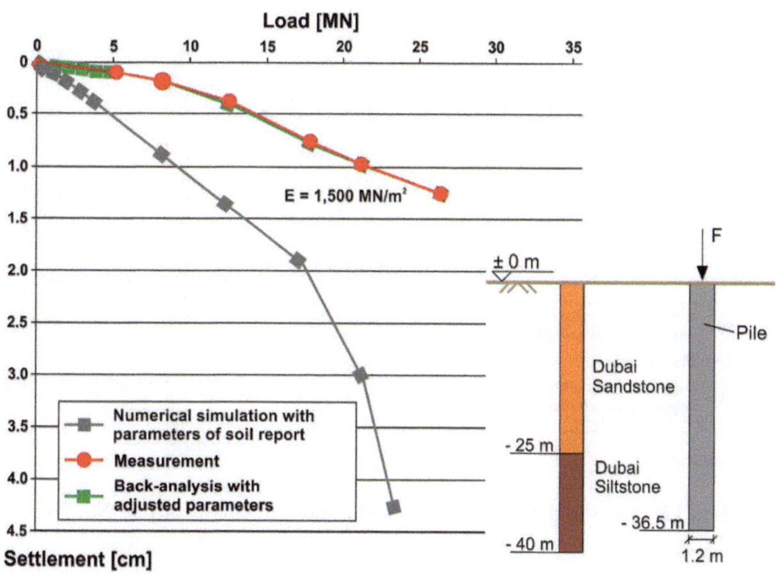

Fig. 10. Measured and calculated load-settlement curves of pile No. 67

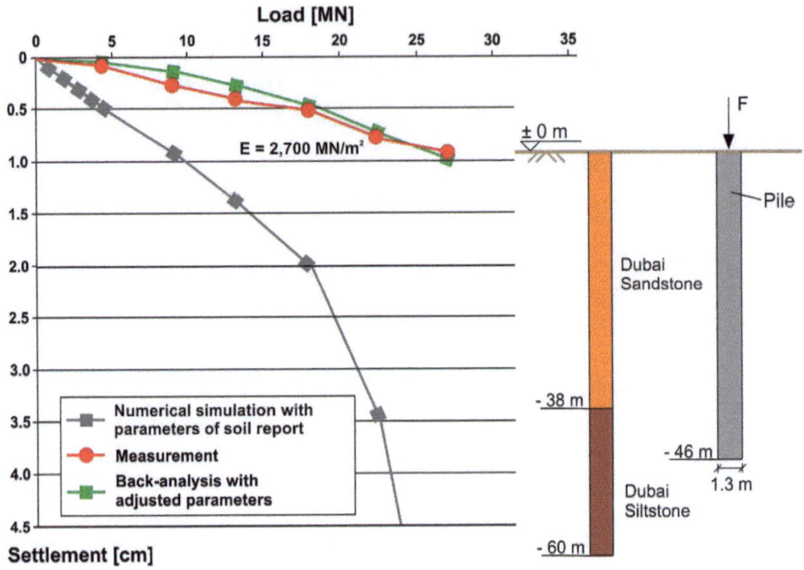

Fig. 11. Measured and calculated load-settlement curves of pile No. 74

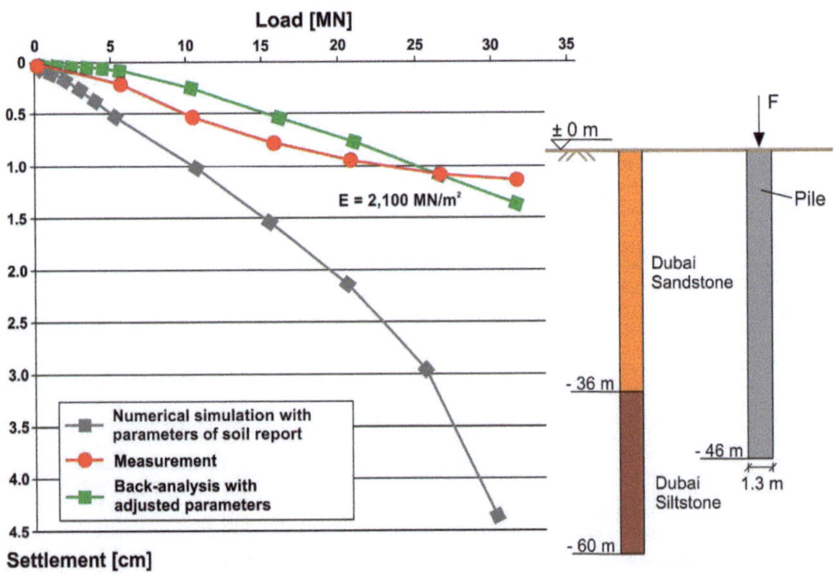

Fig. 12. Measured and calculated load-settlement curves of pile No. 76

4 Conclusions

The results of the scientific research show, that the stiffness of the Dubai Sandstone and the Dubai Siltstone is much higher than estimated in the soil report. The stiffness of the rocky subsoil of Dubai given in the soil report is about 70–80 MN/m^2. The detected stiffness of the scientific research is more than 1,500 MN/m^2. The relation factor is between 21 and 70. A summary is given in Table 1.

Table 1. Comparison of the estimation and of the back-analysis of the stiffness of Dubai Sandstone and Dubai Siltstone

Pile No.	Young's Modulus E [MN/m^2]		Factor
	Estimation $E_{soil\ report}$	Back-analysis $E_{back\text{-}analysis}$	$E_{back\text{-}analysis}/E_{soil\ report}$
67	72	1,500	21
73	78	5,500	70
74	75	2,700	36
76	78	2,100	27

This totally new knowledge is proofed scientifically and will lead to an optimized planning and design of the foundations of large construction projects in Dubai and the whole UAE. This optimized planning and design lead to massive reduction of construction material, money and time and reduces the CO_2 output significantly.

This optimization can be increased in combination with the hybrid foundation system Combined Pile-Raft Foundation (CPRF). The design principles and the application of a CPRF are explained in the CPRF Guideline (ISSMGE 2013) and in Katzenbach et al. (2016).

The advanced new methodology for the identification of stiffness and strength of weak rock can be applied to every place on earth.

Regarding the necessary safety and quality assurance including the observational method practical information is given in Katzenbach et al. (2010) and Katzenbach et al. (2013).

References

ISSMGE International Society of Soil Mechanics and Geotechnical Engineering: Combined Pile-Raft Foundation Guideline (2013)

Katzenbach, R., Bachmann, G., Leppla, S., Ramm, H.: Chances and limitations of the observational method in geotechnical monitoring. In: 14th Danube-European Conference on Geotechnical Engineering, Bratislava, Slovakia, 2–4 June 2010. 13 p

Katzenbach, R., Leppla, S., Weidle, A., Choudhury, D.: Aspects regarding management of soil risk. In: 4th International Seminar on Forensic Geotechnical Engineering, Bengaluru, India, 10–12 January 2013, 12 p

Katzenbach, R., Leppla, S., Choudhury, D.: Foundation Systems for High-Rise Structures. CRC Press Taylor & Francis Group, New York (2016)
Sharif, E.Y., Ahmed, M.J.: Engineering Geology of Dubai. Arab Centre for Engineering Studies (ACES) (2010)
Stipho, A.S.: Soil conditions and foundation problems in the desert regions of the Middle East. In: First International Conference on Case Histories in Geotechnical Engineering, St. Louis, Missouri, USA, pp. 21–25 (1984)

Settlement of Surface Strip Foundation Resting on Soft Clay Subjected to Vertical Cyclic Load

Gobinda Das, Suvendu Kumar Sasmal, Devasish Sahu, and Rabi Narayan Behera(✉)

Department of Civil Engineering, National Institute of Technology, Rourkela, Rourkela, India
das.gvnd@gmail.com, suvendukumarsasmal@gmail.com, civilbydevasish@gmail.com, rnbehera82@gmail.com

Abstract. The nature of dynamic settlement of a shallow foundation is strongly influenced by the response of soil whereas the response of soil largely depends on the cyclic loading. Therefore, it is necessary to study the soil response due to cyclic loading. The present study focuses on investigating the settlement response of a surface strip foundation resting on soft clay subjected to vertical cyclic loading in the form of a rectangular pulse. A strip foundation-soil system is modeled and the settlement response is estimated using numerical programming tool Open System for Earthquake Engineering Simulation (*OpenSees*). Combinations of an allowable vertical static load and rectangular pulse with a frequency of one cycle per second are applied to the footing. The magnitude of the allowable static load and the amplitude of cyclic load are varied to observe the settlement response of the foundation. The effect of variation in the magnitude of static allowable load and increase in the amplitude of cyclic load on the behavior of the foundation is observed. Based on the results obtained from the numerical study, the critical number of load cycles (n_{cr}) after which the settlement becomes negligible for further cyclic loading, is estimated for each case. Furthermore, the effect of frequency of loading on the settlement response of the foundation-soil system for limited cases has been investigated considering two additional frequency of 0.5 Hz and 2 Hz.

1 Introduction

Foundation is the most essential part of any structure and the stability of any structure largely depends on the stability of the foundation on which it rests. The stability of the foundation depends on various factors such as the basic properties of the surrounding soil (cohesion, angle of shearing resistance, unit weight of soil, elastic modulus, shear modulus, Poisson's ratio), the dimensions of the foundation and the nature of load to which it is subjected. A foundation can be subjected to static as well as dynamic loads. Static loads include the weight of superstructure and the weight of the footing. Whereas the dynamic loads may be of various types. These may include slow cyclic loads or may be in the form of severe earthquakes. There are several structures that are subjected to both static as well as dynamic load as a combination. Such structures are the tall structures like tower, chimneys, and others subjected to wind loads, offshore

H. Shehata and B. Das (Eds.): GeoMEast 2018, SUCI, pp. 21–39, 2019.
https://doi.org/10.1007/978-3-030-01923-5_3

platforms, breakwaters and structures subjected to wave loading, structures and slopes subjected to earthquake loading, and foundation subjected to machine vibrations. The dynamic load can be the reason for the reduction of the strength of soil and may result in large settlement of the foundation. Therefore, investigating the behavior of the foundation i.e. settlement is an important aspect while designing a foundation subjected to dynamic load.

The settlement under cyclic load is hugely dependent on the loading and soil conditions. Raymond and Komos (1978) reported that the cyclic load (rectangular pulse) is generally generated from the train wheel passing over a railroad. They found a number of factors such as footing dimension, number of load cycles and amplitude of cyclic load that affect the settlement of any foundation. Das and Shin (1996) studied the settlement of foundation on clayey soil and concluded that the settlement is influenced by the intensity of static load and amplitude of cyclic load. Apart from different experimental techniques there exist a number of numerical methods to accurately predict the settlement of the foundation under cyclic load. When it comes to model a foundation numerically, the Winkler model has always been given importance by various researchers. The nonlinear cyclic settlement response of foundation, using Beam on Nonlinear Winkler Foundation model (BNWF), has been described by various researchers. Harden et al. (2005) described the finite element modelling procedures to estimate the settlement accurately using BNWF model. Gajan et al. (2008) using BNWF model provided the moment-rotation-settlement response of shallow foundation. Winkler model can satisfactorily capture all the cyclic response of a shallow foundation apart from permanent horizontal displacement (Allotey and Naggar 2008a, b). Raychowdhury (2008) found that the BNWF model is capable of predicting experimentally measured footing response in terms of moment, shear, settlement and rotation demands. Allotey and Naggar (2008a, b) reported that the soil structure interaction response can successfully be recorded by the BNWF model.

The experiments conducted by Raymond and Komos (1978), Das and Shin (1996) were unique in their category. However, there is lack of numerical procedures to study how a foundation at railroad behaves under wheel load. Hence, an attempt has been made in the present study to numerically model the foundation system and to study its settlement response under the influence of rectangular pulse. The present study focuses on analyzing the behavior of a strip footing resting on soft clay and subjected to a combination of a static load and vertical cyclic load. The foundation is modeled numerically using Open System for Earthquake Engineering Simulation (*OpenSees*) based on Beam-on-Nonlinear-Winkler foundation approach and the behavior of the foundation under static and cyclic load has been observed. The important factors affecting the cyclic settlement of foundation are highlighted and the critical number of load cycles (n_{cr}) after which the settlement of the foundation becomes negligible for further load cycles are found out for each case. The study will provide a brief idea regarding the settlement pattern of the foundation which will help in predicting the extent of damage and the lifetime of such foundations, which generally undergo a huge number of load cycles.

2 Methodology

In the present study, a shallow strip foundation is modeled using Beam-On-Nonlinear-Winkler foundation (BNWF) model that is capable of capturing the nonlinear behavior of the soil-foundation system. The components of BNWF model are created using numerical programming tool Open System for Earthquake Engineering Simulation (*OpenSees*). The model consists of elastic beam-column elements that capture the behavior of the footing and zero-length elements which capture the soil-footing behavior. A BNWF model requires various input parameters such as soil type (clay or sand), bearing capacity of soil, Poisson's ratio (v), Young's modulus (E), shear modulus (G), vertical and lateral stiffness, radiation damping, tension capacity of soil, distribution and magnitude of vertical stiffness, spring spacing etc. The present model is developed for two-dimensional analyses. The foundation is created by connecting different one-dimensional elastic beam-column elements. Each element joins two nodes having three degrees-of-freedom per node. The zero-length elements are one-dimensional nonlinear inelastic springs that are independent of each other and are modeled using QzSimple1, PySimple1 and TzSimple1 material models, and these spring elements simulate the vertical load-displacement behavior, horizontal passive load-displacement behavior against the side of the footing and horizontal shear-sliding behavior at the base of the footing respectively. The above-mentioned springs are characterized by nonlinear backbone curves resembling a bilinear behavior and are implemented in *OpenSees* by Boulanger (2000). The vertical and lateral stiffness of the springs are calculated from the specified values of shear modulus (G), Poisson's ratio (v) and footing dimensions using the expressions by Gazetas (1991), the same supported by FEMA 356 - 2000. The ultimate bearing capacity (q_{ult}) assigned to the QzSimple1 material is calculated as per Meyerhof (1963) whereas the passive resisting force (p_{ult}) assigned to PySimple1 spring model is obtained as per Rankine (1857). The lateral sliding capacity assigned to the TzSimple1 material is the frictional resistance which is a function of the shear strength of the soil and the footing. The magnitude and distribution of vertical stiffness along the length of the footing is done by assigning suitable values of stiffness intensity ratio ($R_k = K_{end}/K_{mid}$) and end length ratio ($R_e = L_{end}/L$) according to Harden et al. (2005). The spring spacing is taken as a fraction (0.01) of the footing length (L), provided the fact that the smaller the spring spacing, the more accurate the result, when the Winkler model is concerned.

The model foundation has the measurements of 229 mm (length) × 76.1 mm (width) × 38.1 mm (thickness). The depth of embedment is zero since it is a surface foundation. The soil under the foundation is soft clay with a moisture content of 34.4%, moist unit weight of 18.58 kN/m^3 and undrained shear strength of 11.9 kN/m^2. The Young's modulus (E) and Poisson's ratio (v) are taken as 1500 kN/m^2 and 0.3 respectively according to EPRI 1990.

The foundation is subjected to a combination of static safe load and cyclic load. The loading diagram is shown in Fig. 1. First, the static safe load (Q_s) is applied on the center of the foundation. Then, the cyclic load of amplitude (Q_d) is applied in the form of a rectangular pulse load at a frequency of 1 cycle/second (1 Hz). The static safe load is varied from 10% to 45% of the static failure load. The amplitude of cyclic load is varied from 2.5% to 20% of the static failure load of the foundation. Several combinations of both static load and cyclic load are applied to the foundation and the behavior of the foundation to the applied load combinations is observed.

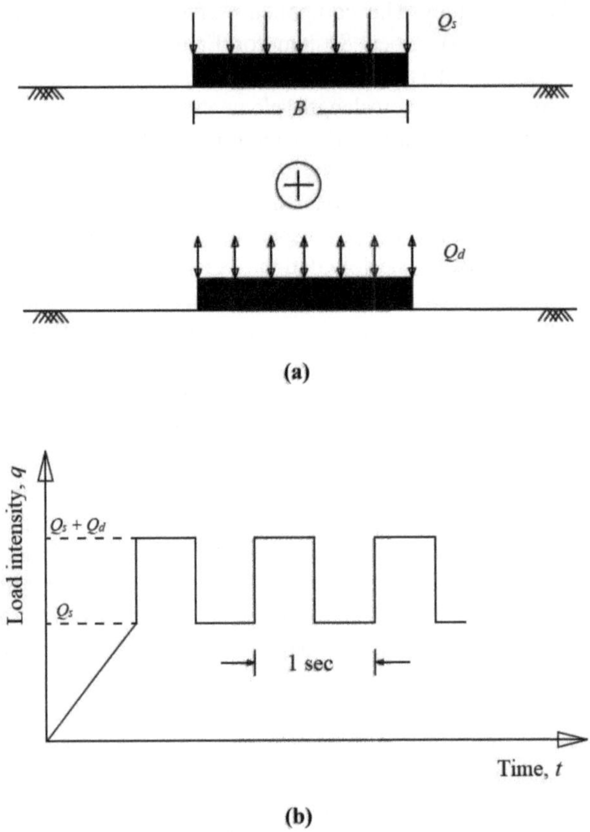

Fig. 1. Details of loads applied on the foundation (a) static + cyclic load, (b) pattern of static + cyclic loading

3 Results and Discussion

The behavior of the foundation towards static load is observed first. The foundation is subjected to a static load equal to its ultimate load (Q_u) and the corresponding settlements are noted. A plot is drawn between the load per unit area (q) and the normalized settlement $(s/B)\%$ to determine the ultimate settlement (s_u) of the foundation. The above-obtained plot is compared with the plot obtained by Das and Shin (1996) and that obtained by using PLAXIS 2D software for the same soil conditions. From Fig. 2, it is observed that both the experimental and numerical results have good agreements, which justifies the ability of *OpenSees* to accurately capture the foundation behavior. Table 1 shows the corresponding ultimate bearing capacity (q_u) and normalized ultimate settlement $[(s_u/B)\%]$ obtained from the above-mentioned approaches.

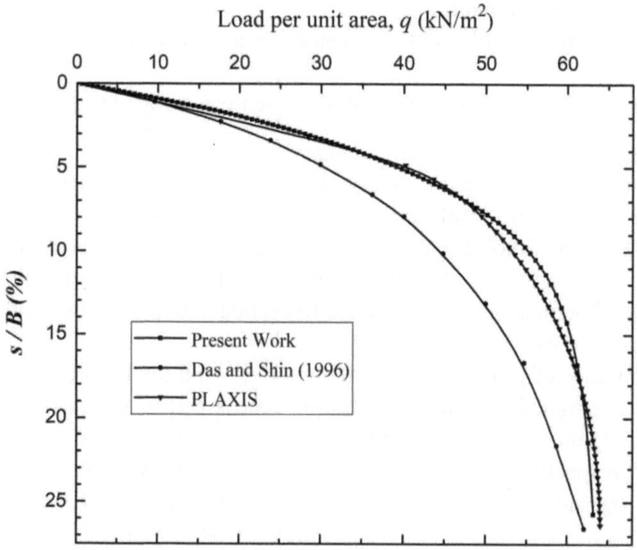

Fig. 2. Static load intensity (q) versus normalized settlement ($s/B\%$)

Table 1. Ultimate bearing capacity (q_u) and normalized ultimate settlement [s_u/B (%)] obtained from different approaches

Different approaches	Ultimate bearing capacity, q_u	Normalized ultimate settlement, $s_u/B(\%)$
Present work	61.5 kN/m^2	17.5
Das and Shin (1996)	58 kN/m^2	17
PLAXIS	62.5 kN/m^2	18.5

The response of the foundation to the cyclic load has been observed in the dynamic analysis where the foundation is subjected to a combination of static and cyclic load. A safe static load (Q_s) is applied to the footing followed by a rectangular pulse load of amplitude (Q_d) for 200000 cycles at a frequency of 1 cycle/second and the settlement for the corresponding number of cycles is noted. Figures 3, 4, 5, 6, 7, 8, 9, 10, 11, 12, 13, 14, 15, 16 and 17 represent the plots between the normalized dynamic settlement $(s_d/B)\%$ (where s_d = settlement due to only cyclic load, B = width of the foundation) and the number of cycles of load (n) for various combinations of Q_s/Q_u and Q_d/Q_u. It is observed that the variation of settlement with increasing load cycles has a definite pattern. The variation of the pattern of $(s_d/B)\%$ with n is shown in Fig. 18 as a straight line approximation. It can be observed that the settlement occurs in three stages:

- A primary rapid settlement: $s_{d(r)}$
- A secondary slower settlement that ceases at a critical number of cycles (n_{cr}) of loading: $s_{d(s)}$
- Beyond $n_{cr,}$ the settlement is very small which can be neglected

From Figs. 3, 4, 5, 6, 7, 8, 9, 10, 11, 12, 13, 14, 15, 16 and 17, it can be observed that the primary rapid settlement occurs within 10–20 cycles whereas the secondary settlement ceases within 20,000 cycles i.e. critical number of load cycles (n_{cr}) lies between 10,000 and 20,000 cycles which is independent of the Q_s/Q_u and Q_d/Q_u combination. When $n \geq n_{cr,}$ the settlement is very less or say negligible. For a given (Q_s/Q_u), the total permanent settlement due to cyclic loading increases with the increase of the amplitude of cyclic load intensity.

The magnitudes of primary rapid settlement $s_{d(r)}$ and total settlement due to cyclic load $s_{d(t)}$ are determined from the straight line approximation for various combinations of Q_s/Q_u and Q_d/Q_u. Figures 19 and 20 show the variations of $s_{d(r)}/s_u$ and $s_{d(t)}/s_u$ against Q_d/Q_u respectively for various Q_s/Q_u. It can be observed that for a particular Q_s/Q_u, both $s_{d(r)}/s_u$ and $s_{d(t)}/s_u$ bear a non-linear relationship with Q_d/Q_u and the rate of increment of $s_{d(r)}/s_u$ and $s_{d(t)}/s_u$ increases with the Q_d/Q_u. Figure 21 shows the plot of $s_{d(r)}/s_{d(t)}$ versus Q_d/Q_u for various Q_d/Q_u. It can be observed that $s_{d(r)}/s_{d(t)}$ bears an approximately linear relationship with Q_d/Q_u.

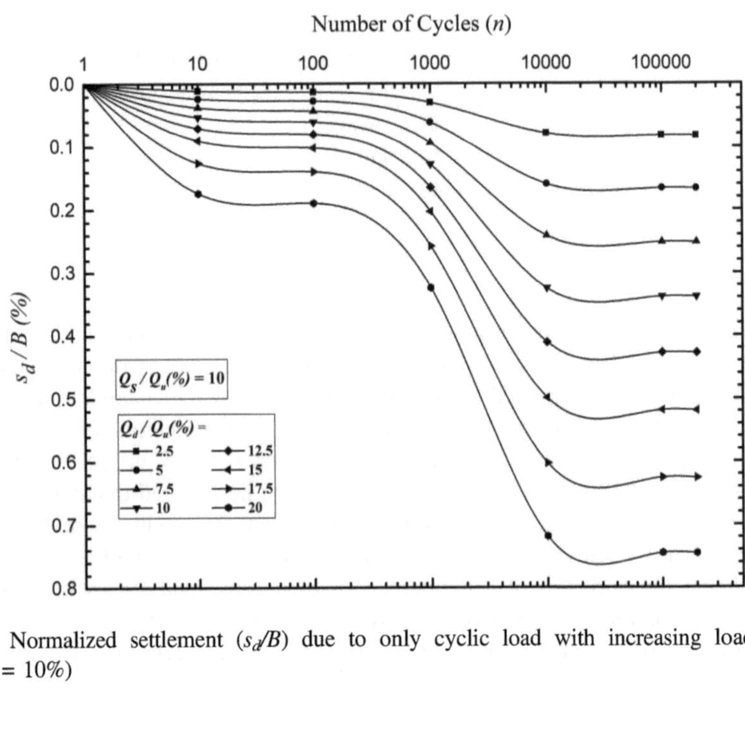

Fig. 3. Normalized settlement (s_d/B) due to only cyclic load with increasing load cycles (Q_s/Q_u = 10%)

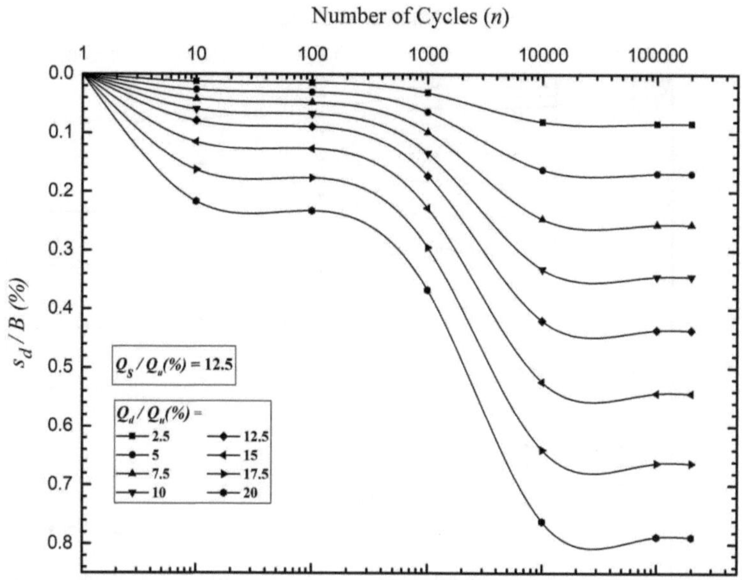

Fig. 4. Normalized settlement (s_d/B) due to only cyclic load with increasing load cycles (Q_s/Q_u = 12.5%)

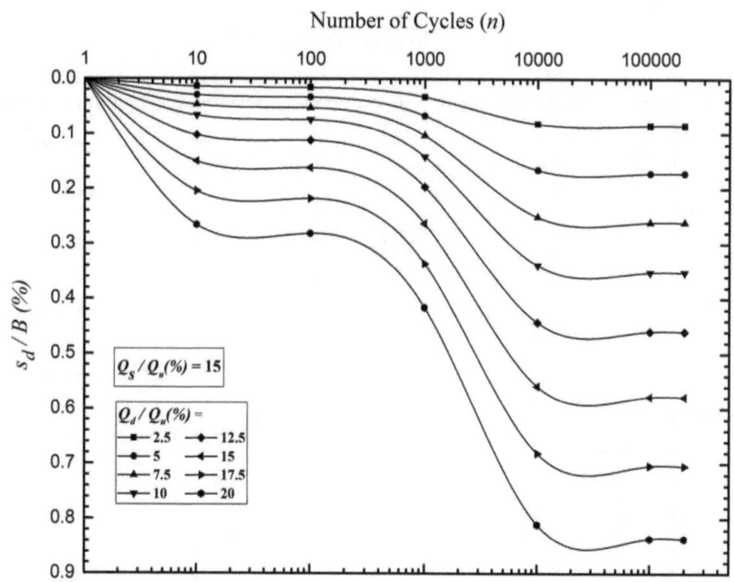

Fig. 5. Normalized settlement (s_d/B) due to only cyclic load with increasing load cycles (Q_s/Q_u = 15%)

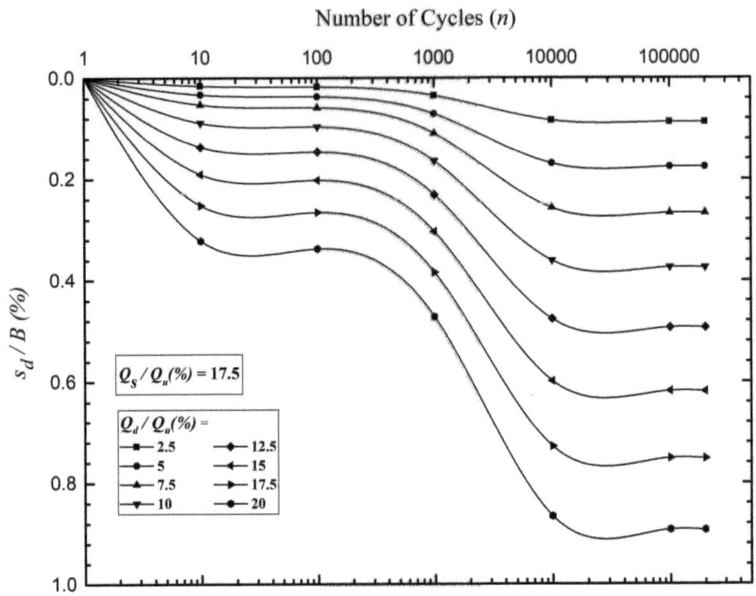

Fig. 6. Normalized settlement (s_d/B) due to only cyclic load with increasing load cycles (Q_s/Q_u = 17.5%)

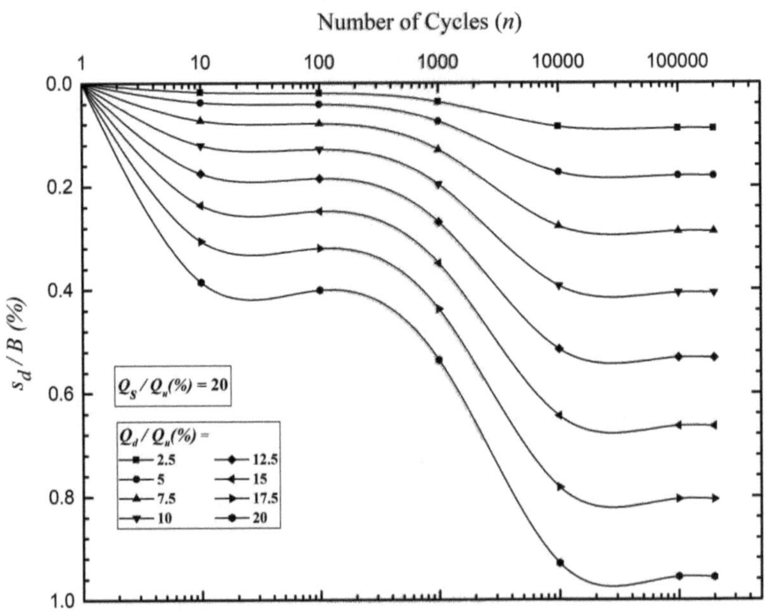

Fig. 7. Normalized settlement (s_d/B) due to only cyclic load with increasing load cycles (Q_s/Q_u = 20%)

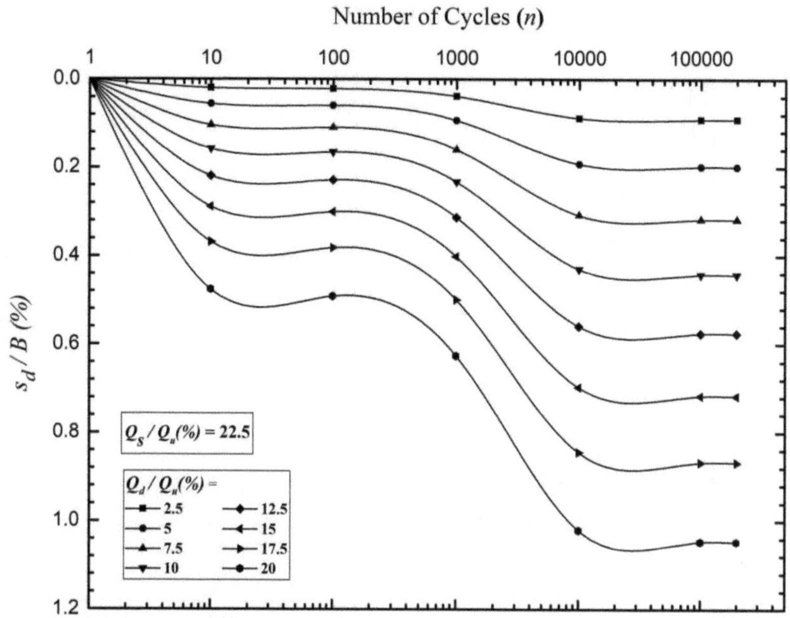

Fig. 8. Normalized settlement (s_d/B) due to only cyclic load with increasing load cycles (Q_s/Q_u = 22.5%)

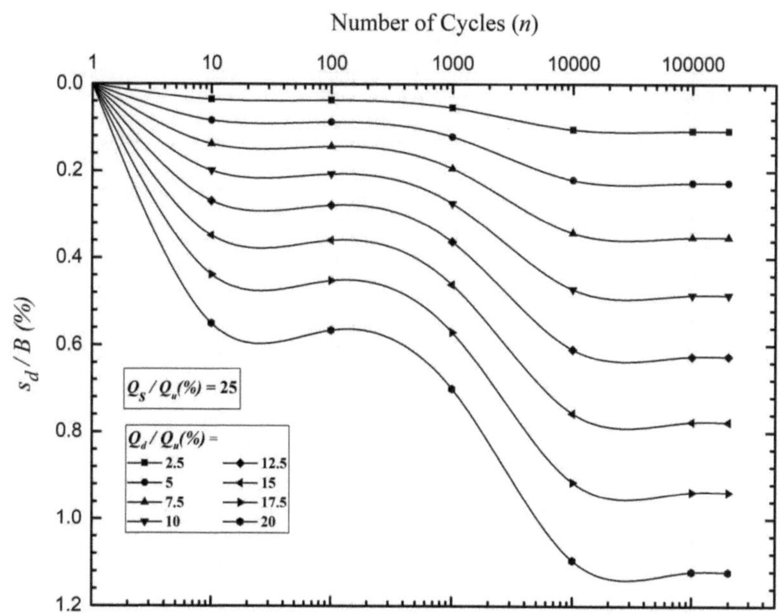

Fig. 9. Normalized settlement (s_d/B) due to only cyclic load with increasing load cycles (Q_s/Q_u = 25%)

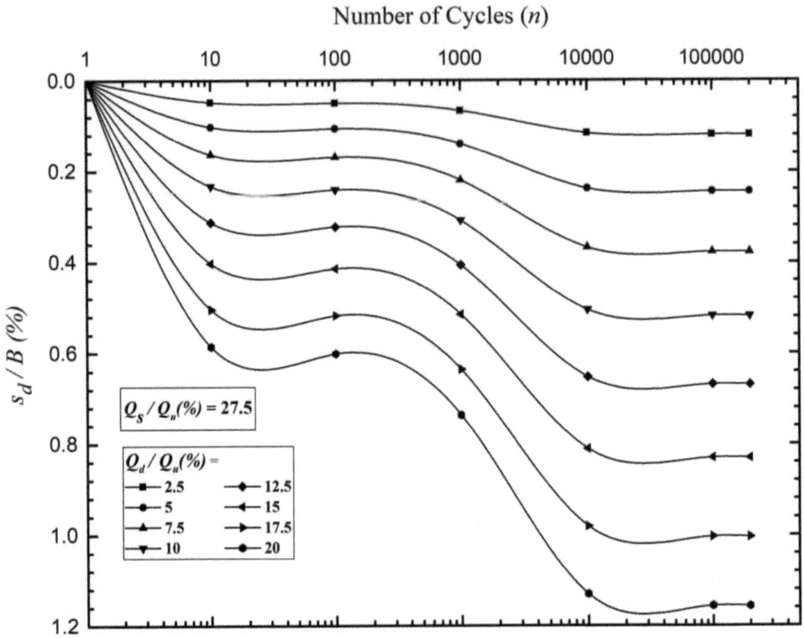

Fig. 10. Normalized settlement (s_d/B) due to only cyclic load with increasing load cycles (Q_s/Q_u = 27.5%)

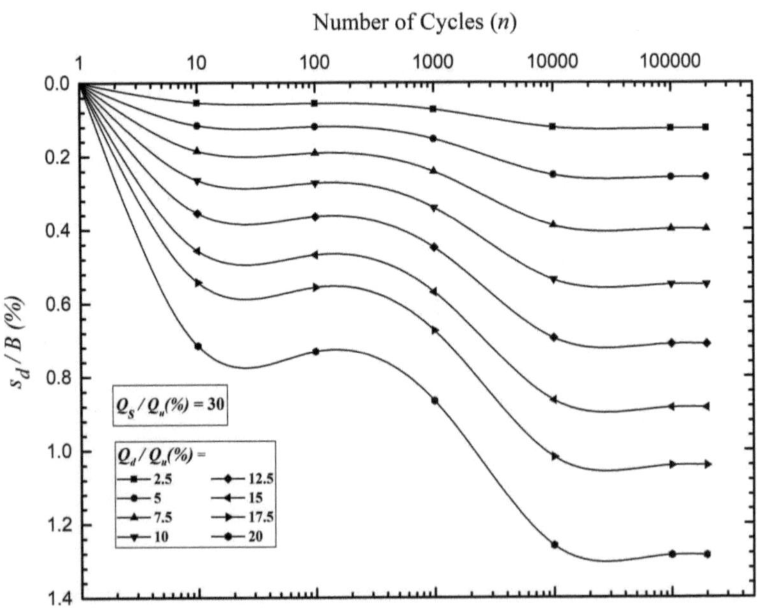

Fig. 11. Normalized settlement (s_d/B) due to only cyclic load with increasing load cycles (Q_s/Q_u = 30%)

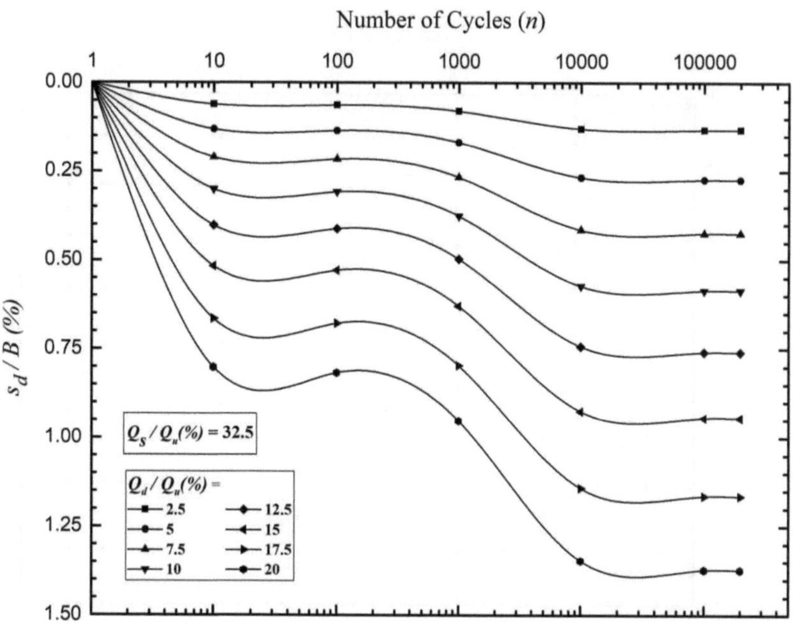

Fig. 12. Normalized settlement (s_d/B) due to only cyclic load with increasing load cycles (Q_s/Q_u = 32.5%)

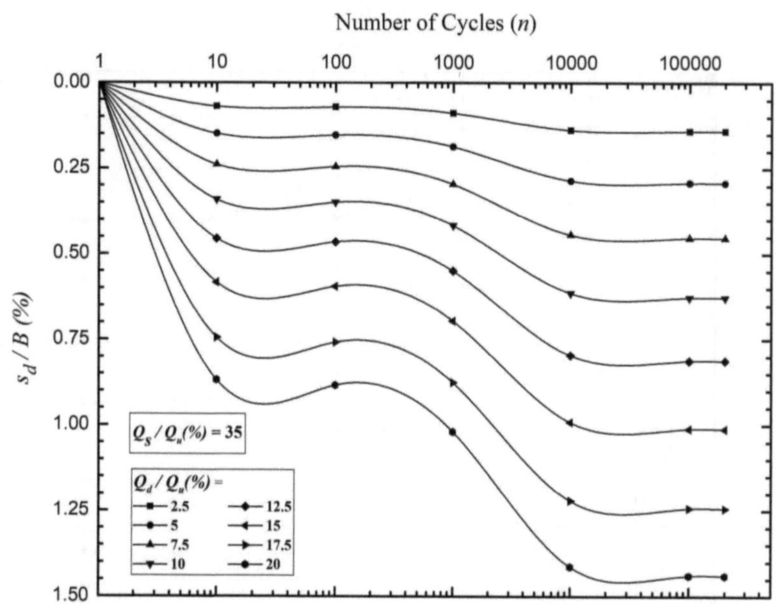

Fig. 13. Normalized settlement (s_d/B) due to only cyclic load with increasing load cycles (Q_s/Q_u = 35%)

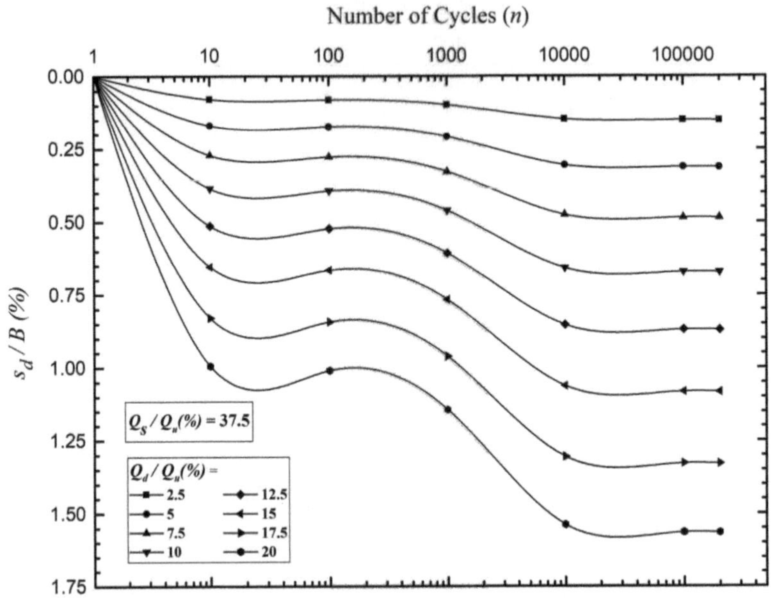

Fig. 14. Normalized settlement (s_d/B) due to only cyclic load with increasing load cycles (Q_s/Q_u = 37.5%)

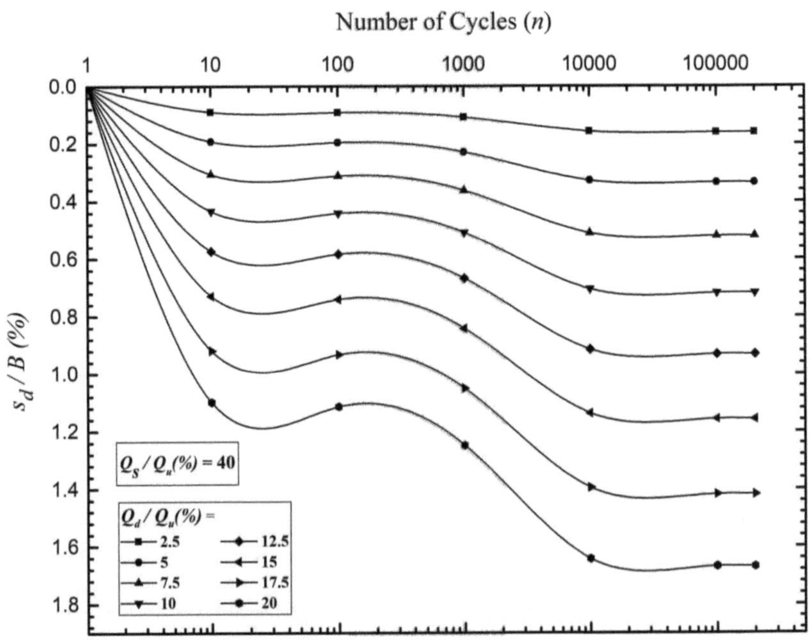

Fig. 15. Normalized settlement (s_d/B) due to only cyclic load with increasing load cycles (Q_s/Q_u = 40%)

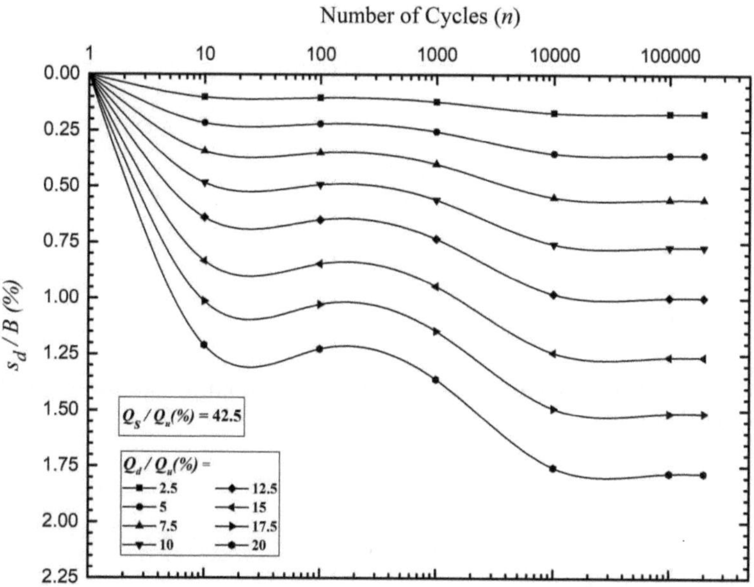

Fig. 16. Normalized settlement (s_d/B) due to only cyclic load with increasing load cycles (Q_s/Q_u = 42.5%)

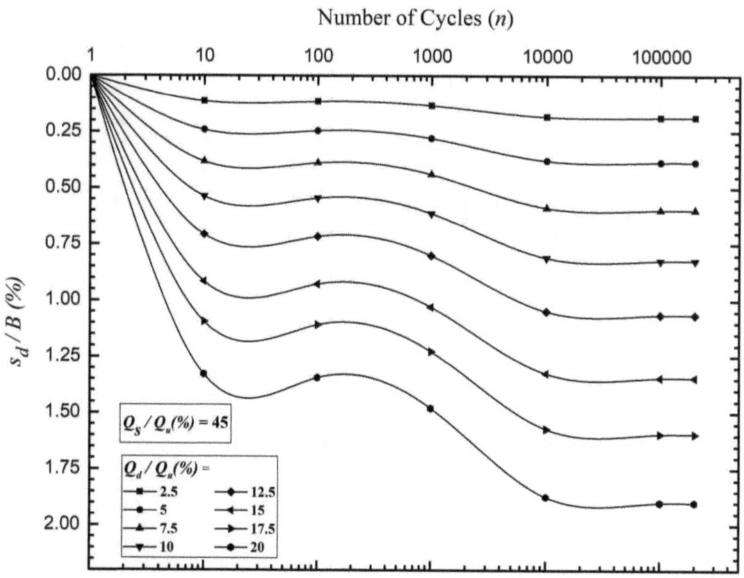

Fig. 17. Normalized settlement (s_d/B) due to only cyclic load with increasing load cycles (Q_s/Q_u = 45%)

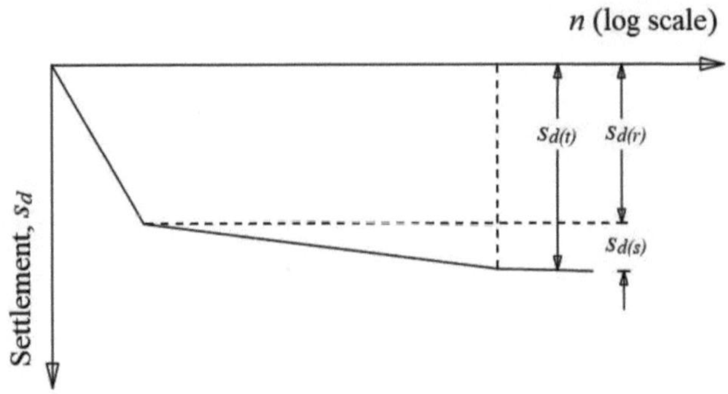

Fig. 18. Nature of variation of dynamic settlement (s_d) versus number of load cycles (n) (Das and Shin 1996)

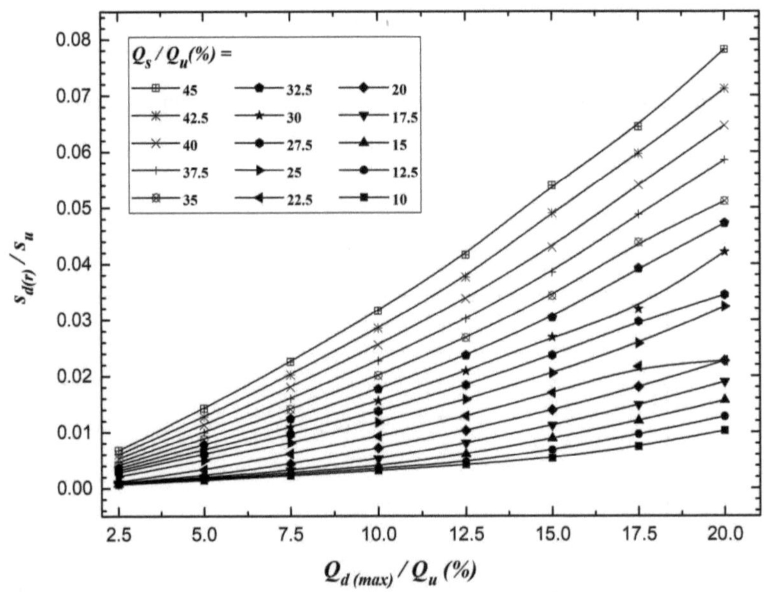

Fig. 19. Variation of ratio of primary rapid settlement to ultimate settlement ($s_{d(r)}/s_u$) with respect to ratio of dynamic to ultimate static load ($Q_{d(max)}/Q_u$)

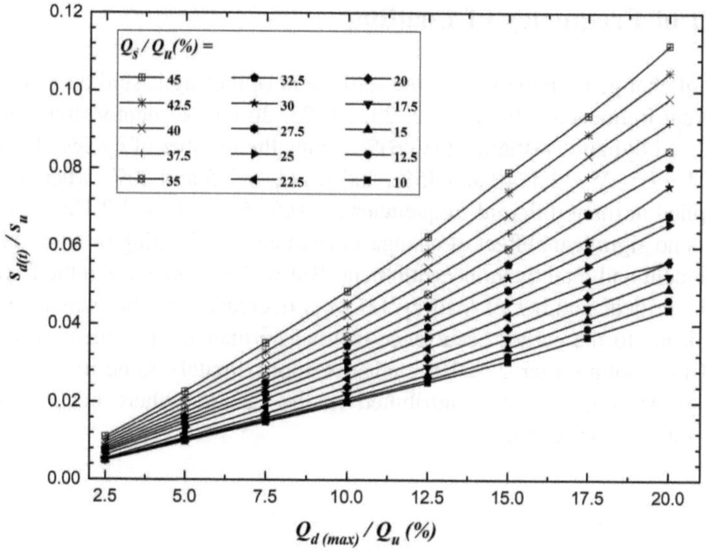

Fig. 20. Variation of ratio of total dynamic settlement to ultimate settlement ($s_{d(t)}/s_u$) with respect to ratio of dynamic to ultimate static load ($Q_{d(max)}/Q_u$)

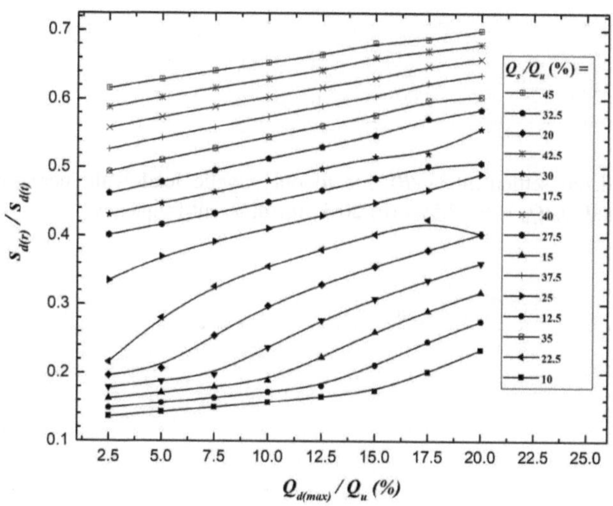

Fig. 21. Plot of ratio of primary rapid settlement to total dynamic settlement ($s_{d(r)}/s_{d(t)}$) versus ratio of dynamic to ultimate static load ($Q_{d(max)}/Q_u$)

4 Effect of Frequency of Loading

The effect of change in frequency on the settlement of footing is studied by considering three different frequencies. Figures 22, 23, 24, 25, 26 and 27 demonstrate the plots of the normalized dynamic settlement $(s_d/B)\%$ versus the number of cycles of load (n) for Q_s/Q_u = 10, 17.5, 25, 32.5, 40 and 45%, and Q_d/Q_u = 2.5 and 20%, where each cyclic load is applied at three different frequencies i.e. 0.5 Hz, 1 Hz and 2 Hz. It is observed that there is no significant effect of change in frequency of loading on both the primary rapid settlement and total dynamic settlement. But with the increase in the frequency of loading, the critical number of load cycles (n_{cr}) decreases i.e. the increase in loading frequency leads to the earlier occurrence of total permanent settlement. The total settlement of the footing after 2×10^5 cycles is approximately same irrespective of the frequency of loading, which is attributed to the fact that there occurs hardly any settlement once n_{cr} is reached.

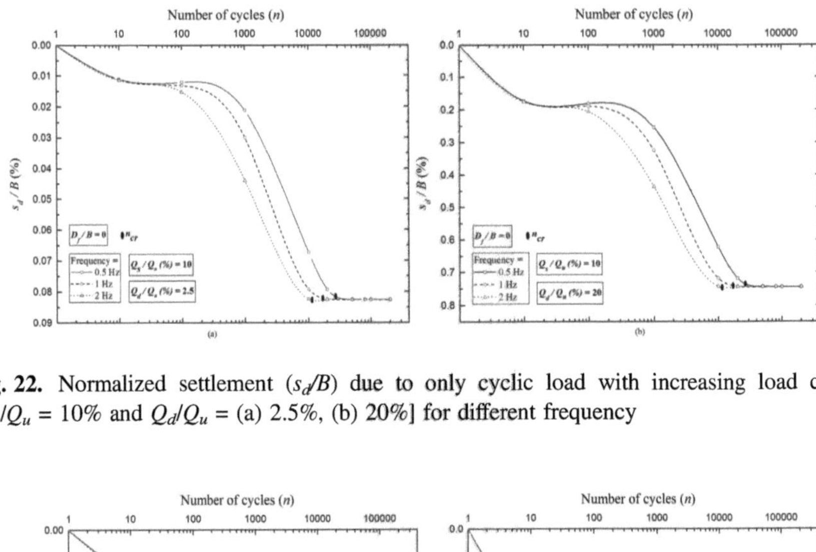

Fig. 22. Normalized settlement (s_d/B) due to only cyclic load with increasing load cycles [Q_s/Q_u = 10% and Q_d/Q_u = (a) 2.5%, (b) 20%] for different frequency

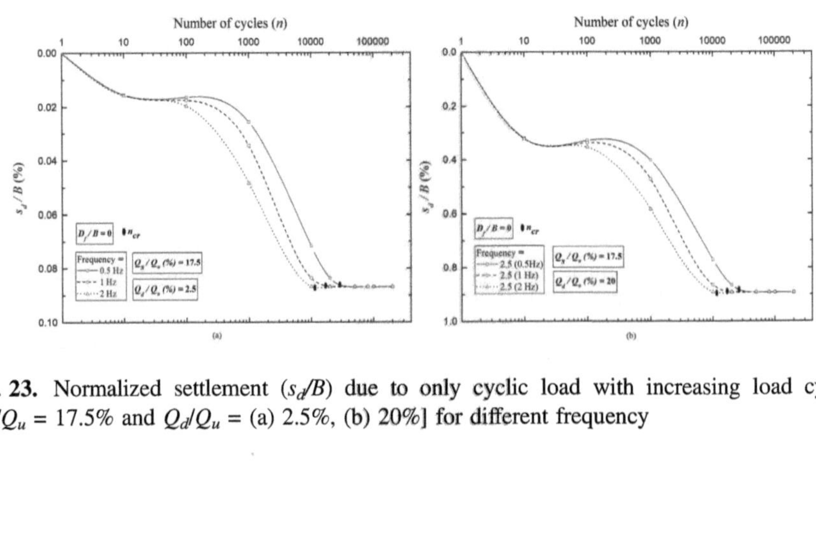

Fig. 23. Normalized settlement (s_d/B) due to only cyclic load with increasing load cycles [Q_s/Q_u = 17.5% and Q_d/Q_u = (a) 2.5%, (b) 20%] for different frequency

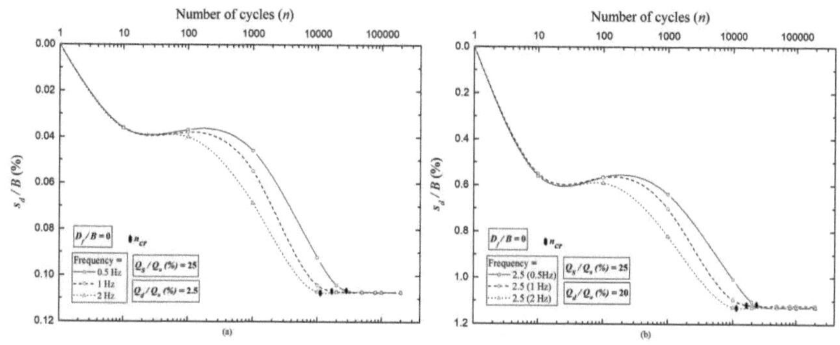

Fig. 24. Normalized settlement (s_d/B) due to only cyclic load with increasing load cycles [Q_s/Q_u = 25% and Q_d/Q_u = (a) 2.5%, (b) 20%] for different frequency

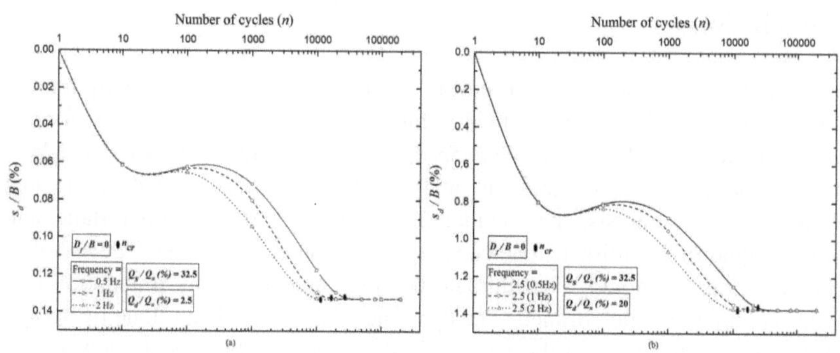

Fig. 25. Normalized settlement (s_d/B) due to only cyclic load with increasing load cycles [Q_s/Q_u = 32.5% and Q_d/Q_u = (a) 2.5%, (b) 20%] for different frequency

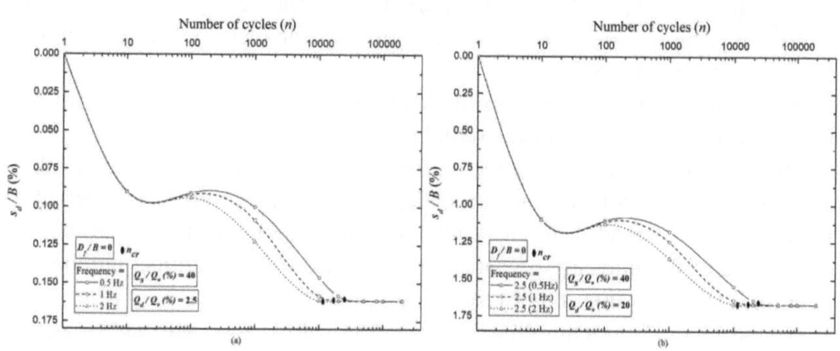

Fig. 26. Normalized settlement (s_d/B) due to only cyclic load with increasing load cycles [Q_s/Q_u = 40% and Q_d/Q_u = (a) 2.5%, (b) 20%] for different frequency

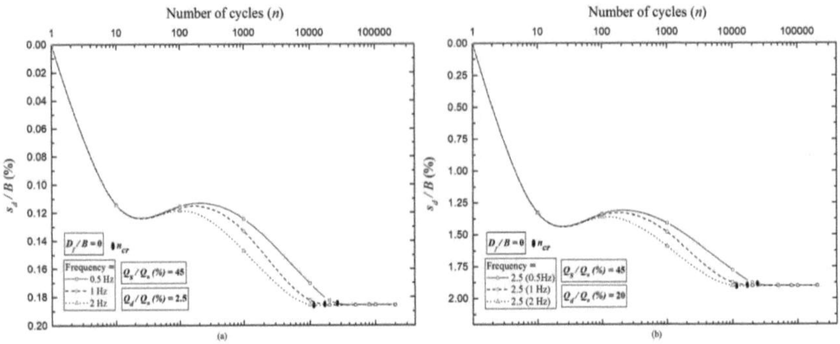

Fig. 27. Normalized settlement (s_d/B) due to only cyclic load with increasing load cycles [Q_s/Q_u = 45% and Q_d/Q_u = (a) 2.5%, (b) 20%] for different frequency

5 Conclusions

The present paper emphasizes on observing the settlement of a shallow strip foundation resting on soft clay subjected to vertical cyclic (rectangular pulse) load. The foundation is modeled numerically using *OpenSees*, based on Beam-on-Nonlinear-Winkler foundation model. The foundation is analyzed for both static and dynamic loading case. In the static analysis, the foundation is subjected to a gradual static load of magnitude equal to the ultimate load capacity of the foundation (as per Meyerhof 1963) and the settlement is noted. The ultimate bearing capacity of the foundation and the ultimate settlement are determined. The dynamic analysis has been done by applying a safe static load depending on the factor of safety and a superimposed low frequency (1 Hz) vertical cyclic load. For various combinations of Q_s/Q_u (10 to 45% @ 2.5%) and Q_d/Q_u (2.5 to 20% @ 2.5%), the settlement with number of cycles are noted. The followings are the important outcomes concluded from the study.

- The load per unit area (q) versus the normalized settlement (s/B) response obtained in this study, has good agreements with the same obtained by Das and Shin (1996) and that using PLAXIS 2D.
- The total settlement due to only cyclic load is composed of two parts. Primary rapid settlement and secondary slower settlement.
- The primary rapid settlement is the response of the foundation for the first 10 to 20 load cycles. The secondary slower settlement takes place up to the arrival of the critical number of load cycle (n_{cr}), after which the settlement becomes negligible.
- Irrespective of the intensity of static and cyclic loading the critical number of load cycle (n_{cr}) takes a value in between 15000 to 20000.
- $s_{d(r)}/s_u$ and $s_{d(t)}/s_u$ bear a non-linear relationship with Q_d/Q_u, whereas $s_{d(r)}/s_{d(t)}$ bears almost linear relationship with Q_d/Q_u.
- Frequency does not affect the rapid and total settlement significantly. However, for a given intensity of static and cyclic loading, the critical number of load cycle (n_{cr}) decreases with the increasing frequency of dynamic loading.

References

Allotey, N., El Naggar, M.H.: An investigation into the Winkler modeling of the cyclic response of rigid footings. Soil Dyn. Earthq. Eng. **28**(1), 44–57 (2008a)

Allotey, N., El Naggar, M.H.: Generalized dynamic Winkler model for nonlinear soil-structure interaction analysis. Can. Geotech. J. **45**(4), 560–573 (2008b)

Boulanger, R.W.: The PySimple1, QzSimple1, and TzSimple1 material documentation. Documentation for the OpenSees platform (2000). http://opensees.berkeley.edu/

Das, B.M., Shin, E.C.: Laboratory model tests for cyclic load-induced settlement of a strip foundation on a clayey soil. Geotech. Geol. Eng. **14**(3), 213–225 (1996)

EPRI: Manual on Estimating Soil Properties for Foundation Design. Electric Power Research Institute, Palo Alto (1990)

Federal Emergency Management Agency (FEMA): Prestandard and Commentary for the Seismic Rehabilitation of Buildings, FEMA (Series) 356, Washington, D.C. (2000)

Gajan, S., Hutchinson, T.C., Kutter, B., Raychowdhury, P., Ugalde, J.A., Stewart, J.P.: Numerical models for the analysis and performance-based design of shallow foundations subjected to seismic loading. Rep. No. 2007/04, Pacific Earthquake Engineering Research Center (PEER), Berkeley, CA (2008)

Gazetas, G.: Formulas and charts for impedances of surface and embedded foundations. J. Geotech. Eng. ASCE **117**(9), 1363–1381 (1991)

Harden, C., Hutchinson, T., Martin, G.R., Kutter, B.L.: Numerical modeling of the nonlinear cyclic response of shallow foundations. Rep. No. 2005/04, Pacific Earthquake Engineering Research Center (PEER), Berkeley, CA (2005)

Meyerhof, G.G.: Some recent research on the bearing capacity of foundations. Can. Geotech. J. **1** (1), 16–26 (1963)

OpenSees (Computer Software): Open system for earthquake engineering simulation, Pacific Earthquake Engineering Research Center (PEER), University of California, Berkeley. http://OpenSees.berkeley.edu/

Raychowdhury, P.: Nonlinear Winkler-based shallow foundation model for performance assessment of seismically loaded structures. Ph.D. dissertation, University of California, San Diego (2008)

Raymond, G.P., Komos, F.E.: Repeated load testing of a model plane strain footing. Can. Geotech. J. **15**(2), 190–201 (1978)

Ultimate Bearing Capacity of Eccentrically Inclined Loaded Circular Foundation on Sand Layer of Limited Thickness Using ANN

B. P. Sethy[1(✉)], C. R. Patra[1], K. Sobhan[2], and B. M. Das[3]

[1] National Institute of Technology, Rourkela, Rourkela, India
barada.jeetu@gmail.com, crpatra19@yahoo.co.in
[2] Florida Atlantic University, Boca Raton, FL, USA
ksobhan@fau.edu
[3] California State University, Sacramento, USA
brajamdas@gmail.com

Abstract. Laboratory model tests were conducted for the ultimate bearing capacity of shallow rough circular surface foundation resting over a sand layer of limited thickness subjected to an eccentrically inclined load. Based on the laboratory model test results, a neural network model is developed to estimate the reduction factor (*RF*). The reduction factor can be used to estimate the ultimate eccentrically inclined load per unit area of the foundation supported by a sand layer of limited thickness from the ultimate bearing capacity of a foundation on a sand layer extending to a great depth under an eccentrically inclined load. A thorough sensitivity analysis was carried out to determine the important parameters affecting the reduction factor. Importance was given on the construction of neural interpretation diagram. Based on this neural interpretation diagram, the direct or inverse relationships that exists between the input and output parameters were determined. Results from the artificial neural network (ANN) were compared with the laboratory model test results and these results are well matched.

Keywords: Eccentrically inclined load · Sand · Limited thickness ANN · Ultimate bearing capacity · Reduction factor

1 Introduction

During the last three decades, a number of laboratory model test results and few field test results have been published that are related to the ultimate bearing capacity of shallow foundation resting over homogeneous sand bed and clay extending to a great depth. The presence of a hard layer within a certain depth below the foundation base can significantly influence the ultimate bearing capacity supported by the soil. In many practical problems, the thin soil layer may be underlain by rigid rock base. In these situations, the ultimate bearing capacity of shallow foundation resting on soil layer of limited thickness is influenced by the lower rigid boundary. Most of the experimental studies were related to centric vertical loading. However, none of the published studies address the effect of load eccentricity, load inclination, and the effect of rigid base

H. Shehata and B. Das (Eds.): GeoMEast 2018, SUCI, pp. 40–53, 2019.
https://doi.org/10.1007/978-3-030-01923-5_4

located at a limited depth on the ultimate bearing capacity of circular foundation on limited thickness of sand layer using ANN. The purpose of this study is to develop a neural network model from the results of laboratory model tests to estimate the reduction factor. Artificial neural network (ANN) is an artificial intelligence system inspired by the behavior of human brain and nervous system. In the present study a feed forward back propagation neural network model has been used to predict the reduction factor of eccentrically inclined loaded circular foundation. Backpropagation neural network is most suitable for prediction problems and Levenberg-Marquadrt algorithm is adopted as it is efficient in comparison to gradient descent backpropagation algorithm (Goh et al. 2005; Hornik et al. 1989). By drawing a neural interpretation diagram relationship between input and output are found out. A prediction model is developed based on the weights of the ANN model. The developed reduction factor is compared with the experimental reduction factor.

2 Analysis and Data

All the laboratory model tests were conducted using a poorly graded sand with effective grain size D_{10} = 0.325 mm, uniformity coefficient, C_u = 1.45, and coefficient of gradation, C_c = 1.15. Relative density D_r of sand is 69% and angle of internal friction, ϕ = 40.9°. Model foundations used for the tests had dimensions of 100 mm diameter. Mild steel plate 30-mm thick was used to make the model foundations. The bottom of the foundation was made rough by applying glue and rolling the steel plate over sand.

One hundred laboratory model tests were conducted. Three parameters *H*/*B*, α/ϕ and *e*/*B* are used as inputs in the ANN model, and the output is the reduction factor *RF* given by

$$RF = \frac{q_{u(H/B,\, \alpha/\phi,\, e/B)}}{q_{u(\alpha/\phi,\, e/B)}} \tag{1}$$

where $q_{u(H/B,\, \alpha/\phi,\, e/B)}$ is the ultimate eccentrically inclined load per unit area of a circular foundation on the surface of a sand layer of limited thickness (Table 1) and $q_{u(\alpha/\phi,\, e/B)}$ is the ultimate eccentrically inclined load per unit area on a sand layer extending to a great depth (Table 2).

Out of 100 tests, 75 tests are considered for training and the remaining 25 are considered for testing. All the inputs and output are normalized in the range of [−1, 1] before training. A feed-forward back-propagation neural network is used with hyperbolic tangent sigmoid function and linear function as the transfer function. The network is trained with Levenberg-Marquardt (LM) algorithm as it is efficient in comparison to gradient descent back-propagation algorithm. The ANN has been implemented using MATLAB V 7.11.0(R2015b).

Table 1. Database used for ANN model and compared with experimental results

Data type	Expt. No.	H/B	α/ϕ	e/B	$q_{u(H/B,\,\alpha/\phi,\,e/B)}$ kN/m^2	$RF_{expt.}$	RF_{ANN}	Deviation (%)
Training	1	0.3	0.000	0	880	7.59	7.91	−4.21
	2	0.3	0.000	0.05	810	7.79	7.84	−0.62
	3	0.3	0.000	0.1	690	7.84	7.77	0.94
	4	0.3	0.122	0	810	7.79	7.44	4.45
	5	0.3	0.122	0.05	745	7.76	7.37	4.97
	6	0.3	0.122	0.15	510	7.29	7.24	0.62
	7	0.3	0.244	0	680	7.08	6.99	1.30
	8	0.3	0.244	0.1	515	6.78	6.86	−1.24
	9	0.3	0.244	0.15	420	6.67	6.80	−1.93
	10	0.3	0.367	0.05	508	6.51	6.48	0.55
	11	0.3	0.367	0.1	427	6.47	6.41	0.95
	12	0.3	0.367	0.15	350	6.60	6.34	4.06
	13	0.3	0.489	0	420	5.92	5.96	−0.77
	14	0.3	0.489	0.05	372	5.90	5.86	0.82
	15	0.3	0.489	0.1	300	5.66	5.75	−1.65
	16	0.5	0.000	0	425	3.66	3.72	−1.52
	17	0.5	0.000	0.05	390	3.75	3.68	1.75
	18	0.5	0.000	0.15	270	3.51	3.62	−3.11
	19	0.5	0.122	0	385	3.70	3.49	5.69
	20	0.5	0.122	0.1	295	3.51	3.43	2.41
	21	0.5	0.122	0.15	245	3.50	3.40	2.96
	22	0.5	0.244	0.05	290	3.30	3.25	1.32
	23	0.5	0.244	0.1	240	3.16	3.22	−2.03
	24	0.5	0.244	0.15	200	3.17	3.19	−0.55
	25	0.5	0.367	0	247	2.98	3.05	−2.45
	26	0.5	0.367	0.05	220	2.82	3.00	−6.42
	27	0.5	0.367	0.1	190	2.88	2.95	−2.38
	28	0.5	0.489	0	186	2.62	2.59	1.00
	29	0.5	0.489	0.05	160	2.54	2.55	−0.28
	30	0.5	0.489	0.15	108	2.54	2.47	2.68
	31	1	0.000	0	194	1.67	1.57	5.84
	32	1	0.000	0.1	144	1.64	1.57	3.93
	33	1	0.000	0.15	110	1.43	1.57	−9.95
	34	1	0.122	0.05	156	1.63	1.56	3.94
	35	1	0.122	0.1	132	1.57	1.56	0.89
	36	1	0.122	0.15	103	1.47	1.55	−5.50
	37	1	0.244	0	140	1.46	1.47	−1.05
	38	1	0.244	0.05	130	1.48	1.44	2.63
	39	1	0.244	0.1	112	1.47	1.40	5.24

(continued)

Table 1. (*continued*)

Data type	Expt. No.	*H*/*B*	α/ϕ	*e*/*B*	$q_{u(H/B,\ \alpha/\phi,\ e/B)}$ kN/m^2	$RF_{\text{expt.}}$	RF_{ANN}	Deviation (%)
	40	1	0.367	0	112	1.35	1.19	11.95
	41	1	0.367	0.05	101	1.29	1.18	9.23
	42	1	0.367	0.15	70.5	1.33	1.16	12.75
	43	1	0.489	0	90	1.27	1.15	9.35
	44	1	0.489	0.1	62	1.17	1.15	1.95
	45	1	0.489	0.15	54	1.27	1.15	9.80
	46	2	0.000	0.05	120	1.15	1.23	−6.55
	47	2	0.000	0.1	104	1.18	1.19	−1.08
	48	2	0.000	0.15	92	1.19	1.17	2.29
	49	2	0.122	0	120	1.15	1.11	3.46
	50	2	0.122	0.05	110	1.15	1.11	3.01
	51	2	0.122	0.1	95	1.13	1.11	1.89
	52	2	0.244	0	112	1.17	1.11	5.13
	53	2	0.244	0.05	99	1.13	1.11	1.63
	54	2	0.244	0.15	74	1.17	1.11	5.79
	55	2	0.367	0	94	1.13	1.11	2.29
	56	2	0.367	0.1	72	1.09	1.11	−1.44
	57	2	0.367	0.15	62	1.17	1.11	5.40
	58	2	0.489	0.05	66	1.05	1.11	−5.63
	59	2	0.489	0.1	54	1.02	1.11	−8.61
	60	2	0.489	0.15	44	1.04	1.11	−6.89
	61	3	0.000	0	119	1.03	1.11	−7.87
	62	3	0.000	0.05	111	1.07	1.11	−3.68
	63	3	0.000	0.1	94	1.07	1.11	−3.60
	64	3	0.122	0	110	1.06	1.11	−4.62
	65	3	0.122	0.05	102	1.06	1.11	−4.15
	66	3	0.122	0.15	74	1.06	1.11	−4.68
	67	3	0.244	0	102	1.06	1.11	−4.15
	68	3	0.244	0.1	82	1.08	1.11	−2.56
	69	3	0.244	0.15	67	1.06	1.11	−4.05
	70	3	0.367	0.05	83	1.06	1.11	−3.99
	71	3	0.367	0.1	70	1.06	1.11	−4.34
	72	3	0.367	0.15	57	1.08	1.11	−2.89
	73	3	0.489	0	74	1.04	1.11	−6.17
	74	3	0.489	0.05	65	1.03	1.11	−7.26
	75	3	0.489	0.1	54	1.02	1.11	−8.61
Testing	76	0.3	0.000	0.15	565	7.34	7.70	−4.92
	77	0.3	0.122	0.1	645	7.68	7.31	4.83
	78	0.3	0.244	0.05	610	6.93	6.93	0.09

(*continued*)

Table 1. (*continued*)

Data type	Expt. No.	H/B	α/ϕ	e/B	$q_{u(H/B,\ \alpha/\phi,\ e/B)}$ kN/m^2	$RF_{\text{expt.}}$	RF_{ANN}	Deviation (%)
	79	0.3	0.367	0	550	6.63	6.54	1.24
	80	0.3	0.489	0.15	245	5.76	5.66	1.86
	81	0.5	0.000	0.1	330	3.75	3.65	2.67
	82	0.5	0.122	0.05	360	3.75	3.46	7.76
	83	0.5	0.244	0	323	3.36	3.28	2.46
	84	0.5	0.367	0.15	150	2.83	2.89	−1.96
	85	0.5	0.489	0.1	135	2.55	2.51	1.57
	86	1	0.000	0.05	170	1.63	1.57	3.74
	87	1	0.122	0	168	1.62	1.56	3.20
	88	1	0.244	0.15	83	1.32	1.35	−2.54
	89	1	0.367	0.1	82	1.24	1.17	6.10
	90	1	0.489	0.05	81	1.29	1.15	10.72
	91	2	0.000	0	128	1.10	1.27	−15.17
	92	2	0.122	0.15	84	1.20	1.11	7.62
	93	2	0.244	0.1	86	1.13	1.11	2.20
	94	2	0.367	0.05	83	1.06	1.11	−3.99
	95	2	0.489	0	75	1.06	1.11	−4.76
	96	3	0.000	0.15	83	1.08	1.11	−2.66
	97	3	0.122	0.1	89	1.06	1.11	−4.44
	98	3	0.244	0.05	95	1.08	1.11	−2.51
	99	3	0.367	0	88	1.06	1.11	−4.37
	100	3	0.489	0.15	43	1.01	1.11	−9.37

3 Results and Discussion

Three inputs and one output parameters were considered in the ANN model. The schematic diagram of the ANN architecture is shown in Fig. 1, which was computed from the database. The number of neurons in hidden layer is varied and the optimum number was taken based on mean square error (mse) value which was maintained at 0.001. In this ANN model there were two neurons evaluated in hidden layer as shown in Fig. 2. Therefore the final ANN architecture as 3-2-1[i.e. 3 (input) – 2 (hidden layer neuron) – 1 (output)].

Mean square error (MSE) is defined as

$$MSE = \frac{\sum_{i=1}^{n} \left(RF_i - RF_p\right)^2}{n} \tag{2}$$

Coefficient of efficiency, R^2 is defined as

Table 2. Database used for ANN model

α/ϕ	e/B	$q_{u\,(\alpha/\varphi,\,e/B)}$ (kN/m^2)
0	0	116
0	0.05	104
0	0.10	88
0	0.15	77
0.122	0	104
0.122	0.05	96
0.122	0.10	84
0.122	0.15	70
0.244	0	96
0.244	0.05	88
0.244	0.10	76
0.244	0.15	63
0.367	0	83
0.367	0.05	78
0.367	0.10	66
0.367	0.15	53
0.489	0	71
0.489	0.05	63
0.489	0.10	53
0.489	0.15	42.5

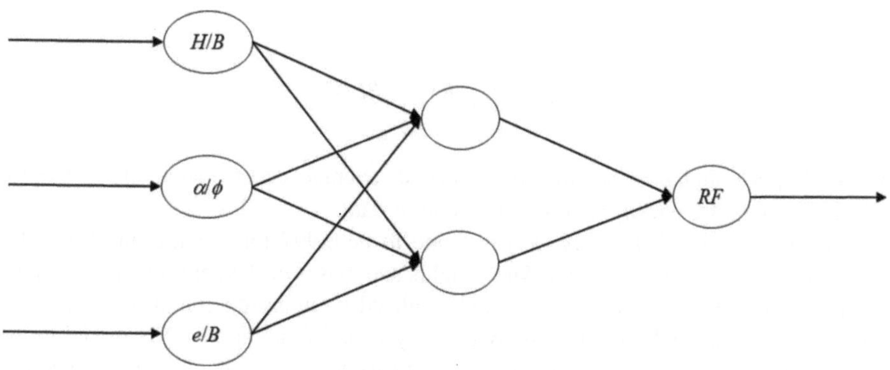

Fig. 1. ANN architecture

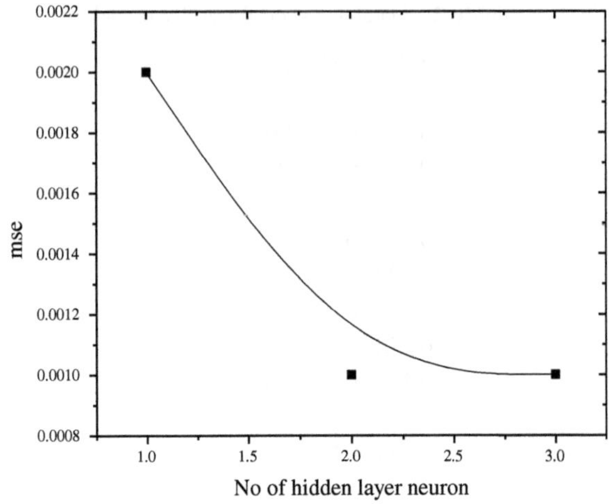

Fig. 2. Variation of hidden layer neuron with mean square error (mse)

$$R^2 = \frac{E_1 - E_2}{E_1} \quad (3)$$

where,

$$E_1 = \sum_{i=1}^{n} \left(RF_i - \overline{RF}\right)^2 \quad (4)$$

and

$$E_2 = \sum_{i=1}^{n} \left(RF_p - RF_i\right)^2 \quad (5)$$

where, $RF_i, \overline{RF}$ and RF_p are the experimental, average experimental, predicted RF values respectively; and n = number of training data.

The coefficient of efficiency (R^2) is found to be 0.997 for training and 0.996 for testing as shown in Figs. 3, and 4. The weights and biases of the network are presented in Table 3. These weights and biases can be utilized for interpretation of relationship in between the inputs and output, sensitivity analysis and framing an ANN model in the form of an equation. The residual analysis was carried out by calculating the residuals in between experimental reduction factor and predicted reduction factor for training data. Residuals can be defined as the difference between the experimental and predicted RF value and is given by

$$e_r = RF_i - RF_p \quad (6)$$

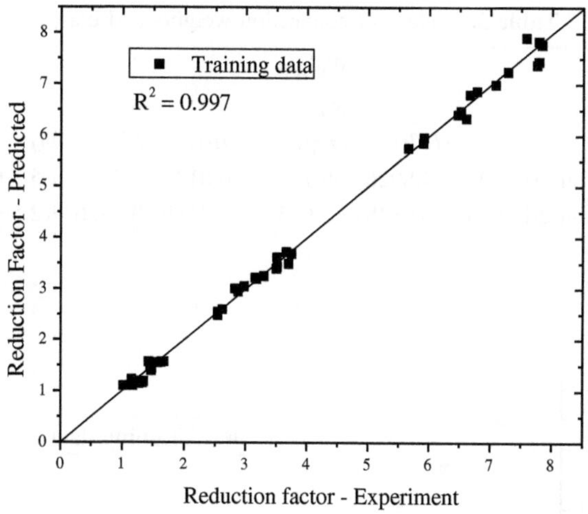

Fig. 3. Correlation between predicted reduction factors with experimental reduction factor for training data

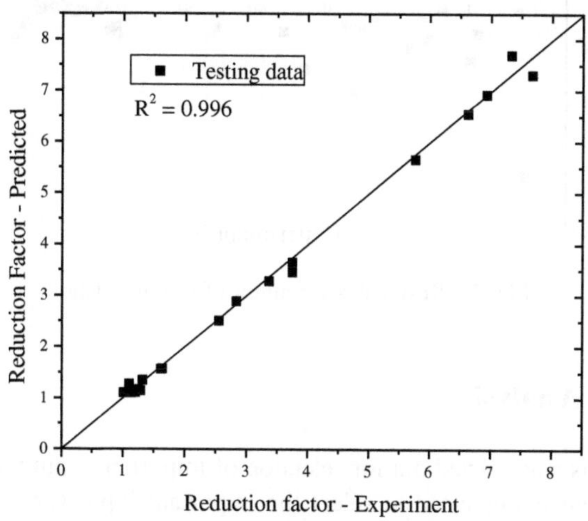

Fig. 4. Correlation between predicted reduction factors with experimental reduction factor for testing data

The residuals are plotted with the experiment number as shown in Fig. 5. It is observed that the residuals are evenly distributed along the horizontal axis of the plot. Therefore it can be said that the network is well trained and can be used for prediction with reasonable accuracy.

Table 3. Values of connection weights and biases

Neuron	Weight					
		w_{ik}		w_k		Bias
	$(H/B)_n$	$(\alpha/\phi)_n$	$(e/B)_n$	RF	b_{hk}	b_0
Hidden neuron 1(k = 1)	−4.8926	−0.1266	−0.014	2.233	−5.1941	1.317
Hidden neuron 2(k = 2)	−6.0138	−3.6338	−0.3389	0.0586	−2.5935	

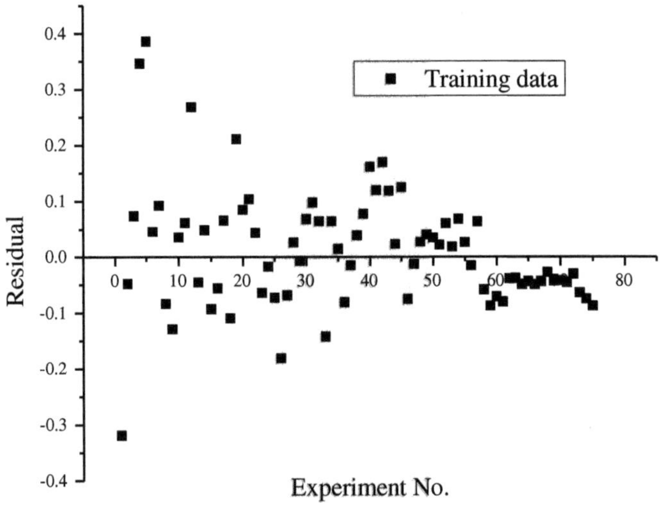

Fig. 5. Residual distribution of training data

4 Sensitivity Analysis

Sensitivity analysis was carried out for selection of important input variables. Different approaches have been suggested to select the important input variables. The Pearson correlation coefficient is one of them in selecting proper inputs for the ANN model. It was approached by Guyon and Elisseeff (2003) and Wilby et al. (2003). Goh (1994), Shahin et al. (2002), Behera et al. (2013), Sahu et al. (2017, 2018), Sethy et al. (2017) have used Garson's algorithm (Garson 1991) in which the input-hidden and hidden-output weights of trained ANN model are partitioned. In Garson's algorithm the absolute values of weights are taken to select the important input variables. It does not provide information on the effect of input variables in terms of direct or inverse relation to the output. Olden et al. (2004) proposed a connection weights approach based on the neural interpretation diagram (NID), in which the actual values of input-hidden and hidden-output weights are taken. Table 4 shows the cross-correlation of the three input parameters with the reduction factor (*RF*) value. From the Table 4 it can be seen that *RF* is highly correlated to *H*/*B* with a values of −0.71 followed by α/ϕ = −0.11 and

$e/B = -0.01$. The relative importance, *quantified through the parameter* S_i of three input parameters as per Garson's algorithm is presented in Table 5. The H/B is found to be the most important input parameters with relative importance value being 78.71% followed by 19.45% for α/ϕ and 1.84% for e/B. As per the connection weight approach (Olden et al. 2004) the relative importance of the present input variables is also presented in Table 5. H/B is also the most important input parameter ($S_i = -11.28$) followed by α/ϕ ($S_i = -0.5$) and e/B ($S_i = -0.05$). The S_i values being negative imply that H/B, α/ϕ, and e/B are indirectly related to RF. In other words increase in H/B, α/ϕ, and e/B leads to decrease in RF and leads to decrease in ultimate bearing capacity.

Table 4. Cross-correlation matrix of input and output for reduction factor

Cross-correlation Matrix				
	(H/B)	(α/ϕ)	(e/B)	RF_{expt}
(H/B)	1	0	0	−0.71
(α/ϕ)		1	0	−0.11
(e/B)			1	−0.01
RF_{expt}				1

Table 5. Relative importance of different inputs as per Garson's algorithm and connection weight approach

Parameters	Garson's algorithm		Connection weight approach	
	Relative importance	Ranking of input as per relative importance	S_i values as per connection weight approach	Ranking of input as per relative importance
H/B	78.71	1	−11.28	1
α/ϕ	19.45	2	−0.5	2
e/B	1.84	3	−0.05	3

5 Neural Interpretation Diagram (NID)

Ozesmi and Ozesmi (1999) proposed neural interpretation diagram for visual interpretation of the connection weight among the neurons. For the present study with the weights as obtained and shown in Table 1, an NID is presented in Fig. 6. The lines joining the input-hidden and hidden output neurons represent the weights. The positive weights are represented by solid lines and negative weights by dashed lines and the thickness of the line is proportional to its magnitude.

It is seen from Table 5 that S_i values for parameters H/B, α/ϕ, and e/B are negative indicating that the parameters are indirectly related to RF values, whereas S_i values for. This is shown in Fig. 6. Therefore, the developed ANN model is not a black box and could explain the physical effect of input parameters on the output.

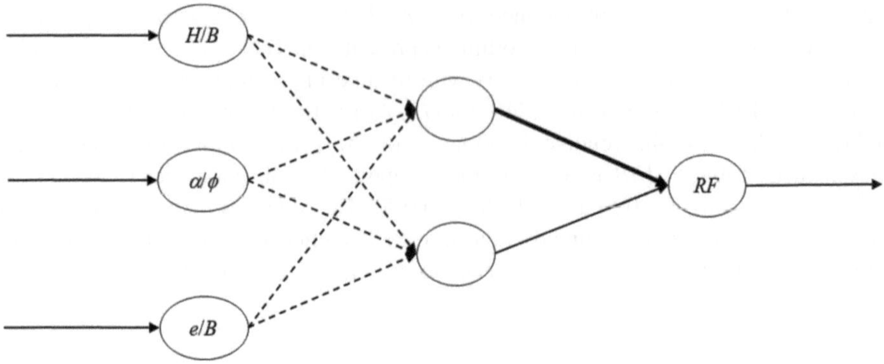

Fig. 6. Neural interpretation diagram (NID) showing lines representing connection weights and effects of inputs on reduction factor (RF)

6 ANN Model Equation for Reduction Factor Based on Trained Neural Network

A model equation is developed using the weights obtained from trained neural network model (Goh et al. 2005). The mathematical equation relating input parameters (*H*/*B*, *α*/*ϕ*, and *e*/*B*) to output given by

$$RF_n = f_n\left\{ b_0 + \sum_{k=1}^{h}\left[w_k f_n\left(b_{hk} + \sum_{i=1}^{m} w_{ik} X_i \right)\right]\right\} \tag{7}$$

where RF_n is the normalized value of *RF* in the range [−1, 1], f_n is the transfer function, *h* is the number of neurons in the hidden layer, X_i is the normalized value of inputs in the range [−1, 1], *m* is the number of input variables, w_{ik} is the connection weight between the ith layer of input and k^{th} neuron of hidden layer, w_k is the connection weight between the kth neuron of hidden layer and single output neuron, b_{hk} is the bias at the kth neuron of hidden layer and b_0 is the bias at the output layer.

The model equation of *RF* of shallow circular foundations on sand layer of limited thickness subjected to eccentrically inclined load was formulated using the values of the weights and biases shown in Table 3 as per the following steps.

Step 1

The input parameters were normalized in the range [−1, 1] by the following expressions

$$X_n = 2\left(\frac{X_n - X_{\min}}{X_{\max} - X_{\min}}\right) \tag{8}$$

Step 2
Calculate the normalized value of reduction factor (RF_n) using the following expressions

$$A_1 = -4.89\left(\frac{H}{B}\right)_n -0.13\left(\frac{\alpha}{\phi}\right)_n -0.01\left(\frac{D_f}{B}\right)_n -5.19 \tag{9}$$

$$A_2 = -6.01\left(\frac{B}{L}\right)_n -3.63\left(\frac{e}{B}\right)_n -0.34\left(\frac{D_f}{B}\right)_n -2.59 \tag{10}$$

$$B_1 = 2.23\left(\frac{e^{A_1} - e^{-A_1}}{e^{A_1} + e^{-A_1}}\right) \tag{11}$$

$$B_2 = 0.06\left(\frac{e^{A_1} - e^{-A_1}}{e^{A_1} + e^{-A_1}}\right) \tag{12}$$

$$C_1 = B_1 + B_2 + 1.32 \tag{13}$$

$$RF_n = C_1 \tag{14}$$

Denormalize the RF_n value obtained from Eq. 14 to actual RF as

$$RF = 0.5(RF_n + 1)(RF_{\max} - RF_{\min}) + RF_{\min} \tag{15}$$

$$RF = 0.5(RF_n + 1)(7.84 - 1.01) + 1.01 \tag{16}$$

Figure 7 Shows the comparison of reduction factor (RF) obtained from Eqs. 16 and 1. It can be seen that the results predicted by using artificial neural network (ANN) are closer to the results obtained from model tests. The deviation between the RF obtained from model tests and those predicted RF is within $\pm 10\%$ except two values as shown in Table 1. The proposed ANN model can be used as an effective tool in predicting the reduction factor (RF) and hence, the ultimate bearing capacity shallow circular foundation on sand layer of limited thickness underlain by rigid rough base subjected to an eccentrically inclined load.

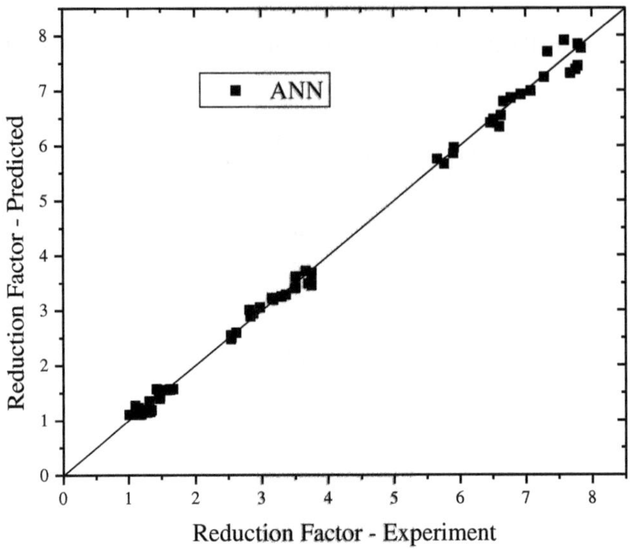

Fig. 7. Comparison of ANN results with experimental *RF*

7 Conclusion

Based on developed neural network model, the following conclusions may be drawn.

1. The errors are distributed evenly along the centerline as per residual analysis. It can be concluded that the network was well trained and can predict the reduction factor (*RF*).
2. Based on Pearson correlation coefficient and Garson's algorithm, it was observed that *H*/*B* is the most important input parameter followed by α/ϕ and *e*/*B*.
3. The developed ANN model could explain the physical effect of inputs on the output, as described in NID. It has been observed that *H*/*B*, α/ϕ, and *e*/*B* is inversely related to *RF*.
4. A model equation is developed based on the trained weights of ANN to calculate the ultimate bearing capacity of circular foundation on sand layer of limited thickness.
5. The deviation between the *RF* obtained from model tests and those predicted from ANN model is within $\pm10\%$

References

Behera, R.N., et al.: Prediction of ultimate bearing capacity of eccentrically inclined loaded strip footing by ANN part 1. Int. J. Geotech. Eng. (2013). https://doi.org/10.1179/1938636212z.00000000012

Garson, G.D.: Interpreting neural-network connection weights. Artif. Intell. Expert **6**(7), 47–51 (1991)

Guyon, I., Elisseeff, A.: An introduction to variable and feature selection. J. Mach. Learn. Res. **3**, 1157–1182 (2003)

Goh, A.T.C., et al.: Bayesian neural network analysis of undrained side resistance of drilled shafts. J. Geotech. Geoenviron. Eng. (2005). https://doi.org/10.1061/(asce)10900241(2005)131:1(84)

Hornik, K., et al.: Multilayer feed forward networks are universal approximators. Neural Netw. **2**, 359–366 (1989)

Olden, J.D., et al.: An accurate comparison of methods for quantifying variable importance in artificial neural networks using simulated data. Ecol Model. (2004). https://doi.org/10.1016/j.ecolmodel.2004.03.0.013

Ozesmi, S.L., Ozesmi, U. An artificial neural network approach to spatial modeling with inter specific interactions. Ecol. Model. (1999). https://doi.org/10.1016/S0304-3800(98)00149-5

Sahu, R., et al.: Use of ANN and Neuro fuzzy model to predict bearing capacity factor of strip footing resting on reinforced sand and subjected to inclined loading. Int. J. Geosynthetics Ground Eng. **3**(3), 29 (2017)

Sahu, R., et al.: Bearing capacity prediction of inclined loaded strip footing on reinforced sand by ANN. In: International Congress and Exhibition "Sustainable Civil Infrastructures: Innovative Infrastructure Geotechnology", pp. 97–109. Springer, Cham (2018)

Sethy, B.P., et al.: Application of ANN and ANFIS for predicting the ultimate bearing capacity of eccentrically loaded rectangular foundations. Int. J. Geosynthetics Ground Eng. **3**(4), 35 (2017)

Shahin, M.A., et al.: Predicting settlement of shallow foundations using neural network. J. Geotech. Geoenviron. Eng. (2002). https://doi.org/10.1061/(asce)1090-0241(2002)128:9(785)

Thoughts on a Simple Means of Estimating Settlement in Thick Soil Layers in Accordance with Hypothesis B

Casey J. Shepheard[1(✉)] and Michael G. Williamson[2]

[1] EPSRC FIBE CDT, University of Cambridge, Cambridge, UK
cjs225@cam.ac.uk
[2] Mott MacDonald, Cambridge, UK

Abstract. Settlement of structures founded on soft soil is a primary concern for geotechnical engineers. Buildings will frequently fail due to reaching the serviceability state due to cracking and or tilting from uneven settlement well before reaching the ultimate limit state. There is ongoing discussion in the geotechnical engineering community on the mechanics of settlement, specifically the point at which secondary consolidation (creep) begins: ought Hypothesis A or Hypothesis B be adopted? This paper addresses the challenges in estimating Hypothesis B, where the strain values in both thick and thin soil samples at the end of primary consolidation are believed to be a function of both the change in void ratio, which is understood well, and creep, which is not. A new empirical-based model, the EOP-Anchor model, is proposed to estimate Hypothesis B strain as a function of settlement time and soil thickness in conjunction with standard laboratory oedometer test data on thin samples.

1 Introduction

The settlement of soft soils is a subject of significant interest to geotechnical engineers, due to the effects ground movements can have on structures (e.g., McMahon and Bolton 2014). Changes in the ground profile may have large effects on a structure founded upon it. These ground movements frequently cause structures to reach their serviceability limit state (e.g., due to cracking, uneven floors etc.) well before the ultimate limit state is thought to have been reached by the soil (Craig 2004, p. 185). Therefore, engineers should have a clear picture of settlement magnitude and rate. The aim of this study is to review the process for estimation of settlements and investigate whether this can be simplified for early analysis of ground movements.

Forces from loading a soil are carried by the soil grains and fluid in the pores (usually assumed to be water). A load to a soil is carried immediately by the water in the pores creating an increase in pore water pressure and thus a state that is not in equilibrium (Terzaghi 1943, article 99), before transferring to the soil grains gradually. the time taken for this excess pore fluid pressure is a function of a number of different variables including the soil permeability and composition.

Settlement is frequently divided into three distinct phases: immediate settlement, primary consolidation settlement and secondary consolidation settlement (Almeida and

H. Shehata and B. Das (Eds.): GeoMEast 2018, SUCI, pp. 54–63, 2019.
https://doi.org/10.1007/978-3-030-01923-5_5

Marques 2013, p. 63). Immediate settlements in soft soils are generally comparatively small and occur in the undrained state; primary consolidation settlements occur while excess pore water pressures created by loading dissipate; and secondary consolidation settlements are considered to be due to creep effects (McMahon and Bolton 2014). Creep is strain at constant stress. As immediate settlement is not due to consolidation, this paper focuses on primary and secondary consolidation processes and settlements.

Primary consolidation is the process by which the soil re-establishes the equilibrium state by expelling water from the pores until the excess force is dissipated. Depending on the permeability and thickness of the soil layer in question, this can take anywhere between a few minutes in the lab to months or years in the field (Degago et al. 2011). The consolidation process results in the reduction in volume of the soil mass (Craig 2004, p. 227). As a soil element in the field is generally considered to be laterally confined, this volume change is generally considered only in terms of vertical consolidation settlements.

Creep in soils is a mechanical process and occurs through rearrangement of the soil particles themselves and can occur under constant effective stress over a long period, contributing significantly to settlement (e.g., Alonso et al. 2000). Additionally part of the creep process is likely also attributed to the decay and volume change of individual of soil particles, particularly in organic soils. Traditional Terzaghi theory does not take into account the contribution of creep processes in consolidation, leading to this being classed as secondary consolidation, a separate phenomenon to primary consolidation (Craig 2004, p. 257).

2 Hypotheses A and B

While it is widely accepted that consolidation includes both components of primary and secondary consolidation, there exists in the geotechnical community continuing debate on the contribution of creep processes. Effectively, the argument can be summarised by the question of whether secondary consolidation begins after or during primary consolidation, with each school of thought respectively titled Hypothesis A or Hypothesis B. Both theories have been apparently supported by large quantities of laboratory data (Degago et al. 2011), and numeric models using both hypotheses have been used in research (Svanø et al. 1991; Kutter and Sathialingam 1992; Nash and Ryde 2001).

Standard Terzaghi consolidation theory, supported by researchers such as Mesri and Choi (1985) and Wahls (1962), suggests that the primary and secondary phases of consolidation occur in succession, and therefore the creep processes do not begin until all excess pore water pressures have dissipated to infinitesimal levels. Hence, the strain (change in void ratio) at the end of primary consolidation (EOP) (Fig. 1) is argued to be unique for a particular soil (e.g., Degago et al. 2009, 2011; Watabe and Leroueil 2015). This theory is dubbed Hypothesis A and suggests that the results from laboratory tests on a thin sample (industry standard is 20 mm thick) may be translated directly to the field scale (Degago et al. 2009) (Fig. 2).

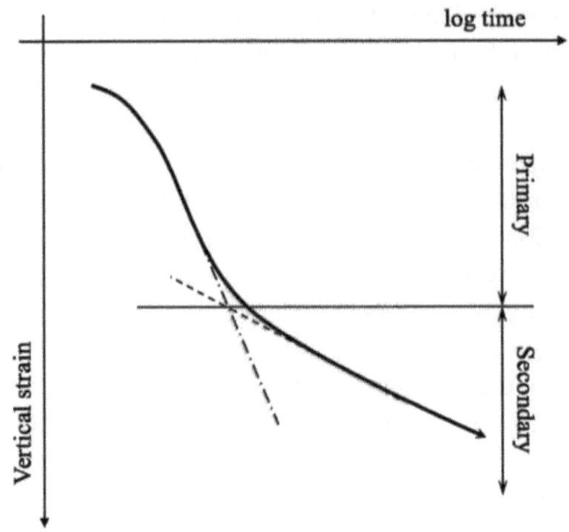

Fig. 1. Typical traditional consolidation curve for a soil exhibiting creep behaviour, with Casagrande construction for primary to secondary consolidation transition

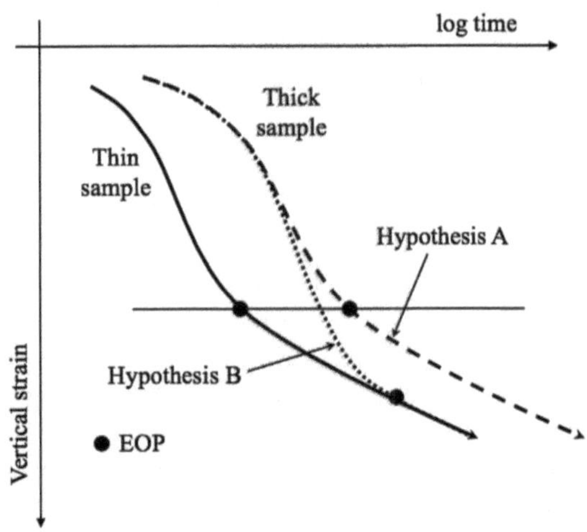

Fig. 2. Comparison of consolidation curves according to creep hypotheses A and B (after Ladd et al. 1977)

However, the counter-theory Hypothesis B has been proposed by Leroueil, Degago and many others (Berre and Iversen 1972; Aboshi 1973; Kabbaj et al. 1988; Degago et al. 2009, 2011). This argues that creep processes do take place while pore pressures are dissipating. Taking this approach, the thickness of the soil layer being consolidated will have an effect on the EOP (defined in this instance as the point at which excess pore pressure is equal to zero) strain: a thicker soil layer will require a longer period to complete primary consolidation, hence increased creep effects, and a larger EOP strain (Degago et al. 2009) (Fig. 2). The Hypothesis B approach suggests that scaling up lab results to field-scale soil layers, which can be multiple orders of magnitude thicker, will likely give settlement predictions too low. In particular, for soils in which creep plays a significant role in consolidation (i.e. organic soils, soft clays and loose gravels) this assumption of direct comparison can be incorrect and may lead to expensive, or even dangerous, design errors. Finite element comparisons of Hypotheses A and B settlement on a soil layer 4 m thick with a vertical drain by Feng and Yin (2018) showed that the Hypothesis B settlement can be around 30% higher.

In addition to simple comparison of observed settlements to Hypothesis A, work by Degago et al. (2011) has re-evaluated results of some lab tests that did appear to support Hypothesis A and concluded that it is often actually possible to apply Hypothesis B theory to this data. It has also been argued that Hypothesis A violates some of the rules of continuum mechanics (Degago et al. 2013). Therefore, if it is assumed that Hypothesis B is the more correct theory, the question of why Hypothesis A is still used in the majority of settlement analysis may be posed. The answer is simple, current methods of analysing Hypothesis B are much more involved and complex than for Hypothesis A. In particular, the commonly employed isotache and soft soil creep (SSC) formulations, while proved to be good models of consolidation settlement, are highly complex and require numerous input parameters.

3 Standard Hypothesis B Model: The Isotache Formulation

The isotache (Šuklje 1957) and SSC (Stolle et al. 1999) formulations consider strain rate as a function of void ratio and effective stress at a specific time (Degago et al. 2009). Yuan and Whittle (2013) evaluated the isotache and SSC formulations and proved them to be essentially equivalent. Hence further discussion of this method of analysis will focus on the isotache formulation.

The isotache concept is best illustrated with the aid of Fig. 3. The parallel dashed lines are individual isotaches, each representing a particular strain rate (Degago et al. 2011). 'Isotache' effectively translates to 'equal rate' (Watabe and Leroueil 2015). The gradient of these lines is equal to the primary consolidation index, C_c. With each increase in effective stress, a soil sample will follow a consolidation path reaching EOP (full black dot mark) on the same isotache as it began, but at the new effective stress. Thicker samples will exhibit slower strain rates than thin samples.

The isotache concept is of use to engineers as it yields both the magnitude and rate of strain due to a loading increment, accounting for both primary and secondary consolidation. However these models must be calibrated and numerical analysis (usually FE) necessary to model the response which requires numerous to model the response of thick soil samples/field scale to loading (e.g., Degago et al. 2009, 2011; Yuan and Whittle 2013). Hence, there is arguably an appetite in the geotechnical community for a significantly simpler means of estimating creep settlement in thick layers.

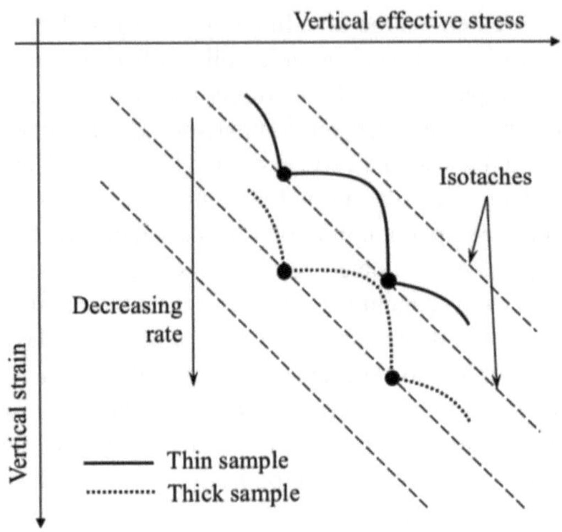

Fig. 3. Stress-strain effects on samples of different thickness according to the isotache concept

4 A New Hypothesis B Model: The EOP-Anchor Estimation

It can be seen in the sketch Fig. 2 that the Hypothesis B consolidation curves for the thick and thin samples eventually meet and extend as one secondary consolidation line. It is at this intersection of the curves that Hypothesis B defines the point at which primary consolidation ends for the thick sample (Šuklje 1957). After this point, where creep is the only settlement mechanism, it is believed that samples of the same material will continue to settle identically, irrespective of the thickness. Simply, after this EOP consolidation point for the thick sample any more settlement will occur at the same rate and the strain will be the same magnitude as in a thin sample. Additionally, the gradient of the secondary consolidation line is believed to be effectively constant with a gradient of C_α.

This concept has some interesting implications for the estimation of settlement magnitude in soils and could be a useful tool for engineers when conducting first pass analyses of serviceability. As the secondary consolidation component of the consolidation curve for a thin sample also applies to a thick sample, any strains after the EOP point (i.e., strains induced by creep settlement only) for a thick sample may be simply found through the results of classic thin sample oedometer tests. Of course, the assumption of a constant C_α gradient implies infinite settlement. As design life of most structures is limited to less than 150 years, this limitation is negligible.

Thus, it follows that if either the time or strain at the EOP point for a thick soil sample or layer is known, through use of a sister thin sample consolidation curve from regular tests, the respective strain or time may be found. Of course, at present knowing one coordinate generally implies knowledge of both.

This study proposes the use of an empirical relationship between soil layer thickness and the time for primary consolidation to complete. Through use of this proposed relationship in conjunction with a standard thin sample oedometer test curve to find the strain at the end of primary consolidation. This would give geotechnical engineers a simple means of estimating the expected settlement in a given layer through Hypothesis B theory.

Further value could also be yielded from this method, as a simple primary-creep consolidation line could be drawn between the end of immediate settlement strain point (which may be calculated or found experimentally) to the EOP point to illustrate (generally conservatively) the settlement while primary consolidation is still taking place. A simple straight-line consolidation curve can be constructed. Figure 4 illustrates the graphical construction of this proposed new Hypothesis B model, dubbed the EOP-Anchor model for its use of the EOP point of a thick soil layer as the point about which the settlement properties are derived.

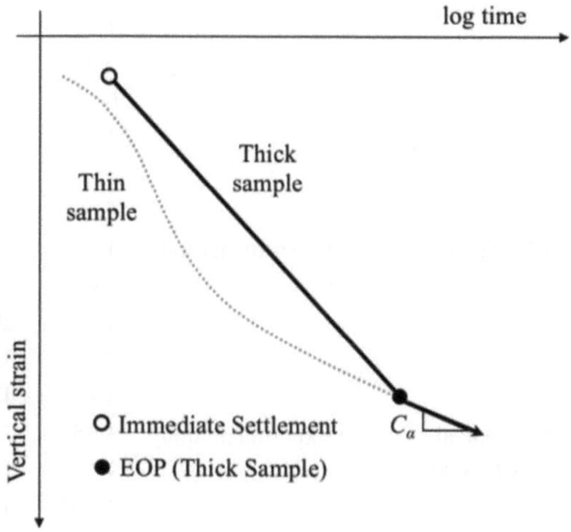

Fig. 4. Construction of the proposed EOP-Anchor model

4.1 Regression Analysis for Empirical Thickness-Time Model

The proposed EOP-Anchor model is founded upon an empirical relationship between soil layer thickness and the time it takes to reach EOP consolidation. For the calculation of this relationship, a preliminary database of 40 experimental settlement data points has been collected from the literature (Table 1). The t_{EOP} values have been found from the figures in the references. The database is used to perform regression analysis to investigate the possibility of a reliable relationship being found.

It should be noted that a study by Aboshi (1973) on samples of different thicknesses concluded that the Šuklje (1957) isotache theory is not strictly correct, and that while EOP strain did increase with sample thickness, it did not plot onto the creep component of the consolidation curve. Aboshi (1973) concluded that this may be due to effects of stress history or duration of primary consolidation on the clay structure. However, as the analysis in this study is intended only to be used in simple first-pass analysis, and the observed predicted-measured difference was less than 3%, this disparity is beyond the scope of this paper.

Table 1. Sample thickness vs time to EOP database

Source	Soil type	Data type	h_i (cm)	t_{EOP} (min)	ε_{EOP}
Watabe et al. 2012	Osaka bay clay, marine	Experimentation	2	3.987	0.13
			2	4.221	0.81
			2	1.75	12.73
			2	5.206	7.77
			2	33.72	7.04
			2	49.2	10.94
Li et al. 2004	Alluvial clay	Experimentation	1	35.583	8.19
			3.5	445.38	8.06
			6	2695	7.98
			1	16.111	5.66
			3.5	211.29	5.51
			6	1337.4	5.62
			1	1422.3	6.39
			3	1136.2	6.39
			5	1177.2	6.05
Aboshi 1973	Marine clay	Experimentation	1	21.45	7.03
			2.4	127.76	7.44
			10	2093.4	7.75
			20	8242.5	7.90
			50	48984	8.63
Berre and Iversen 1972	Postglacial marine clay	Experimentation	1.75	71.34	1.16
			7.5	1429.1	2.68
			15.01	4932.7	2.65
			45	37550	4.32
			1.75	599.5	5.37
			7.5	1629.3	5.37
			15.01	4613.5	5.86
			45	39750	6.21
Imai and Tang 1992	Yokohama bay mud	Experimentation	0.5	29.64	
			1	87.76	
			2	195.12	
			3.5	643.3	
Konovalov and Bezvolev 2005	Postglacial marine lithorine	Experimentation	1	34.437	6.23
			2.5	307.49	7.16
			5	1423.8	7.78
			10	4674	8.22
Kabbaj et al. 1988	Grey silty clay	Field	180	636480	0.07
			270	466560	0.03
			320	720000	0.11
Bjerrum 1967	Drammen clay - layered	Field	1100	8646645.6	5.73

Regression analysis is used in this study to establish and assess the correlation between EOP strain and soil layer thickness. Limited information is available to quantify secondary consolidation; as such, regression analysis can be a powerful tool to inform predictions. Some care must be taken, however, in using such results, as scatter may be disguised and smoothed within a broad data range. Where data is available, it may be of use to perform a multiple linear regression analysis to further refine the results.

Four formulations of simple regressions have been calculated to investigate the proposed relationship: linear, logarithmic, exponential, and power. Where data points are taken from soils at depth, the thickness has been taken as the distance to surface.

The linear fit was found to have the strongest relationship, but for layers under 170 mm gave negative time values. Hence the best fit power equation (from which only positive time values are yielded) is:

$$t_{EOP} = 0.4536h_i^{1.8259}\ R^2 = 0.84,\ n = 40,\ p < 0.001 \quad (1)$$

where t_{EOP} is the time at EOP in minutes and h_i is the initial specimen depth in mm. The coefficient of determination, R^2, value of 0.84 implies that 84% of the variability of the data can be explained by this relationship (Montgomery et al. 2007, p. 294). While the high R2 value is encouraging some critical thinking should be applied; the data is over many log-cycles, with both axes in logarithmic scale, which generally leads to high coefficients of determination where a strong trend exists, regardless of the proposed fit. Of course, it is well known already that thicker layers do take longer to finish primary consolidation. The predictive power of Eq. 1 is illustrated in Fig. 5.

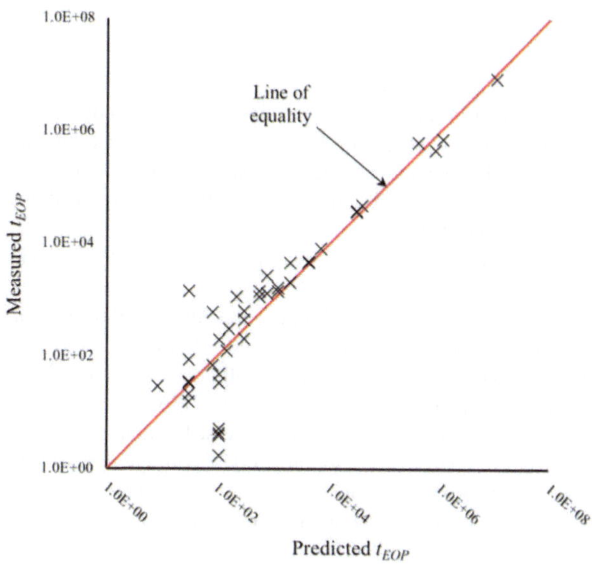

Fig. 5. Predictive power of Equation 1

Some error is incorporated by the data from very thick soil layers as few data points could be found at this higher end. When only the samples 20 cm or less are studied, the coefficient of determination value increases to 0.89. This is not a large change in the coefficient of determination and thus the EOP-Anchor theory is still supported.

Of course, further validation with data points from the field is necessary. Further, only clays have been included in this study, as these were available in the literature; yet granular material, such as that used in land reclamation, can also exhibit creep. Some further refinement of the relationship will likely be necessary once more data for deeper soil layers and other materials is included in the database. Multiple linear regression analysis could be performed to include in particular the effects of material properties such as Atterberg limits, grain size and C_{α}.

5 Conclusions

A summary of the theory behind and ongoing debate on creep settlement theories Hypotheses A and B has been presented in this paper. Despite clearly violating the principles of continuum mechanics, Hypothesis A is still most commonly used in design due to the comparative simplicity.

A simple method of estimating Hypothesis B settlement for thick soil layers has been proposed, the EOP-Anchor model. This new model should be competitive with the simplicity of Hypothesis A, as it requires only application of a thickness-time empirical relationship in conjunction with a simple standard oedometer test (which would also be performed for Hypothesis A analysis). Further expansion of the collected settlement database for the calculation of the empirical relationship will allow for further refinement of the EOP-Anchor model for use in more accurate design.

References

Aboshi, H.: An experimental investigation of the similitude in the consolidation of a soft clay, including the secondary creep settlement. In: Proceedings of the 8th International Conference on Soil Mechanics and Foundation Engineering, Moscow, vol. 4, no. 3, p. 88 (1973)

Almeida, M.S.S., Marques, M.E.S.: Design and Performance of Embankments on Very Soft Soils. CRC Press, London (2013)

Alonso, E.E., Gens, A., Lloret, A.: Precompression design for secondary settlement reduction. Géotechnique **50**(6), 645–656 (2000)

Berre, T., Iversen, K.: Oedometer tests with different specimen heights on a clay exhibiting large secondary compression. Géotechnique **22**(1), 53–70 (1972)

Bjerrum, L.: Engineering geology of Norwegian normally-consolidated marine clays as related to settlements of buildings. Géotechnique **17**(2), 81–118 (1967)

Craig, R.F.: Craig's Soil Mechanics, 7th edn. Spon Press, London (2004)

Degago, S.A., Grimstad, G., Jostad, H.P., Nordal, S.: The non-uniqueness of the end-of-primary (EOP) void ratio-effective stress relationship. In: Proceedings of the 17th International Conference on Soil Mechanics and Foundation Engineering, Alexandria 1, pp. 324–327 (2009)

Degago, S.A., Grimstad, G., Jostad, H.P., Nordal, S., Olsson, M.: Use and misuse of the isotache concept with respect to creep hypotheses A and B. Géotechnique **61**(10), 897–908 (2011)

Degago, S.A., Grimstad, G., Jostad, H.P., Nordal, S.: Misconceptions about experimental substantiation of creep hypothesis A. In: Proceedings of the 18th International Conference on Soil Mechanics and Geotechnical Engineering, Paris, pp. 215–218 (2013)

Feng, W., Yin, J.-H.: A new simplified Hypothesis B method for calculating the consolidation settlement of ground improved by vertical drains. Int. J. Numer. Anal. Meth. Geomech. **42**(2), 295–311 (2018)

Imai, G., Tang, Y.-X.: A constitutive equation of one-dimensional consolidation derived from inter-connected tests. Soils Found. **32**(2), 83–96 (1992)

Kabbaj, M., Tavenas, F., Leroueil, S.: In situ and laboratory stress-strain relationships. Géotechnique **38**(1), 83–100 (1988)

Konovalov, P.A., Bezvolev, S.G.: Analysis of results of consolidation tests of saturated clayey soils. Soil Mech. Found. Eng. **42**(3), 81–85 (2005)

Kutter, B.L., Sathialingam, N.: Elastic-viscoplastic modelling of the rate-dependent behaviour of clays. Géotechnique **42**(3), 427–441 (1992)

Ladd, C.C., Foott, R., Ishihara, K., Schlosser, F., Poulous, H.G.: Stress deformation and strength characteristics. State-of-the-art report. In: Proceedings of the 9th International Conference on Soil Mechanics and Foundation Engineering, Tokyo 2, pp. 42–429 (1977)

Li, S., Shirako, H., Sugiyama, M., Akaishi, M.: Time effects on one-dimensional consolidation analysis. In: Proceedings of the School of Engineering of Tokai University, Series E, vol. 29, pp. 1–8 (2004)

McMahon, B.T., Bolton, M.D.: Centrifuge experiments on the settlement of circular foundations on clay. Can. Geotech. J. **51**(6), 610–620 (2014)

Mesri, G., Choi, Y.K.: The uniqueness of the end-of-primary (EOP) void ratio-effective stress relationship. In: Proceedings of the 11th International Conference on Soil Mechanics and Foundation Engineering, San Fransisco, vol. 2, pp. 587–590 (1985)

Montgomery, D.C., Runger, G.C., Hubele, N.F.: Engineering Statistics, 4th edn. Wiley, Hoboken (2007)

Nash, D.F.T., Ryde, S.J.: Modelling consolidation accelerated by vertical drains in soils subject to creep. Géotechnique **51**(3), 257–273 (2001)

Šuklje, L.: The analysis of the consolidation process by the isotaches method. In: Proceedings of the 4th International Conference on Soil Mechanics and Foundation Engineering, London, vol. 1, pp. 200–206 (1957)

Svanø, G., Christiansen, S., Nordal, S.: A soil model for consolidation and creep. In: Proceedings of the 10th European Conference on Soil Mechanics and Foundation Engineering, Florence, vol. 1, pp. 269–272 (1991)

Terzaghi, K.: Theoretical Soil Mechanics. Wiley, New York (1943)

Wahls, H.E.: Analysis of primary and secondary consolidation. J. Soil Mech. Found. Div. ASCE **88**(SM6, Part 1), 207–234 (1962)

Watabe, Y., Leroueil, S.: Modelling and implementation of the isotache concept for long-term consolidation behaviour. Int. J. Geomech. **15**(5), A4014006 (2015)

Yuan, Y., Whittle, A.J.: Examination on time-dependent soil models in one-dimensional consolidation. Springer Ser. Geomech. Geoengin. 159–166 (2013)

Upgrading of the Technologies of Soil Preparation and Construction of Foundations for Structures of Oil and Gas Industry

Talal Awwad[1,2(✉)], Andrei Gruzin[3], and Vladimir Gruzin[4]

[1] L.N.Gumilyov Eurasian National University, Astana, Kazakhstan
dr.awwad.gfce@gmail.com
[2] Department of Geotechnical Engineering, Damascus University, Damascus, Syria
[3] Omsk State Technical University, Omsk, Russian Federation
[4] S.Seifullin Kazakh Agro-Technical University, Astana, Kazakhstan

Abstract. The progressive technologies, which are aimed at reinforcement of soils, play a special role in complex conditions of construction of foundations for laying aboveground main pipelines. Initially, it is explained by ever-changing loads in the technological chain of the pipeline transport of hydrocarbons as well as the variety of soils in green deposits. First and foremost, these requirements include the limitations associated with values of admissible absolute and relative settlements of foundations of the technological chain structures. Based on the abovementioned, a research for innovative approaches to meet the requirements of the reliability of the "soil-foundation" system for these structures is quite relevant.

In practice, if it is impossible to reduce the values of loads, it is proposed to control a required value of soil settlement by changing a form of load through foundation geometry.

There was conducted comprehensive research, which included the simulation of compressive stresses of soil depending on a geometry of a pile foundation, and plate model experiments. Based on its results and taking into account physical-mechanical soil properties there have been selected technologies of preparation and rational parameters of consolidation of soil for the structures of oil and gas industry (Innovative patents ## 28833, 28834, 29158, 29424).

Pre-consolidation of subsoil and construction materials allows increasing stability (reliability) of foundations of engineering facilities of main pipelines during a long period of their exploitation.

Moreover, application of the obtained research results for construction of foundations of closely located structures of the technological chain allows reducing their mutual influence on each other at the soil level. It provides the admissible values of absolute and relative soil settlements excluding additional financial costs.

H. Shehata and B. Das (Eds.): GeoMEast 2018, SUCI, pp. 64–72, 2019.
https://doi.org/10.1007/978-3-030-01923-5_6

1 Introduction

Development of oil and gas greenfields as well as reconstruction of the existing technological infrastructure of pipeline transport of hydrocarbons, as a rule, faces the necessity of providing stability of soils under buildings and structures during a whole period of their exploitation (Awwad et al. 2018; Abramenkov and Gruzin 1999; Abramenkov et al. 2002, 2012; Awwad and Al-Asali 2014; Gruzin et al. 2009; Awwad et al. 2017; Awwad and Donia 2016). The task of reducing settlements of subsoils at the sites of hydrocarbon pipeline transport under construction and reconstruction is nowadays solved by using technologies of local consolidation with a help of special machine tools or by a rational deployment of the existing technological equipment.

Traditionally in practice, if it is impossible to reduce the values of loads, it is proposed to control a required value of soil settlement by changing a form of load through foundation geometry.

A research for innovative approaches to meet the requirements of the reliability of the "soil-foundation" system for these structures is quite relevant.

2 Simulation of Vertical Compressive Stresses in Soil Caused by Concentrated Forces Applied to a Foundation

In order to study characteristics of an influence of foundation base geometry on a spatial distribution of vertical compressive stresses in a soil half-space, it is supposed to use the solution of the problem of action of a vertical concentrated force applied to the surface of an elastic half-space. It was elaborated by Boussinesq in 1885. This approach allows defining a stress tensor caused by a concentrated force at any point of a half-space. A value of soil settlement caused by compressive stresses emerging due to an external load P in the coordinate system $x - y - z$ has the greatest importance for construction and exploitation of soils and foundations for facilities of pipeline transport of hydrocarbons (Stavnitser 1969; Ukhov 2007):

$$\sigma_z = k \cdot \frac{P}{z^2}, \tag{1}$$

where z is a depth of a point where stress σ_z is defined;

k is a coefficient depending on the mutual location of the point under consideration in a half-space and the points of application of the concentrated force P on the surface of an elastic half-space.

The coefficient k is calculated according to the formula:

$$k = \frac{3}{2\pi} \cdot \frac{1}{\left[1 + \left(\frac{r}{z}\right)^2\right]^{5/2}}, \tag{2}$$

where r is a distance from the location of application of the concentrated force P to the point where a vertical compressive stress σ_z is defined.

In case of application of a load, which is evenly distributed over a surface of limited dimensions, to the ground surface of a soil half-space, a stress in any point of the soil bulk can be defined in accordance with the principle of independence of action of forces as a sum of stresses caused by concentrated loads. They substitute the action of a load evenly distributed in elementary sites and applied in their centre of gravity (see Fig. 1).

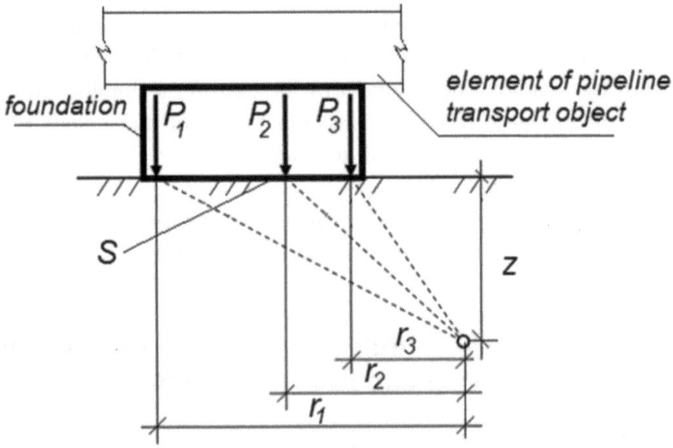

Fig. 1. Definition of vertical compressive stresses caused by action of several concentrated forces

Therefore, using the principle of superposition one can define a value of a vertical compressive stress σ_z in each point of a soil half-space caused by the action of several concentrated forces $P_1, P_2, \ldots, P_n$, applied to the soil ground surface:

$$\sigma_z = k_1 \cdot \frac{P_1}{z^2} + k_2 \cdot \frac{P_2}{z^2} + \ldots + k_n \cdot \frac{P_n}{z^2} = \frac{1}{z^2} \cdot \sum_{i=1}^{n} k_i \cdot P_i. \quad (3)$$

A specialized software "Soil 3D" was developed to analyze the influence of the identified existing forms and values of load on a spatial distribution of vertical compressive stresses and search the new rational ones.

Based on the algorithm of the specialized software "Soil 3D" for calculating values and spatial distribution of vertical compressive stresses σ_z in a soil half-space, dependency (3) has been utilized. For calculations, there have been adopted the following input values: a load value P, an area of loading S, a form of load in the plan - a rectangle with the different ratio between its side lengths and an equilateral n-angle.

During numerical modelling with a help of the software "Soil 3D" there has been studied the influence of the load transferred via a foundation base of the rectangular shape on the spatial distribution of vertical compressive stresses σ_z depending on the ratio of its side lengths (Abramenkov 2002). There have been studied rectangular loads with the following side ratio 1:1, 2:1, 5:1 and 10:1 (see Figs. 2, 3, 4 and 5). The results of the numerical modelling are presented in graphs. By using conditional colours the spatial distribution of the fields of vertical compressive stresses of different intensity has been visualized.

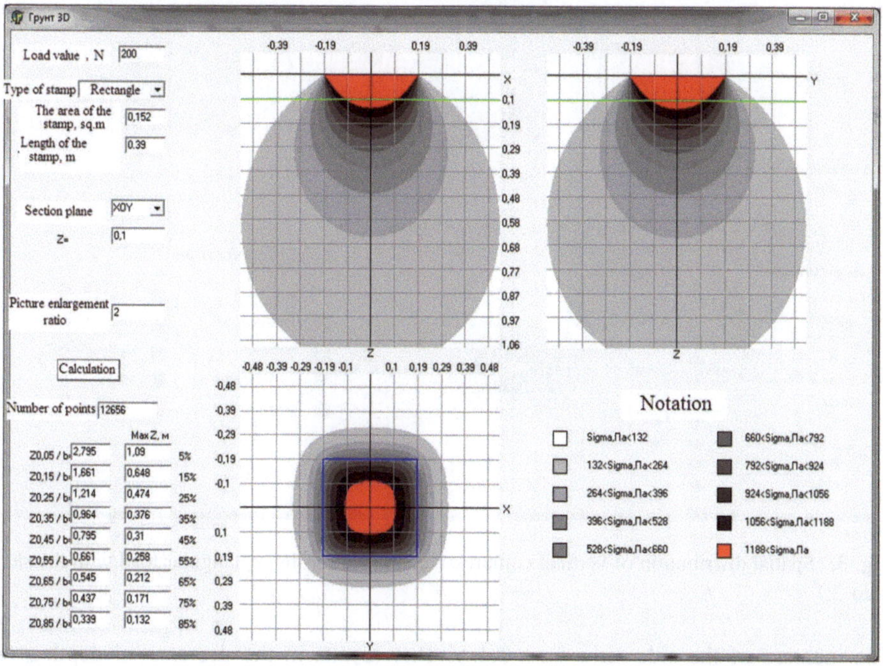

Fig. 2. Spatial distribution of vertical compressive stresses for the rectangular load with the side ratio 1:1

The software "Soil 3D" allowed visualizing the spatial distribution of vertical compressive stresses σ_z both in vertical and horizontal planes at a given distance from the centre of load application. Their spatial distributions are presented in the top left and right graphical fields of the window of the software "Soil 3D". The horizontal distribution of vertical compressive stresses σ_z is given in the bottom graphical field of the window of the software "Soil 3D".

The location of the horizontal plane in the top left and right graphical fields of the window of the software "Soil 3D" is indicated by a green line. The form of the load is presented in the bottom graphical field of the window of the software "Soil 3D" by a dark blue line. In order to make comparative analysis more convenient, there have been selected the same values of square areas of the foundation base *S*, the same values of load *P* and the same values of a location depth in the horizontal plane for all values of a side ratio of the load in the form of a rectangle.

There have been adopted the same values of load *P* and loading area *S* for all shapes. Based on the obtained results of numerical modelling (see Figs. 2, 3, 4 and 5), in case of applying a load in the form of a rectangle, a change of a side ratio leads to considerable changes of spatial distribution of vertical compressive stresses σ_z.

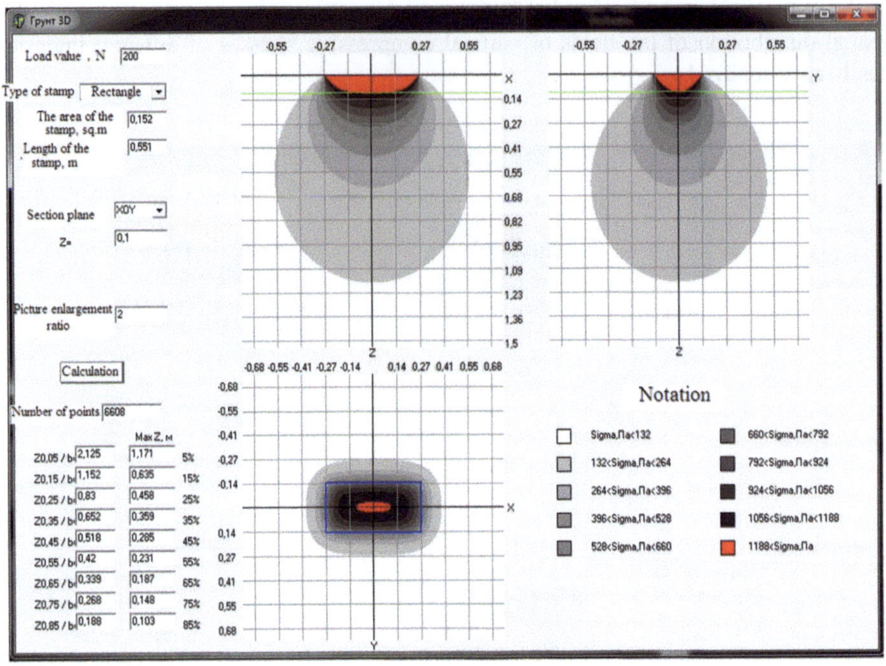

Fig. 3. Spatial distribution of vertical compressive stresses for the rectangular load with the side ratio 2:1

The increase of the side ratio from 1:1 to 10:1 results in the decrease of depth z of maximum distribution vertical compressive stresses σ_z. It comprises 15% of the stress in the base of load σ_n, namely, from 0.648 m to 0.473 m or 1.37 time (see Fig. 6). As regards vertical compressive stresses σ_z, which amount to 85% of the stress in the base of load σ_z, there is the decrease of depth z from 0.132 m to 0.022 m or 6 times.

Therefore, in terms of the reduction of the depth z of compressive stress σ_z distribution in the soil bulk, it is preferable to construct a foundation having the rectangular shape in plan with a maximum possible ratio of lengths of foundation sides. It is built in the technological chain of pipeline transport of hydrocarbons in a construction site (Innovative patent #29425). It is explained by reducing the depth z of compressive stress σ_z distribution in the soil bulk. One should mention that the following phenomena will have considerable influence on structural features and parameters of cross-sections of foundations of facilities for pipeline transport of hydrocarbons in the rectangular shape (including by a value of the ratio of its sides):

- Physical-mechanical properties of soil milieu in which such a foundation is being constructed;
- Technical characteristics and sizes of facilities for pipeline transport of hydrocarbons;
- Construction of junctions of pipeline transport together with transport railway or highway facilities, or other engineering technological structures.

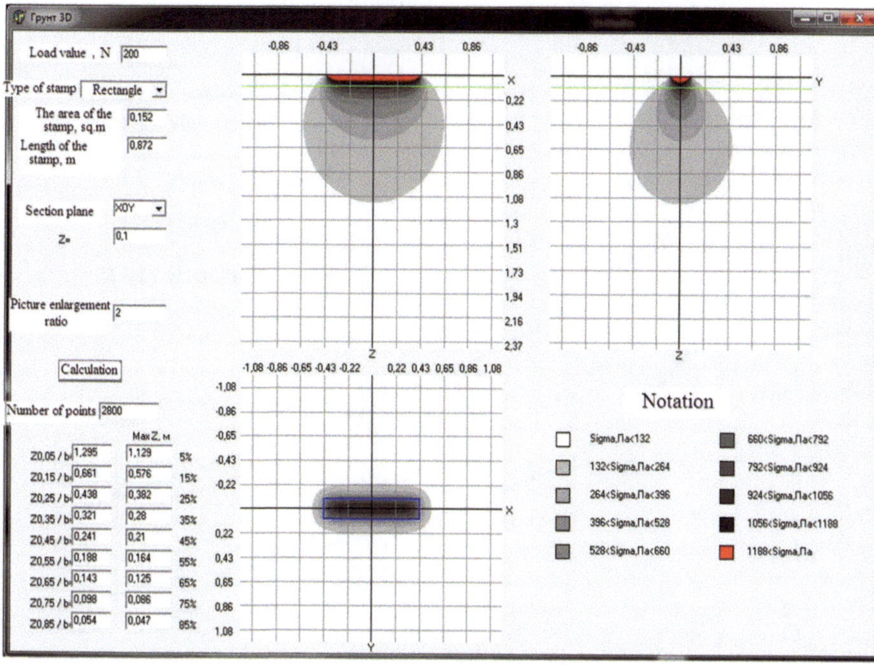

Fig. 4. Spatial distribution of vertical compressive stresses for the rectangular load with the side ratio 5:1

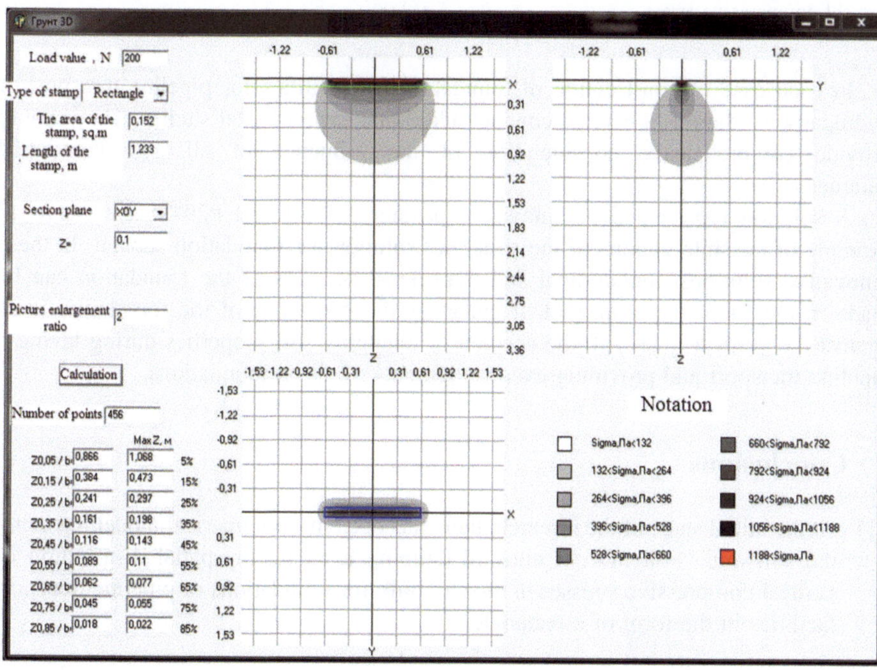

Fig. 5. Spatial distribution of vertical compressive stresses for the rectangular load with the side ratio 10:1

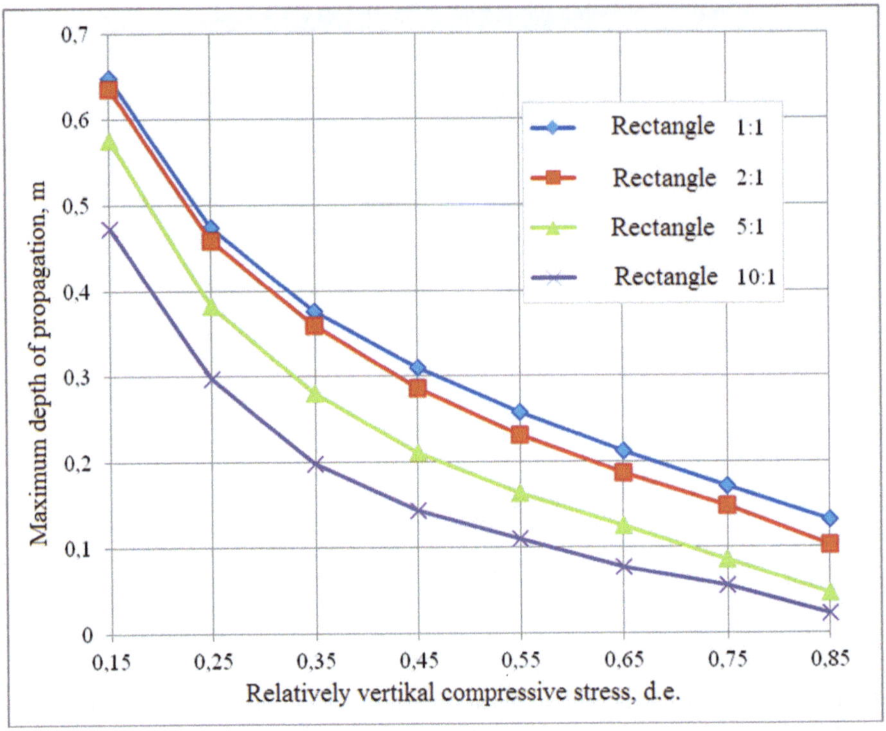

Fig. 6. The influence of a shape of the rectangular load on the maximum depth of distribution of vertical compressive stress

In each case of construction of foundations of facilities for pipeline transport of hydrocarbons, there is a need to conduct additional experimental studies. They should provide comprehensive consideration of the influence of all abovementioned parameters.

A solution represented in innovative patents #27504 and #29425 can serve an example of possible options of the structural solution for foundation subsoil. In these innovative solutions, longitudinal and transversal sections of the foundation can be made in the form of a trapezoid. It enhances the resistance of the foundation itself against the vertical load with the account of changing soil properties during laying a pipeline transport and providing even settlements of all its foundations.

3 Conclusions

(1) At the initial stage of the research, there has been made numerical modelling using the software "Soil 3D". It allowed defining a nature of spatial distribution of vertical compressive stresses in the soil bulk for foundations of pipeline transport facilities in the form of a rectangle.

(2) There have been found out the characteristics of the influence of foundation base geometry on the spatial distribution of vertical compressive stresses in a soil half-space. Based on the results of the numerical modelling there have been defined the values and nature of spatial distribution of vertical compressive stresses. There were also determined the features of their dependence on a shape of the base of foundation, which transfers loads from the structural elements of the facilities of the technological chain of pipeline transport of hydrocarbons.
(3) Based on the received data there have been determined geometrical parameters of a shape of foundations. It exerts the effective influence of the value and nature of spatial distribution of compressive stresses in the soil bulk of the foundations of pipeline transport facilities for hydrocarbons.
(4) It has been stated that application of the obtained research results for construction of the foundations of closely located elements of the technological chain facilities will allow reducing their mutual influence on each other at the soil level and provide the admissible values of absolute and relative soil settlements without additional financial costs. It will also ensure rational deployment of technological equipment in the infrastructural facilities of pipeline transport of hydrocarbons (Innovative patents of the Republic of Kazakhstan #27504).
(5) There was conducted comprehensive research, which included the simulation of compressive stresses of soil depending on the geometry of a foundation taking into account physical-mechanical soil properties. According to its results, there have been selected technologies of preparation and rational parameters of consolidation of soil for the structures of oil and gas industry (Innovative patents of the Republic of Kazakhstan ##28833, 28834, 29158, 29424).

References

Abramenkov, D.E., Gruzin, A.V., Gruzin, V.V., Nuzhdin, L.V.: Technology and Mechanical Tools for Construction of Foundations on Consolidated Soils: Monograph. Bolashak Baspa, Karaganda (2002). 264 p

Abramenkov, D.E., Gruzin, A.V., Gruzin, V.V.: Mechanical Tools and Technology of Construction: Monograph. Palmarium Academic Publishing, Saarbrucken (2012). 327 p. Edited by Abramenkov, E.A.

Abramenkov, E.A., Gruzin, V.V.: Mechanical Tools for Preparation of Soils and Construction of Foundations. The Publishing House of Novosibirsk State University of Architecture and Construction, Novosibirsk (1999). 215 p

Awwad, T., Al-Asali, B.E.: Efficiency of improving the specifications of soil lenses that are formed near the tunnel during the stages of its investment. In: Proceedings of the TC207 ISSMGE Conference, Soil-Structure Interaction, Underground Structures and Retaining Walls. Advances in Soil Mechanics and Geotechnical Engineering, vol. 4, pp. 111–118 (2014)

Awwad, T., Donia, M.: The efficiency of using a seismic base isolation system for a 2D concrete frame founded upon improved soft soil with rigid inclusions. Earthq. Eng. Eng. **15**(1), 49–60 (2016)

Awwad, T., Donia, M., Awwad, L.: Effect of a stiff thin foundation soil layer's depth on dynamic response of an embankment dam. Procedia Eng. **189**(2017), 525–532 (2017)

Gruzin, A.V., Gruzin, V.V., Abramenkov, E.A.: Soil Milieu in the Conditions of Static and Dynamic Loading: Monograph. The Publishing House of Omsk State Technical University, Omsk (2009). 140 p

Stavnitser, L.R.: Deformation of subsoils of structures due to impact loads. The publishing house for special literature in the field of construction (1969). 284 p

Awwad, T., Gruzin, V., Kim, V.: Sustainable reconstruction in conditions of dense urban development. In: Weng, M.-C., et al. (eds.) Current Geotechnical Engineering Aspects of Civil Infrastructures, Sustainable Civil Infrastructures. Springer International Publishing AG, part of Springer Nature (2018)

Ukhov, S.B.: Soil mechanics, soils and foundations. Vyshaya shkola (2007). 566 p

Seismic Soil Structure Interaction of a Midrise Frame Structure

Sahar Ismail(✉), Fouad Kaddah, and Wassim Raphael

Department of Civil and Environmental Engineering, Université Saint-Joseph de Beyrouth, Beirut, Lebanon
sahar.ismail@net.usj.edu.lb,
{fouad.kaddah,wassim.raphael}@usj.edu.lb

Abstract. Earthquakes have occurred for millions of years and will continue in the future affecting millions of lives, thousands infrastructure and costing billions of dollars. Seismic soil structure interaction is the process in which the structure rested on the ground and subjected to an earthquake is affected by the response of the soil medium beneath it having its own characteristics. Thus, soil structure interaction response is dictated by the interaction between the structure, foundation and underlying soil or rock. To clarify the contribution of the seismic soil structure interaction, 3D time history finite element models of 15 story concrete frame building were performed using Abaqus under Kocaeli's Mw = 7.5 strong ground motion. The effects of soil boundary limits, mat thickness and soil linearity were investigated. To demonstrate the importance of modeling correct soil structure interaction problems, the numerical analysis was carried out for three cases: (1) a fixed-base structure, (2) a structure rested on silty sandy soil, and (3) a structure supported by raft foundation and rested on silty sandy soil. The results, presented in terms of the structural lateral deflection and base shear versus time, show that soil structure interaction effects should be evaluated when the maximum lateral deflection is obtained at top of the structure regardless when it occurred as well as at maximum absolute lateral deflection at each level. Moreover, excluding soil structure interaction effects may result in inadequate structural safety for the frame building rested on silty sandy soil and considering linear soil properties may result in non-economical structural design.

1 Introduction

Soil structures interaction (SSI) is a dynamic interaction problem between the structure, foundation and underlying soil or rock. It is the process in which the structure rested on the ground and subjected to an earthquake is affected by the response of the soil medium beneath it having its own characteristics. SSI affects the dynamic response of a flexible supported structure founded on soft soil by an increase in the structure's fundamental period, damping ratio, lateral displacement and a reduction or an increase in the structure's base shear depending on the soil spectral period response compared to a fixed-base structure (Kramer 1996).

Seismic SSI has been considered to be beneficial to structures founded on soft soils. Therefore, many seismic codes neglect SSI effect as a mean to improve the factor of

H. Shehata and B. Das (Eds.): GeoMEast 2018, SUCI, pp. 73–88, 2019.
https://doi.org/10.1007/978-3-030-01923-5_7

safety. Mylonakis and Gazetas (2000) discussed this subject by considering only inertial interaction in non- liquefiable soils. They found that using average spectral motions, like in all seismic codes, without considering proper normalization of periods may lead to unsafe designs. In addition, the authors proved that the concept that ductility demands of fixed-based structures decrease the structural period is not always true where an increase in structural period was obtained for structures founded on soft soils. As a result, the effect of SSI on structures founded on soft soils has been studied by Spyrakos et al. (1989), Veletsos and Prasad (1989), Safak (1995), Krawinkler et al. (2003), Galal and Naimi (2008), El Ganainy and El Naggar (2009), Iida (2012), Farghaly and Ahmed (2013), Hokmabadi et al. (2012a), Fatahi and Tabatabaiefar (2014), Tabatabaiefar et al. (2014a, b), Shehata et al. (2015), Nguyen et al. (2016, 2017) and many other researchers.

Recently, using finite element and finite difference software, the seismic SSI behavior of mid-rise building frames rested on soft soils has been considered by several researchers who using the direct method of analysis concluded that SSI must be considered in design. Using SAP 2000 and ETABs, Farghaly and Ahmed (2013) found that SSI increases the building's story displacement that gets amplified with the building height. Shehata et al. (2015) showed that seismic codes underestimate the structure period that is function of SSI. Moreover, the story number amplifies the SSI effects described by higher story displacement responses mostly affected at the lower stories. The story shear response depends on the foundation and soil type: as the soil stiffness decreases, the story displacement and drift increase. Using Flac2D, Hokmabadi et al. (2012a) found that although soil-structure interaction has less effect on piles than shallow foundation it decreases structure's base shear and interstory drift compared to the fixed base case. Fatahi and Tabatabaiefar (2014) verified that the equivalent linear method can predict the base shear of midrise buildings resting on soft soils considering SSI. However, this method underestimates the inelastic response (interstory drift and lateral deflection) compared to fully nonlinear method leading to underestimating the performance level. Tabatabaiefar et al. (2014a) showed that SSI can be neglected in sand but not in clay. The increase in SSI effects lead to an increase in structural displacement response that depending on the earthquake PGA may change the structure performance level from life safe to near collapse level. In addition, Tabatabaiefar et al. (2014b) proved that flexible base case's lateral deflection is amplified compared to the fixed base case while the base shear is reduced. The reason behind the reduction in base shear is the reduction in lateral distortions. This reduction leads to an increase in structure interstory drift that could affect the structure performance level. Finally, using Abaqus, Nguyen et al. (2016) showed that SSI interaction amplifies the lateral deflection of the structure and decreases the shear forces absorbed by the structure during the earthquake excitation. As the foundation size decreases, the lateral deflection, maximum interstory drifts and rocking experienced by the structure increase. Thus, the inertial forces decrease, the spectral acceleration gets reduced significantly due to the increase in structure's natural frequency. Moreover, Nguyen et al. (2017) found that increasing the length of piles caused the structure to undergo more lateral deflection which led to an increase in interstory drifts thus influenced the total shear forces absorbed by the structure during an earthquake.

Unlike previous studies, 3D time history finite element models of a 15 story midrise concrete frame building are performed using Abaqus under Kocaeli's $M_w = 7.5$ strong ground motion. The effects of soil boundary limits, mat thickness and soil linearity are investigated. Thus, linear and nonlinear soil models are considered in comparing the behavior of (1) a fixed-base structure, (2) a structure rested on silty sandy soil, and (3) a structure supported by raft foundation and rested on silty sandy soil. The results are presented in terms of structural lateral deflection, interstory drift and base shear versus time. This study demonstrates the importance of modeling correct SSI problems to aid engineers obtaining economic and safe designs.

2 Finite Element Model

2.1 Problem Definition

The finite element software, Abaqus (2017), was used to analyze the seismic behavior of 15 story mid-rise structures under Kocaeli's $M_w = 7.5$ strong ground motion. Figure 1 shows the variation of velocity with respect to time of Kocaeli earthquake. The 3D time history models were performed for linear and nonlinear cases for three cases: (1) a fixed-base structure (S-model), (2) a structure rested on silty sandy soil (SS-model), and (3) a structure supported by raft foundation and rested on silty sandy soil (SSF-model). The finite element models, consisting of the frame structure formed of beams, columns and slab, the soil and raft were hit by the earthquake load (V_x) at the bottom. The effects of soil horizontal boundary limits, soil linearity and raft thickness were varied in the analyses. The 3D models were simulated using the direct method of analysis: the entire structure-foundation-soil was modeled at the same time and the input earthquake load was specified at bedrock.

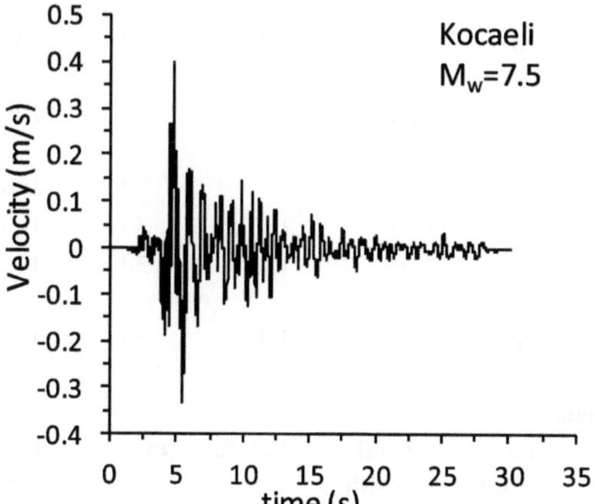

Fig. 1. Velocity with respect to time of Kocaeli earthquake

2.2 Model Characteristics

In this study, typical mid-rise concrete building frame of 15 stories was considered. The simulated frame structure is formed of three bays in both horizontal directions with a total width of 15 m and total story height of 3 m as detailed in Table 1. Dead (gravity load) and live loads were applied to the structural model in term of uniformly distributed load over the floors with a value of 2.5 kN/m^2. The inelastic behavior of structural elements was modeled using elastic-perfectly plastic material by defining the concrete yield stress and a structural damping ζ of 5% was adopted. Rayleigh modal coefficients $\alpha = 0.24$ and $\beta = 0.008$ were used and calculated based on the first and second structure's mode frequencies (f_i and f_j in rad/s):

Table 1. Finite element model configuration

Reference name	Number of stories	Raft dimension			Soil dimensions		
		Length (m)	Width (m)	Thickness (m)	Height (m)	Length (m)	Width (m)
SS-136-136	15	-	-	-	30	136	136
SS-136-76	15	-	-	-	30	136	76
SS-136-30	15	-	-	-	30	136	30
SS-106-106	15	-	-	-	30	106	106
SS-106-76	15	-	-	-	30	106	76
SS-106-30	15	-	-	-	30	106	30
S	15	-	-	-	-	-	-
SS-L	15	-	-	-	30	106	106
SSF-e-1 m-L	15	20	20	1	30	106	106
SSF-e-1.25 m-L	15	20	20	1.25	30	106	106
SSF-e-1.5 m-L	15	20	20	1.5	30	106	106
SS	15	-	-	-	30	106	106
SSF-e-1 m	15	20	20	1	30	106	106
SSF-e-1.25 m	15	20	20	1.25	30	106	106
SSF-e-1.5 m	15	20	20	1.5	30	106	106
Structures Dimensions							
Number of Bays	**Story Height (m)**	**Bay Width (m)**	**Total Width (m)**	**Column (cm)**		**Beam (cm)**	**Slab (m)**
3	3	5	150	50 * 50		50 * 50	5 * 5 * 0.25

$$\zeta = \frac{1}{2f_n}\alpha + \frac{f_n}{2}\beta,\ \alpha = 2\zeta_i \frac{f_i f_j}{f_i + f_j} \text{ and } \beta = 2\zeta_i \frac{1}{f_i + f_j} \text{ (Chopra 2011)} \quad (2)$$

The fixed base structure (S-model) was then rested on a silty sandy soil medium (SS-models) having a shear wave velocity of 63 m/s. The properties of the concrete and silty sandy soil are listed in Table 2 noting that soil damping was not accounted for in this study. The soil model length and width used in studying the effects of the varied parameters are detailed in Table 1, while the vertical distance to the soil boundary was extended 30 m to the bedrock. Finally, the structure and the soil were connected to a 20 m width raft (SSF-models) with different thicknesses (Fig. 2 and Table 1); noting that every time history model took around 60 h to be completed using fast computational facilities at Université Saint-Joseph de Beyrouth.

Table 2. Properties

Silty Sandy Soil	**E (MPa)**	**v**	**γ_d (kN/m^3)**	**e**	**$\varphi°$**	**$\psi°$**	**c (kPa)**
	20	0.3	18.98	0.4	35	10	50
Concrete	**E (GPa)**	**v**	**ρ (kg/m^3)**	**σ_y (MPa)**	**f'$_c$ (MPa)**	**α**	**β**
	20	0.2	2400	30	30	0.24	0.008

The beams (50–50 cm) and columns (50–50 cm) were built using 2-node linear beam elements B31 with around 9400 elements. Moreover, the slab (5–5 m) was built using 4-node doubly curved shell, reduced integration, hourglass control; finite membrane strains elements with around 100 elements. Whereas eight node linear brick, reduced-integration, hourglass control continuum solid elements C3D8R were used to model the soil with around 32000 elements and the raft with 4800 elements. Beams and floors slabs as well as raft bottom surface and soil top surface were tied in the numerical models. Moreover, embedded columns were inserted in the soil in structure-soil (SS) models and in the raft in structure-raft-soil (SSF) models. In static analysis, the bottom soil boundary was assumed fix to accommodate rigid conditions (rock) while movement along the soil boundaries was restrained axially (Fig. 2). In dynamic analysis, the soil wall boundaries were free to move in the horizontal directions but restrained to move in the upward direction to simulate absorbed boundary condition.

3 Numerical Results and Analyses

3.1 Effect of Soil Boundary Limits

To study the effect of artificial horizontal soil boundary limits, the soil was modeled as a semi-infinite medium in soil-structures models (SS-models). These models were formed of 15 story frame concrete structures and 30 m height silty sandy soil blocks as detailed in Fig. 2 and Table 1. Ghosh and Wilson (1969) showed that to obtain

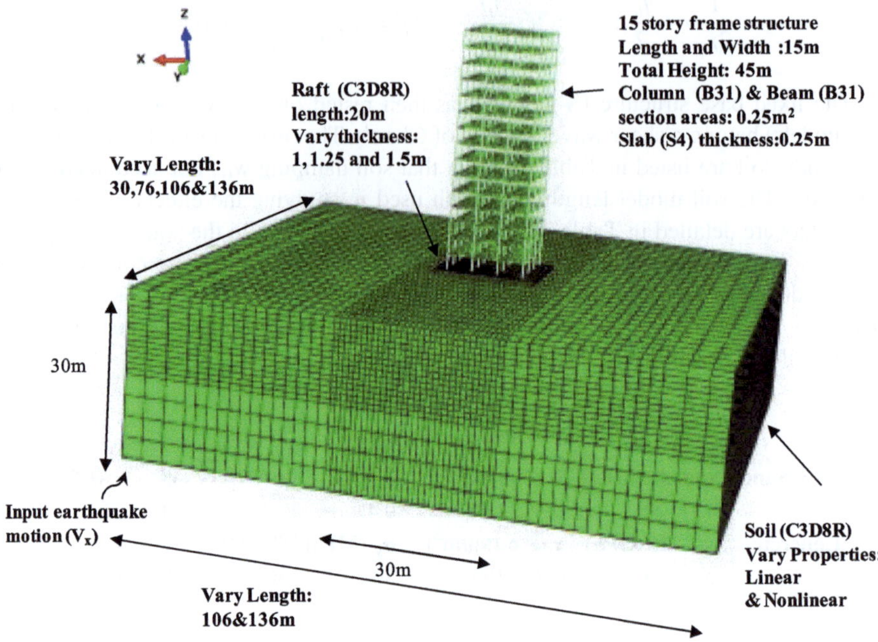

Fig. 2. Finite element model geometry, mesh distribution and varied parameters used in the numerical simulations

insignificant reflexive wave effects, the distance from the center of the foundation to the soil horizontal boundary should be 3–4 times the foundation radius and to the vertical boundary 2–3 times the foundation radius. Moreover, Rayhani and El-Naggar (2008) showed that most ground motion amplification occurs within the first 30 m of the soil profile and increasing the soil boundary from 5 to 10 times the width of the structure leads only to 5% difference in the results. Thus, based on these recommendations, soil boundary limits effects were investigated and the lengths (x-direction) and widths (y-direction) of the soil block (Fig. 2) were varied between 30, 76, 106 and 136 m corresponding to 2B, 5B, 7B and 9B respectively with B being the width of the structure. Moreover, Hokmabadi et al. (2012b) proved that lateral deflection results should be presented when the maximum is obtained at top of the structure (max top) regardless when it occurred while other studies present the lateral deflection results at maximum absolute lateral deflection value at each level. Therefore, the variation of lateral deflection and interstory drifts at maximum absolute value and at maximum lateral deflection at top (at maximum top) of the structure regardless of the time at each level for SS models under the influence of Kocaeli earthquake excitation (Fig. 1) were plotted in Figs. 3 and 4 for the effect of varying soil boundary limits noting that lateral deflection at each story was measured relative to the base of the structure.

The curves on Fig. 3(a, b) indicate that SS-136-30 and SS-106-30 models corresponding to 9B and 7B (x-direction) in lengths and 2B (y-direction) in widths respectively present unrealistic lateral deflection results compared to the other

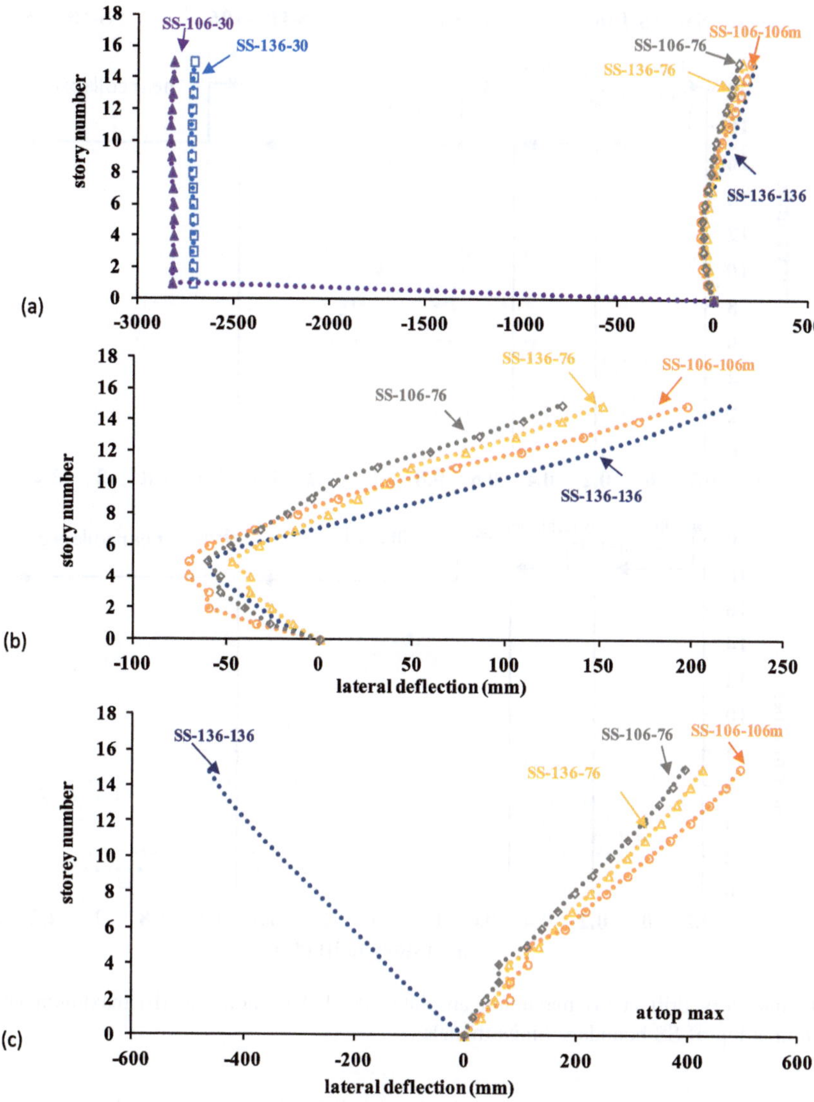

Fig. 3. (a & b) Maximum absolute lateral deflection and (c) maximum lateral deflection at top of SS boundary limits models

SS-models results. This is due to the reflexive wave's effect. Thus, the remaining SS-models were plotted in Figs. 3(b, c) and 4. All these models follow the same deformation pattern (they present the same deformation curve shape) at maximum absolute lateral deflection at each level: as the story level increases, the lateral deflection increases then reaches a peak at the 5th story, changes its direction and re-increases in a linear manner till the top of the structure (15th level) while at maximum top, the curves

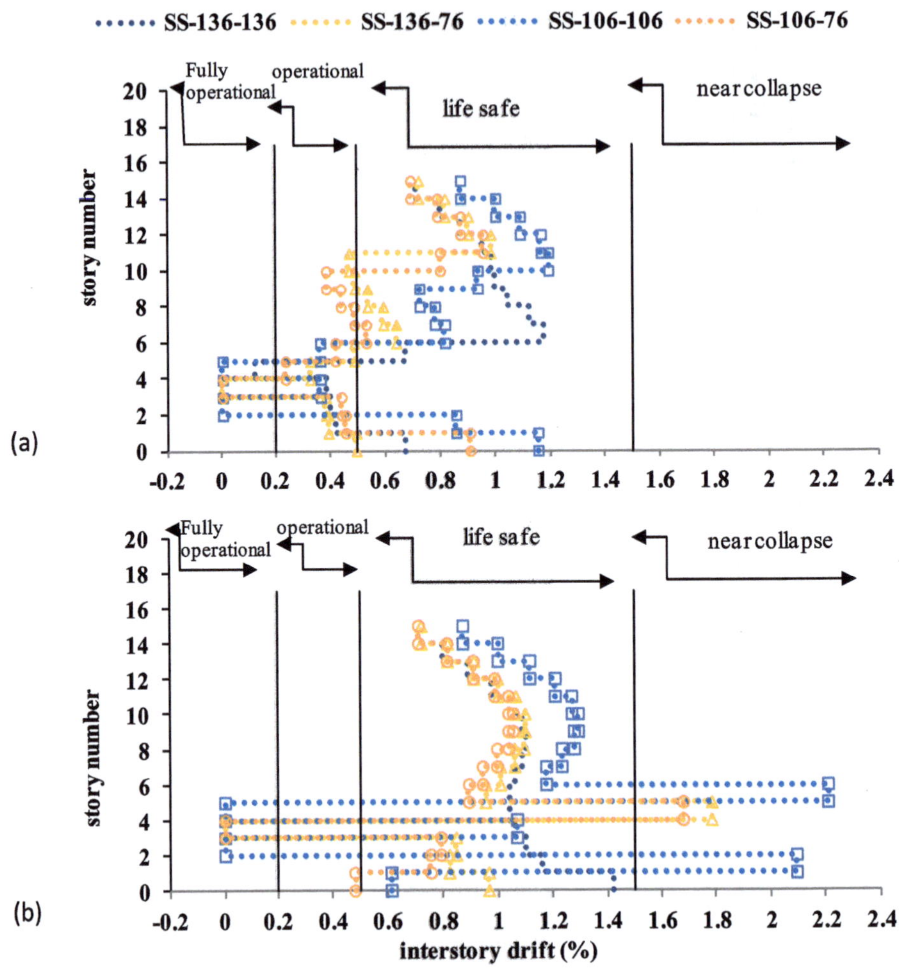

Fig. 4. Interstory drift at (a) maximum absolute lateral deflection and (b) maximum lateral deflection at top of SS boundary limits models

follow a more realistic pattern where the lateral deflection increases when moving to higher stories.

To express the level of damage of an earthquake on a structure based on safety and risk assessment, performance levels are used. Interstory drift, one of the most common damage parameters used, is the difference between the deflections of two stories when the building is subjected to an earthquake load normalized by the story height:

$$drift = (d_{i+1} - d_i)/h \tag{1}$$

where d_{i+1} and d_i are the deflections at i+1 and i levels respectively and h is the story height. Thus, performance levels are classified as: (1) fully operational if interstory drift <0.2%, (2) operational if interstory drift <0.5, (3) life safe if interstory drift <1.5%,

(4) near collapse if interstory drift <2.5% and (5) collapse if interstory drift >2.5% (BSSC 1997, 2009). As shown in Fig. 4, even though the 4 models present the same performance level category "life safe" at maximum absolute value, performance levels of SS-136-136 and SS-106-106 models are shifted to higher values and reach "near collapse" category at maximum top noting that in most building codes the maximum allowable performance level category is "life safe".

The maximum absolute lateral deflection and thus, interstory drifts are greater in the square soil blocks models having the same lengths and widths: SS-136-136 and SS-106-106 than in the rectangular soil blocks models: SS-136-76 and SS-106-76 noting that 76 m is the widths of the blocks in the y-direction while the input earthquake excitation is in the x-direction. Therefore, the average difference in maximum absolute lateral deflection at each level is 20, 10, 16 and 7% while at maximum top is 24, 5, 24 and 5% between SS-136-136 and SS-136-76, SS-136-136 and SS-106-106, SS-106-106 and SS-106-76 and SS-136-76 and SS-106-76 respectively. Therefore, since SS-106-106 presents 5% higher lateral deflection at maximum top while it presents 18% higher peak value at the 5th level and a 10% lower value at the 15th level than that of SS-136-136 and the variation of base shear is almost the same for these 2 models with time (Fig. 5), soil blocks having 106 m (7B) lengths and widths were used in the remaining analysis in this paper as detailed in Table 1.

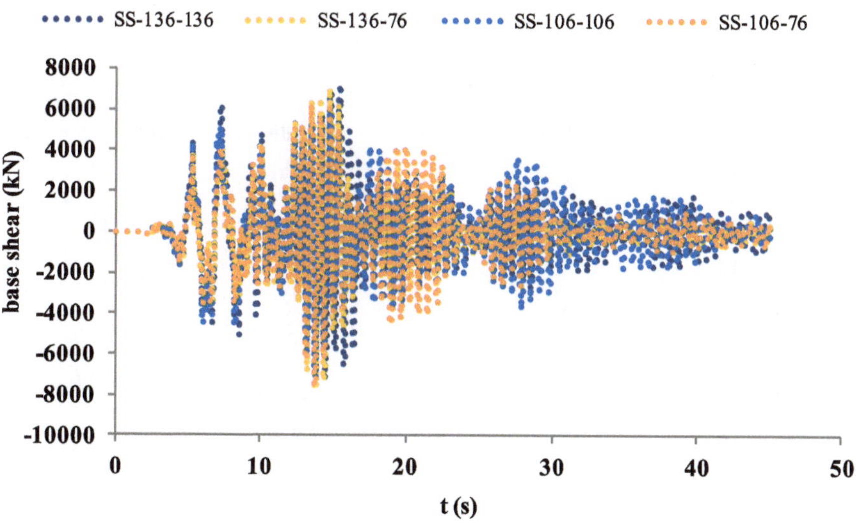

Fig. 5. Variation of base shear with time of SS boundary limits models

3.2 Effect of Soil Linearity and Mat Thickness

Previous studies, using the direct method of analysis, compared the effect of SSI of a structure founded on raft or pile foundation and soil to a fixed based structure adopted in most building codes (Hokmabadi et al. 2012a; Fatahi and Tabatabaiefar 2014; Tabatabaiefar et al. 2014a, b; Shehata et al. 2015; Nguyen et al. 2016, 2017, etc). In this study, the effect of SSI was considered in various situations: linear and nonlinear silty sandy soil (Table 2) and 1, 1.25 and 1.5 m mat thicknesses for three model cases as shown in Fig. 2. These models included: (1) a fixed-base structure (S-model), (2) a structure rested on silty sandy soil (SS-model), and (3) a structure supported by 20 m width raft foundation and rested on silty sandy soil (SSF-model). The variation of lateral deflection and interstory drifts at maximum absolute value and at maximum lateral deflection at top of the structure regardless of the time at each level under the influence of Kocaeli earthquake excitation for the effect of soil linearity and mat thickness were plotted in Figs. 6, 7 and 8.

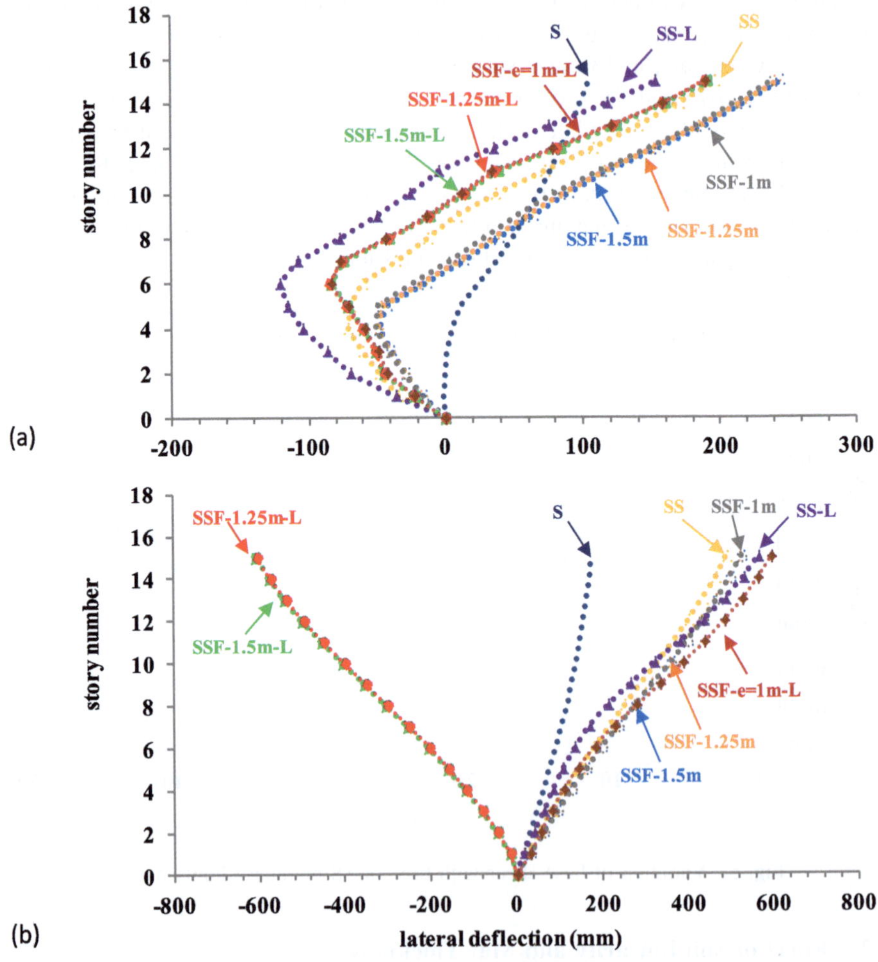

Fig. 6. (a) Maximum absolute lateral deflection and (b) maximum lateral deflection at top for linear and nonlinear S, SS and SSF Model cases

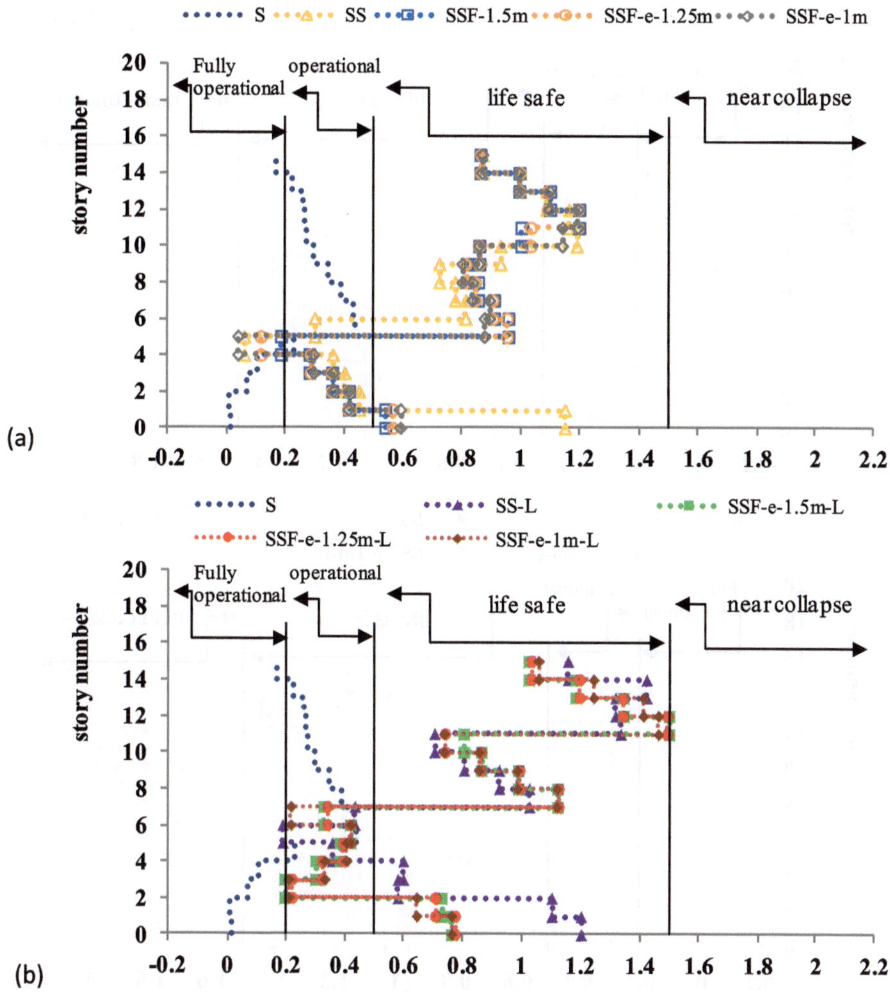

Fig. 7. Interstory drifts at maximum absolute lateral deflection of (a) nonlinear and (b) linear S, SS and SSF Model cases

The results show that mat thicknesses values used in this paper slightly affected the SSF-models lateral deflection and base shear results. This is due to the high mat thicknesses values adopted where a minimum mat thickness of 1 m (0.7 N with N being the number of stories) was used. However, lateral deflections are greater in SSF than in SS and S-models. The amount of energy the raft and soil absorbed led to the increase in lateral deflection. SSF-models lateral deflection curves are similar to SS-models curves: they increase with moving to higher stories at maximum top while peaks are obtained at the 5th and 15th story levels at maximum absolute values. In addition, at maximum absolute value, lateral deflection values are greater at lower stories (<5th level) and higher at higher stories in SS-models compared to SSF-models.

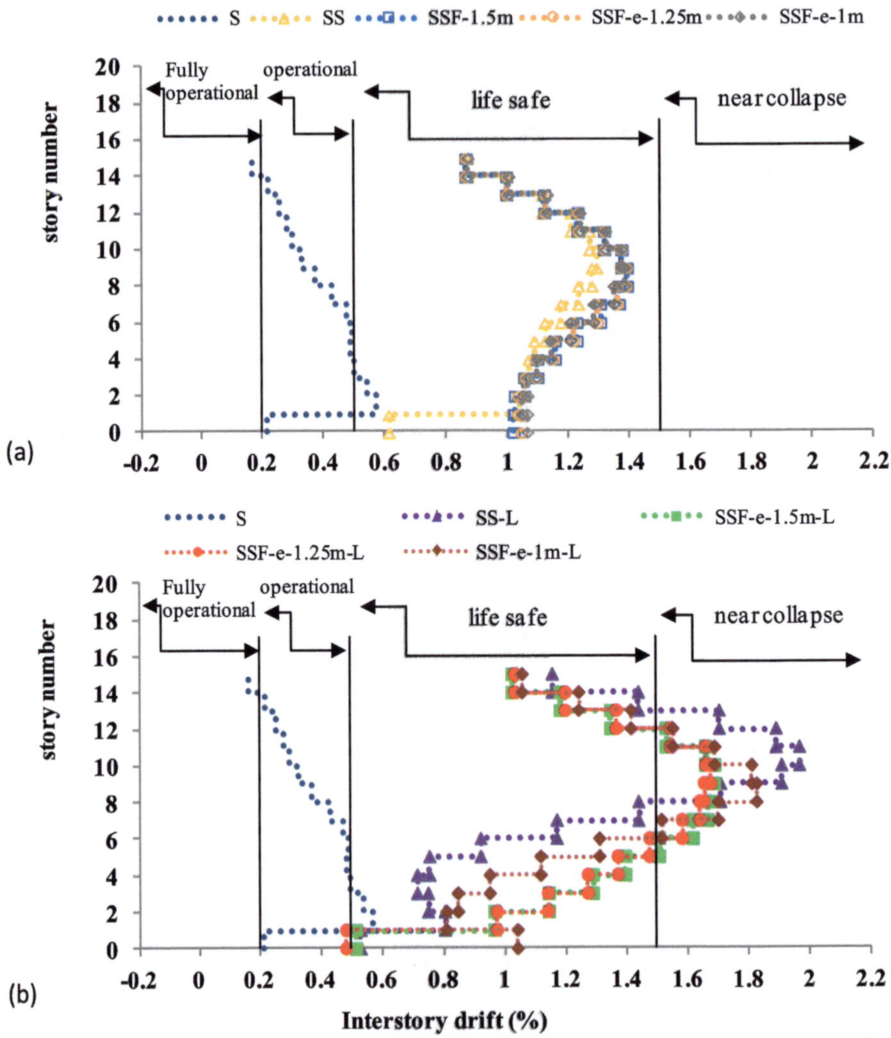

Fig. 8. Interstory drifts at maximum top lateral deflection of (a) nonlinear and (b) linear S, SS and SSF Model cases

The results are in accordance with Shehata et al. (2015) who showed that structure's story number amplifies the SSI effects described by higher story displacement responses mostly affected at the lower stories.

Since soil properties affect earthquake motion characteristics at the base of the structure through amplification and SSI (Rayhani and El-Naggar 2008), lateral deflections were greater at lower stories and lower at higher stories at maximum absolute value while they increased with the number of stories at maximum top in linear models compared to nonlinear models. As shown in Fig. 6, at maximum top, SSF models lateral deflection values are 7% and 300% higher in linear and nonlinear cases than SS and S-models values respectively. Also, at maximum absolute values at the 15th level, SSF models lateral deflection values are 25% and 200% higher in linear and 20% and 250% higher in nonlinear cases than SS and S-models values while at the 5th level, SSF models lateral deflection values are 38% lower and 600% higher in linear and 60% lower and 350% higher in nonlinear cases than SS and S-models values respectively.

Although both SS and SSF-models (flexible structures) are within "life safe" category in nonlinear soil cases, their performance levels are shifted to "near collapse" category in linear soil cases while S-model presents a much lower performance level value: "operational" level. SSF-models average interstory drifts values are 1 and 7% higher in nonlinear cases and 10 and 7% higher in linear cases than SS-models values at maximum absolute values and maximum top respectively.

In a similar manner, flexible structures (SS and SSF-models) base shears are almost the same (Fig. 9) and are 2 times as much as fixed base structure's (S-model) base shear in linear cases and 1.4 in nonlinear cases. Therefore, structure's base shear was amplified in both soil linearity models. This is related to the soil spectral response induced by Kocaeli earthquake excitation: each seismic excitation provokes different levels of shear strains in the soil block thus, induces different soil damping and modulus values (Kramer 1996). In addition, at the 5th level, level shear force is amplified in linear and nonlinear cases compared to the level shear value in fixed base structure. The ratio of level shear force of flexible to fixed base structures at the 5th level is 1.3 and 1.5 for SS and SSF-models respectively in nonlinear cases and 1.8 in SS and SSF-models in linear cases as shown in Fig. 10. Thus, in nonlinear cases, level shear force value at the 5th level is lower than that in fixed base structure while in linear cases, it is higher for SS and SSF-models.

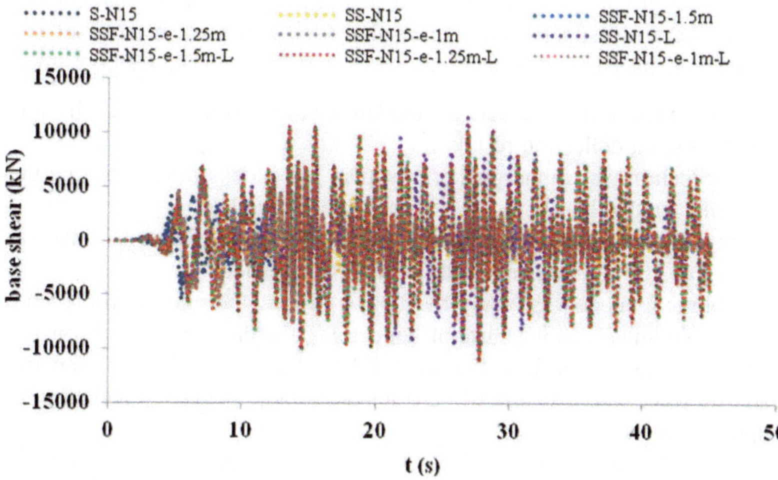

Fig. 9. Variation of base shear with time for linear and nonlinear S, SS and SSF Model cases

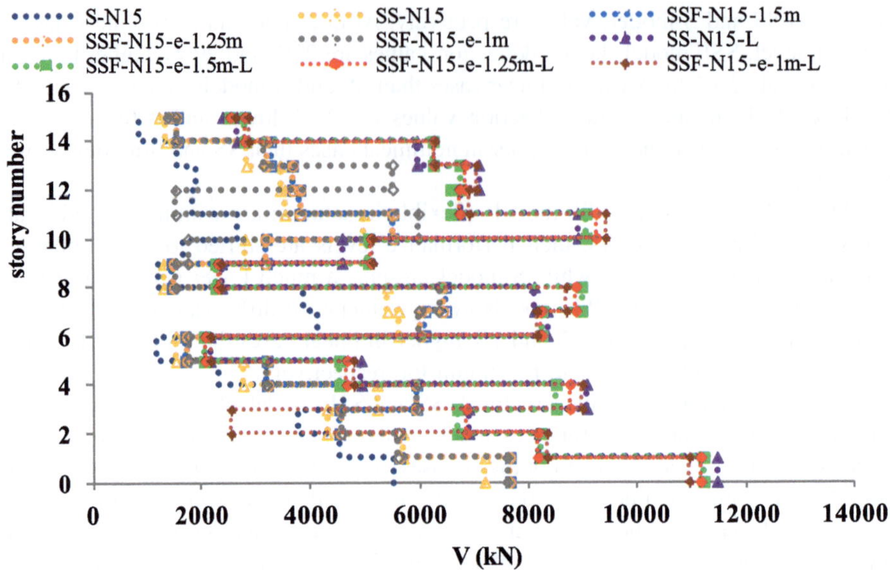

Fig. 10. Variation of level shear force at maximum absolute value for linear and nonlinear S, SS and SSF Model cases

4 Conclusions

3D time history linear and nonlinear finite element models of 15 story concrete building frame rested on silty sandy soils were performed using Abaqus under Kocaeli's Mw = 7.5 strong ground motion. The effect of soil boundary limits, mat thickness and soil linearity were investigated and the results were presented in terms of structural lateral deflection, interstory drift and base shear versus time. The results showed that:

- Due to reflexive wave's effect, SS-models having 30 m (2B) widths presented unrealistic lateral deflection results.
- Lateral deflection was found greater in square soil blocks models having the same lengths and widths than in rectangular soil blocks models.
- Increasing the soil boundary limits form 7 to 9 times the widths of the structure leads only to 10% difference in lateral deflection results.
- All flexible base models (SS and SSF-models) follow the same deformation pattern where at maximum absolute lateral deflection at each level, deflection increases with time to reach a peak at the 5th level then changes direction and increases linearly till top of the structure while at maximum top, deflection increases with the number of stories.
- SSI effects should be evaluated at structure's maximum top lateral deflections regardless of the time because performance levels may be shifted to a higher category compared to the maximum absolute lateral deflection results.
- Lateral deflection was slightly affected by mat thicknesses values tested in this paper due to the high values used.

- At top of the structure, SSF models presented 25% in linear and 20% in nonlinear soil model cases higher lateral deflection than SS models. Nevertheless, flexible structures (SS-SSF models) gave higher story lateral deflections and structures base shear values than fixed base structure (S-model). Thus, higher interstory drifts that changed the performance level category and went beyond the safe limit given by structures code "life safe".
- At the 5th level, level shear force is amplified in linear and nonlinear cases compared to the level shear value in fixed base structure. Moreover, in nonlinear cases, level shear force at the 5th level is lower than the value in fixed base structure while in linear cases, it is higher for SS and SSF-models.
- Linear cases presented higher story lateral deflection, interstory drifts and structures base shear than nonlinear cases. Thus, designing based on linear soil model properties may result in non-economical structural design.

Therefore, even though mat thickness values used slightly altered lateral deflection and base shear results in this study, engineers should model SSI problems in 3D using nonlinear SSF-models that lead to safe and economic results.

References

Abaqus: [Computer software]. Dassault Systèmes SIMULIA Corporation, Minneapolis (2017)

BSSC: NEHRP Guidelines for the seismic rehabilitation of buildings, 1997 Edition, Part 1: Provisions and Part 2: Commentary. Federal Emergency Management Agency (1997)

BSSC: NEHRP Recommended seismic provisions for new buildings and other structures, Federal Emergency Management Agency (2009)

Chopra, A.K.: Dynamics of Structures: Theory and Applications to Earthquake Engineering, 4th edn. Prentice Hall, Englewood Cliffs (2011)

El Ganainy, H., El Naggar, M.H.: Seismic performance of three-dimensional frame structures with underground stories. Soil Dyn. Earthq. Eng. (2009). https://doi.org/10.1016/j.soildyn.2009.02.003

Farghaly, A., Ahmed, H.: Contribution of soil-structure interaction to seismic response of buildings. KSCE J. Civ. Eng. (2013). https://doi.org/10.1007/s12205-013-0261-9

Fatahi, B., Tabatabaiefar, S.: Fully nonlinear versus equivalent linear computation method for seismic analysis of mid-rise buildings on soft soils. Int. J. Geomech., ASCE, 04014016 (2014). https://doi.org/10.1061/(ASCE)GM.1943-5622.0000354

Galal, K., Naimi, M.: Effect of conditions on the response of reinforced concrete tall structures to near fault earthquakes. Struct. Des. Tall Spec. Build. (2008). https://doi.org/10.1002/tal.365

Ghosh, S., Wilson, E.L.: Dynamic stress analysis of axi-symmetric structures under arbitrary loading. Report no. EERC 69–10, University of California, Berkeley (1969)

Hokmabadi, A., et al.: Effect of soil-pile-structure interaction on seismic response of moment resisting buildings on soft soil. In: At the 3rd International conference on New Development in Soil Mechanics and Geotechnical Engineering, 28–30 June 2012, Near East University, Nicosia, North Cyprus (2012a)

Hokmabadi, A.S. et al.: Recording interstorey drifts of structures in time-history approach for seismic design of building frames. Aust. J. Struct. Eng. (2012b). https://doi.org/10.7158/13287982.2012.11465107

Iida, M.: Three-dimensional finite-element method for soil-building interaction based on an input wave field. Int. J. Geomech., ASCE, 430–440 (2012). https://doi.org/10.1061/(ASCE)GM.1943-5622.0000232

Kramer, S.L.: Geotechnical Earthquake Engineering. Prentice Hall, Englewood Cliffs (1996)

Krawinkler, H., et al.: Seismic drift and ductility demands and their dependence on ground motions. Eng. Struct. (2003). https://doi.org/10.1016/S0141-0296(02)00174-8

Mylonakis, G., Gazetas, G.: Seismic soil-structure interaction: beneficial or detrimental. J. Earthq. Eng. (2000). https://doi.org/10.1080/13632460009350372

Nguyen, Q.V., et al.: The effects of foundation size on the seismic performance of buildings considering the soil-foundation-structure interaction. Struct. Eng. Mech. (2016). https://doi.org/10.12989/sem.2016.58.6.1045

Nguyen, Q.V., et al.: Influence of size and load-bearing mechanism of piles on seismic performance of buildings considering soil-pile-structure interaction. Int. J. Geomech., ASCE (2017). https://doi.org/10.1061/(ASCE)GM.1943-5622.0000869

Rayhani, M., El Naggar, M.: Numerical modeling of seismic response of rigid foundation on soft soil. Int. J. Geomech., ASCE, 336–346 (2008). https://doi.org/10.1061/(ASCE)1532-3641(2008)8:6(336)

Safak, E.: Detection and identification of soil-structure interaction in buildings from vibration recordings. J. Struct. Eng., ASCE, 889–906 (1995). https://doi.org/10.1061/(ASCE)0733-9445(1995)121:5(899)

Shehata, E., et al.: Evaluation of soil-foundation-structure interaction effects on seismic response demands of multi-story MRF buildings on raft foundations. Int. J. Adv. Struct. Eng. (2015). https://doi.org/10.1007/s40091-014-0078-x

Spyrakos, C., et al.: Assessment of computational practices in dynamic soil-structure interaction. J. Comput. Civ. Eng., ASCE, 143–157 (1989). https://doi.org/10.1061/(ASCE)0887-3801(1989)3:2(143)

Tabatabaiefar, S., et al.: An empirical relationship to determine lateral seismic response of mid-rise building frames under influence of soil-structure interaction. Struct. Des. Tall Spec. Build. (2014a). https://doi.org/10.1002/tal.1058

Tabatabaiefar, S., et al.: Numerical and experimental investigations on seismic response of building frames under influence of soil-structure interaction. Int. J. Adv. Struct. Eng. (2014b). https://doi.org/10.1260/1369-4332.17.1.109

Veletsos, A.S., Prasad, A.M.: Seismic interaction of structures and soils: Stochastic approach. J. Struct. Eng., ASCE, 935–956 (1989). https://doi.org/10.1061/(ASCE)07339445(1989)115:4(935)

Slope Stability Analysis Under External Static Surcharge

Soumia Merat, Lynda Djerbal(✉), and Ramdane Bahar

LEEGO, University of Science and Technology Houari Boumediene, 32 El Alia, 16111, BP Algiers, Algeria
ldjerbal@yahoo.fr

Abstract. In the assessment of slopes, factor of safety values still remain the primary indexes for determining how close or far slopes are from failure. Limit equilibrium and finite elements methods are among the most popular methods of slope analysis. Both methods were used to analyze homogeneous and inhomogeneous slopes. However, Limit equilibrium methods are relatively simple and suffer from many assumptions compared with finite element analysis. Recently, as computational techniques advance, more and more attention has been paid to the slope stability evaluations using finite element methods (FEM). This article demonstrates a finite element approach to analyze the response of homogeneous and layered slopes. A detailed parametric analysis is presented to study the effect of surcharge (it can be an imposition of building load….etc.) on the stability of the slope. In regard with these results it is seen that the factor of safety increases with increase in soil properties, groundwater depth, slope length, distance from the crest of slope. However, it decreases with increase in surcharge intensity, increase of weak soil layer thickness and increase of slope angle and slope height.

Keywords: Slope stability · Surcharge · Finite element method
Factor of safety · Layered slope

1 Introduction

Slope stability analysis is one of the most important areas of interest in geotechnical engineering. The embankment slope can be failure due to its geotechnical properties, geological structure conditions, other internal factors and the various external conditions such as depth of water, seismic loads, surcharge, etc. The evaluation of stability of the natural slopes becomes very essential for the safe design, especially when the slopes are situated close to residential areas or when structures are built on these slopes. The objective of our work is the study of the stability of the slope the under static loads combining with various parameters using the software Plaxis^{2D} 2015 for finite element analysis.

Many slope stability analysis methods and tools have also been developed over the last 50 years. The main interest of slope stability analysis is typically to determine a factor of safety (FS) against slope failure (Wei et al. 2010). Each stability analysis method differs from the others in some of the basic assumptions, but most of these

H. Shehata and B. Das (Eds.): GeoMEast 2018, SUCI, pp. 89–102, 2019.
https://doi.org/10.1007/978-3-030-01923-5_8

methods will give similar factors of safety/slip surfaces for normal cases, which are sufficient for normal engineering uses (Cheng and Lau 2008).

Recently, as computational techniques advance, more and more attention has been paid to the slope stability evaluations using finite element methods (FEM) which has become popular for the analysis of slopes and has been successfully implemented to calculate Factor of Safety for more than a decade (Dawson et al. 1999; Griffiths and Lane 1999; Hammah et al. 2004).

The paper deals with the assessment of slope stability under static load taking into account a number of influencing parameters such as: ground water level (GWL), slope geometry, soil properties, distance to crest…etc. The effect of each factor is discussed in this paper. Consequently, the objectives of the present study are: **(i)** to investigate the effect of slope geometry on the factor of safety, **(ii)** to show the effect of ground water position and surcharge intensity, **(iii)** to evaluate the influence of homogeneous and a layered soil (in term of mechanical properties) on the stability of slope, **(iv)** to assess the effect of the position of surcharge on the factor of safety.

2 Slope Stability Analyses by FEM and Boundary Conditions

FEM is a powerful numerical tool for solving many engineering problems and mathematical physics. Due to rapid development of computer technology, FEM has gained increasing popularity over the traditional methods in geotechnical engineering (Moni and Sazzad 2015).

The geometric model (Fig. 1) is incorporated in the Plaxis^{2D}. After incorporating the model, the properties of soil is assigned for the specified interface and the material model is set to be the Mohr–Coulomb model. The two lateral boundaries were allowed to move only in the vertical directions (roller type: restrained against horizontal movement) whereas the bottom boundary was completely restrained and fully blocked from movement (Closed: restrained in both horizontal and vertical direction).

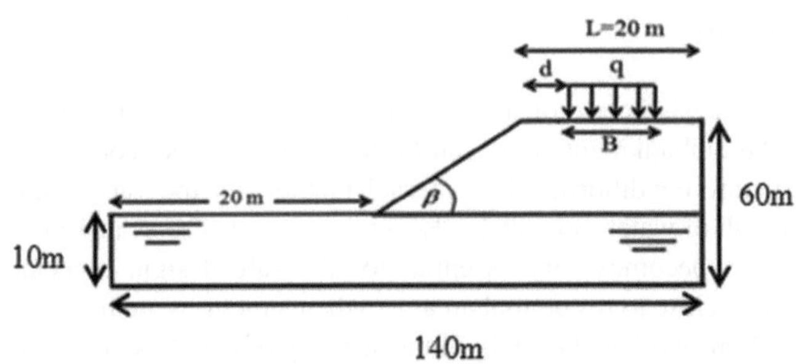

Fig. 1. Geometry of the numerical model of slope considered in the present study

In this paper for Finite Element Method analysis, meshes are generated and set to be medium with 167 elements and 1435 nodes. After generating the meshes, and assigning related properties, the stability analysis is performed then a surcharge is added on the terrain of slope (Griffiths and Lane 1999).

3 Geometry and Soil Parameters

In this study, a slope with an inclination angle of 26.56° (2/1) to the horizontal plane and a height of 50 m is considered (Fig. 1). The groundwater table is assumed to be at the ground surface. The soil type was specified as «Marl» of Azazga. The soil properties related to the unsaturated soil were taken as given by van Genuchten (1980) (Sun et al. 2015).

The surface soil layer behavior was modeled as linear, elastic-perfectly plastic the material by using Mohr–Coulomb model and analyzed using soil parameters giving by undrained tests. The FEM analysis using Mohr-Coulomb material model requires parameters such as modulus of elasticity, poison's ratio, angle of internal friction and cohesion.

The properties of soil used in the present study are presented in Table 1. Two types of soils are considered. Soil-1 (Marl) is used in the analysis of a slope of homogeneous soil. Whereas soil-2 (clayey loam) is used for thin layer of slope of layered soil. Both Soil-1 and Soil-2 are used in the analysis of a slope of layered soil.

Table 1. Summary of soils parameters of Azazga soil layer used in the FEM

Soil layer	Soil model	γunsat (kN/m^3)	γsat (kN/m^3)	ν	C (kpa)	φ (°)	K (m/min)	E′ (Kpa)
Soil 1 (Marl)	Mohr-Coulomb	18	19	0.33	165	14	Kx = 10^{-5} Ky = 10^{-5}	4 * 10^6
Soil 2 Clayey loam	Mohr-Coulomb	15.8	17.8	0.33	55	15	Kx = 10^{-3} Ky = 10^{-3}	4.038 * 10^5

ν: Poisson ratio, C: Cohesion, φ': Friction angle, K_x and K_y: Horizontal and vertical hydraulic conductivity respectively, E′: Young modulus.

4 Parametric Analysis

A parametric study is carried out to analysis the effect of distributed surcharge (***q***) on the stability of the slope in terms of the factor of safety. The distribution of weight along slope and loading the top of a slope may have great influence on stability (Noroozi and Hajiannia 2015).

There are many factors affect the stability of the slope. Usually a slope fails by the change in location of surcharge, magnitude of surcharge, variation of sloping angle, etc. and this paper focuses on the level water table, geometry of the slope (length, height, slope angle…etc.), variation of mechanical properties (cohesion and friction angle), surcharge range, surcharge distribution width, distance to crest (Fig. 2).

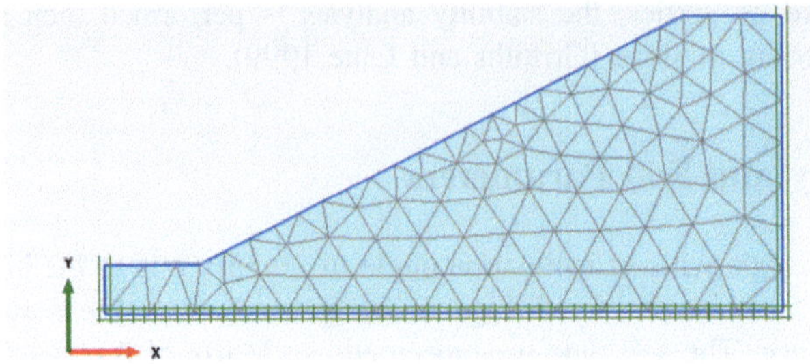

Fig. 2. Model slope with triangular medium mesh

4.1 Effect of Ground Water Depth

Before applying a surcharge to the model, we tried to understand the effect of ground water level on the slope stability; we had decreased the ground water level by a step of 5 m. The preliminary model indicated, as obvious, that the stability of the slope is found to increase if the water table is located at deeper locations: for a depth close to the soil surface, the factor of safety seems to be smaller (at 5 m depth, FS took a value **1.56**). However, it increases when the ground water level is located at a deeper layer (ex: at 50 m depth, it equals to **2.08**) and it's linearly proportional as shown in Fig. 3.

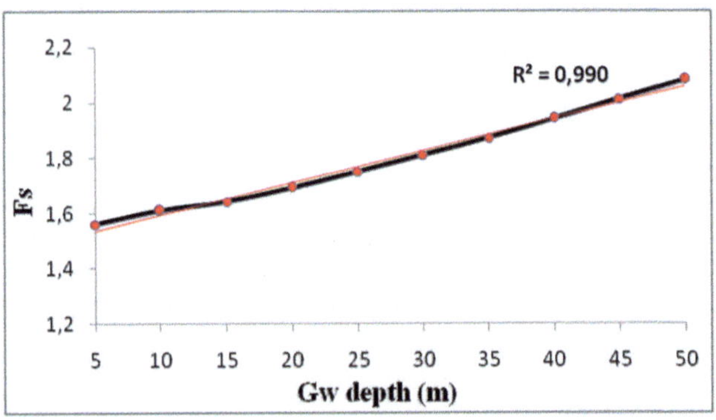

Fig. 3. Variation of factor of safety (FS) for different values of ground water depth

4.2 Effect of Ground Water Level and Surcharge

A combined study used to show the effect of ground water level and surcharge on the slope stability. It is noted that the factor of safety appears to be proportional with ground water depth (polynomial second order) for the different applied surcharges where the safety factor is higher for lower surcharge values and deeper ground water locations, and it will be smaller if the case is the opposite (unfavorable) as shown in Fig. 4. For q = 100 KN/m^2 and ground water table located at 50 m depth, FS is equal to 1.9, while it took the value 1.01 if q = 700 KN/m^2 and the ground water table is found at 5 m depth.

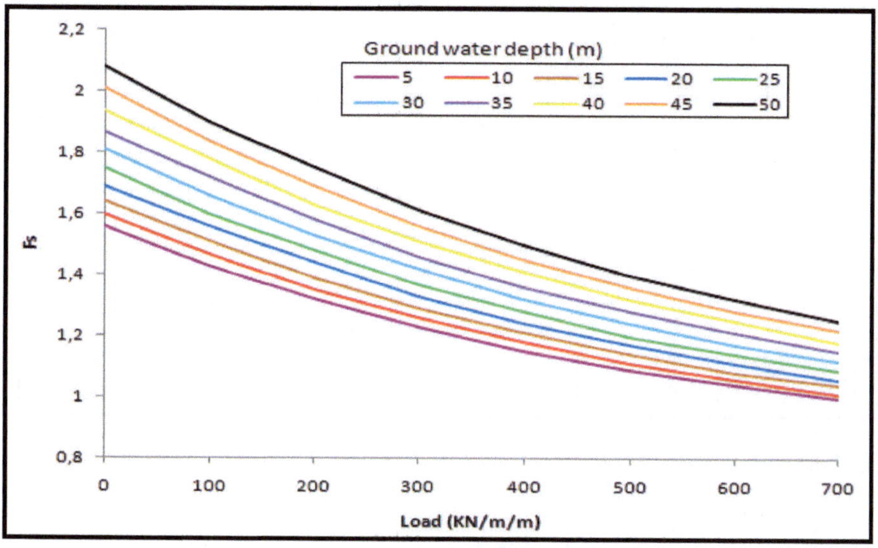

Fig. 4. Variation of factor of safety (FS) with surcharge (q) for different ground water depth

4.3 Effect of Soil Properties (Cohesion and Friction Angle)

In addition to the effect of ground water level, the mechanical properties of soil layer have an important influence on slope stability, i.e., a soil with high mechanical properties (ex: cohesion, friction angle) represents more stable case (with a high safety factor). In other words, the greatest factor of safety values are a combination of more favorable parameters of cohesion and friction angle (Chaudhary et al. 2016). The initial factor of safety for soil-1 (2.08) is greater than safety factor of soil-2 (1.29) for initial values of cohesion and friction angle because the soil-1 had stronger mechanical properties (C' and φ') than soil-2.

As an example, when the soil 2 took the initial values both of cohesion and friction angle ($C' = 55$ kPa; $\varphi' = 15°$), FS is found to be 1.189. If the cohesion value is doubled and tripled to take the following values ($2C' = 110$ kPa; $3C' = 165$ kPa), FS is found to be 1.617 and 2.045 respectively. As expected, greater the cohesion, greater is the safety factor. It can be seen from the figure that there is significant increase in the safety factor with increase in cohesion (Fig. 5). If the friction angle is doubled and tripled to take the following values ($2\varphi' = 30°$; $3\varphi' = 45°$), FS is found to be 1.93 and 2.827 respectively. The soil friction angle has higher influence on the factor of safety than cohesion.

For different values of cohesion and friction angle, a comparison between a model without load and a loaded one reveals that the surcharge decreases the stability of the slope as shown in Figs. 6 and 7. It is also observed that the slope stability also varies linearly both with cohesion and friction angle (Fig. 8).

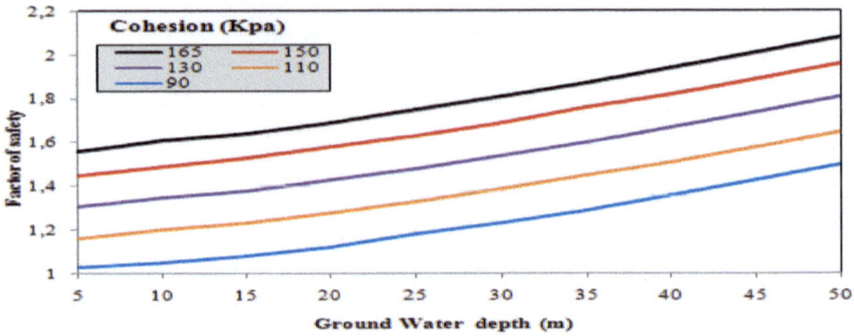

Cohesion	90	110	130	150	165
R^2	0.977	0.986	0.988	0.99	0.99

Fig. 5. Variation of factor of safety (Fs) for different values of Cohesion and ground water depth

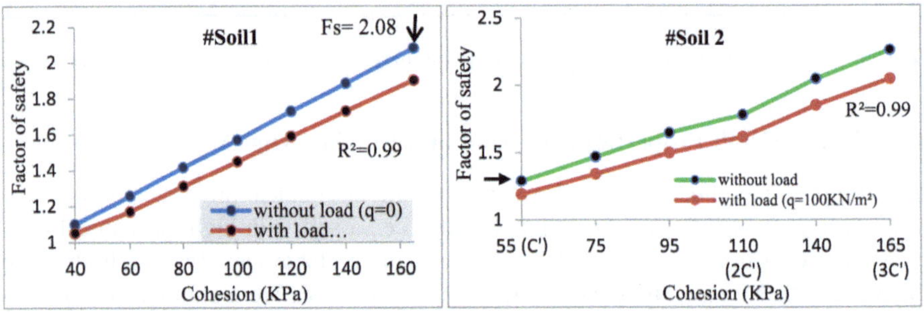

Fig. 6. Variation of factor of safety (FS) with and without surcharge (q) for different values of Cohesion

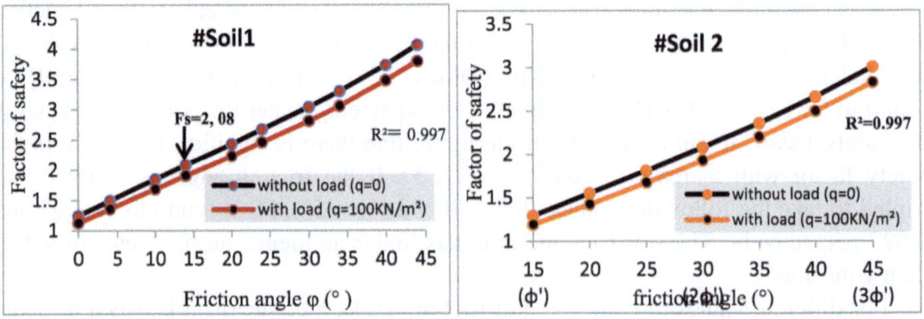

Fig. 7. Variation of factor of safety (FS) with and without surcharge (q) for different values of friction angle

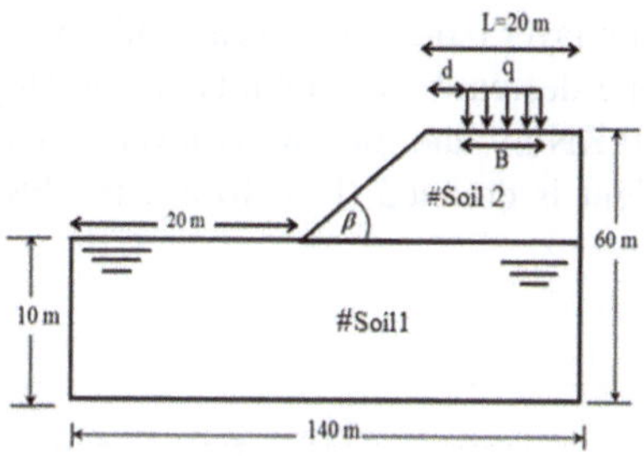

Fig. 8. Layered slope considered in the present study

4.4 Effect of Layered Slope

From the Table 1, it is noted that the soil-1 represents high properties in comparison with soil-2, for that, the slope failure (critical stability) is reached first if the layer takes the soil-2 properties during the application of a surcharge (FS equals to 1 when applying 400 kPa as a load), i.e. counting from this threshold the slope will be unstable and the failure may occur. The slopes that are quite stable can become unsafe, with FS < 1 when the surcharge exceeds a threshold value. In other words, limiting the surcharge can improve the safety of the slope significantly. However, with the same loading condition but with Soil-1, FS is found to increase (1.5) because of the higher soil properties (see Fig. 9) which means that the surcharge threshold is different.

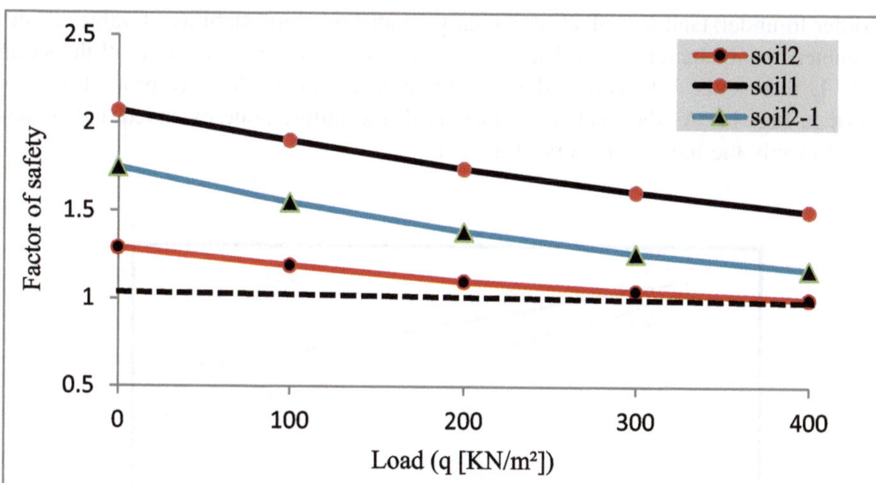

Fig. 9. Variation of factor of safety for the different soil cases

A layered slope as shown in Fig. 9 gave a higher degree of stability (in comparison with a slope only with layer of soil-2. It means that the additional layer with soil-1 properties played a role as a support for strengthening the slope stability.

Because the clayey loam layer (soil-2) is a weak soil layer, the increase in the layer thickness in a layered slope decreases the stability of the slope as shown in Fig. 10. Under a surcharge q = 100 KN/m^2 and for a weak layer height H = 20 m, FS is found to be 1.86. When the height is doubled (H = 40 m), the FS decreases from 1.86 to 1.406.

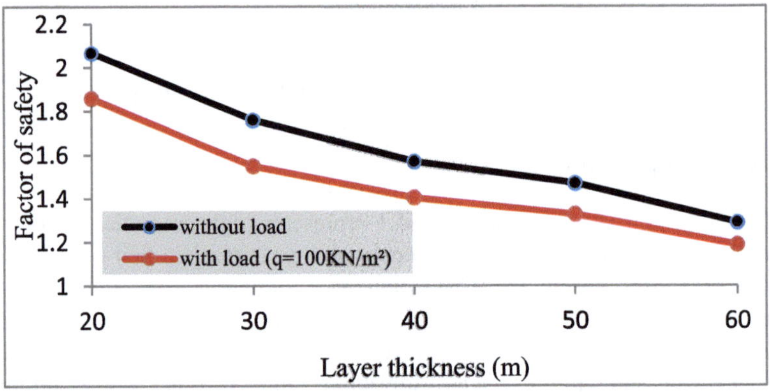

Fig. 10. Variation of factor of safety for the different layer thicknesses

4.5 Effect of Surcharge Width

In order to understand the effect of surcharge width on slope stability, 4 values of the parameter **B** with Increment of 5 m. It is noted that a distributed load on all the slope width **L** represents a delicate and unfavorable case and it's linearly proportional as shown in Fig. 11. As the load width increased, the failure zone extended and propagated towards the toe of the slope (Fig. 12).

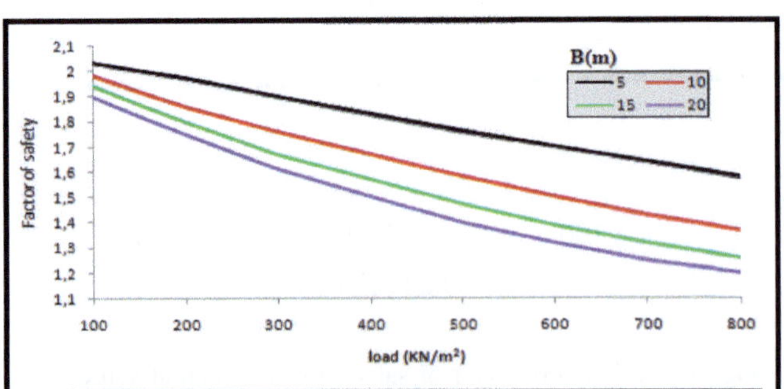

B (m)	5	10	15	20
R^2	0.999	0.99	0.98	0.972

Fig. 11. Variation of factor of safety (FS) with surcharge (q) for different surcharge width values

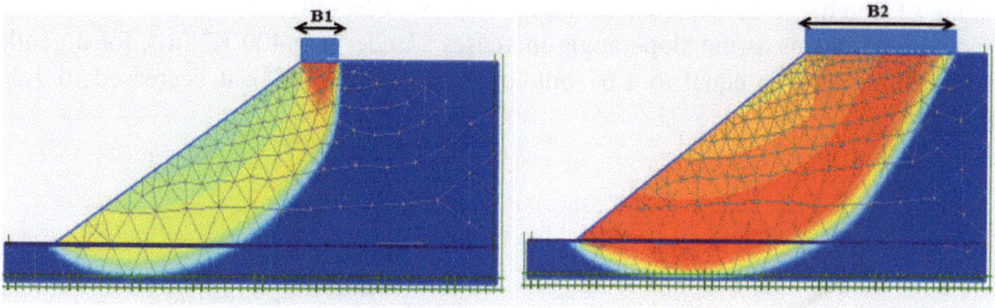

Fig. 12. Propagation and extension of the failure zone (B1 < B2)

4.6 Effect of the Slope Geometry

The geometry of the material has a significant effect in the analysis of the slope. The slope angle and slope height are another important variables (Noroozi and Hajiannia 2015).

4.6.1 Effect of Slope Height

For the same model as shown in Fig. 1 fixing the slope angle (26.56°) and varying the slope height (10 m-20 m-30 m-40 m and 50 m). The results show that the factor of safety decreases gradually when the slope height increases. For a slope with 20 m height, FS is equal to 3.28, However, when the height is doubled to be 40 m, the Factor of Safety decreased to 2.25 (Fig. 13).

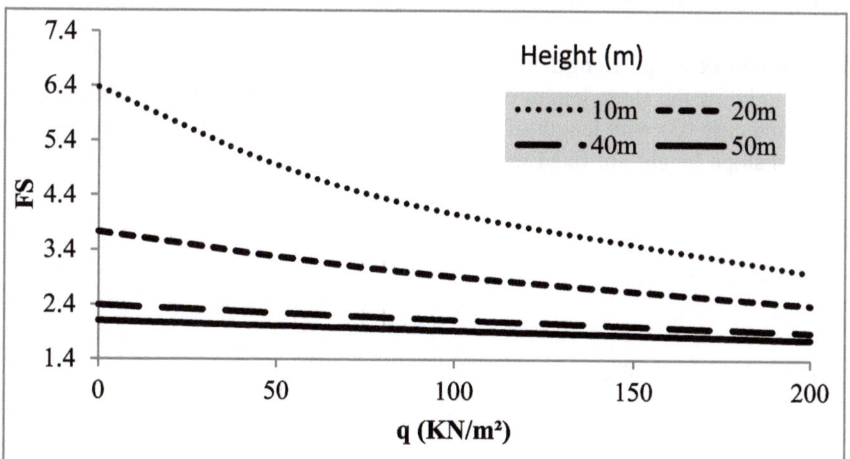

Fig. 13. Variation of factor of safety (FS) with surcharge (q) for different slope heights

4.6.2 Effect of Slope Angle

The slope height is fixed, it took a 50 m as a value and the slope angle is varied (26.56°; 33.69°; 45°). The effect of slope angle on the factor of safety for different values of surcharge is depicted in Fig. 14. It is observed that the factor of safety gradually decreases as the slope angle increases.. Under q = 400 KN/m^2, for a gentle slope (26.56°), FS is equal to 1.67 but for a steep slope (45°), it decreased to 1.36 (Fig. 14).

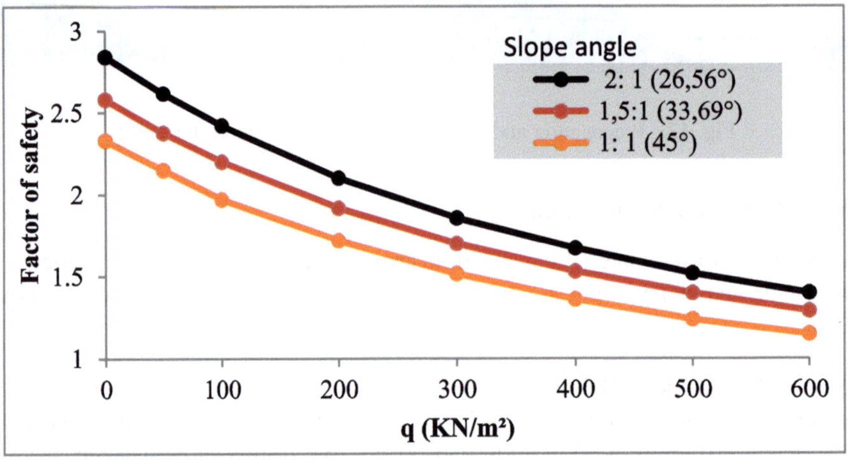

Fig. 14. Variation of factor of safety (FS) with surcharge (q) for different slope angles

It is noted that the factor of safety appears to be proportional with slope angle (polynomial second order) for the different applied surcharges

4.6.3 Effect of Slope Length

In order to know the slope length effect, we had chosen three different cases (case 01, case 02 and case 03) with length of 20 m, 40 m and 80 m respectively while the slope angle and slope height are fixed as shown in Fig. 15.

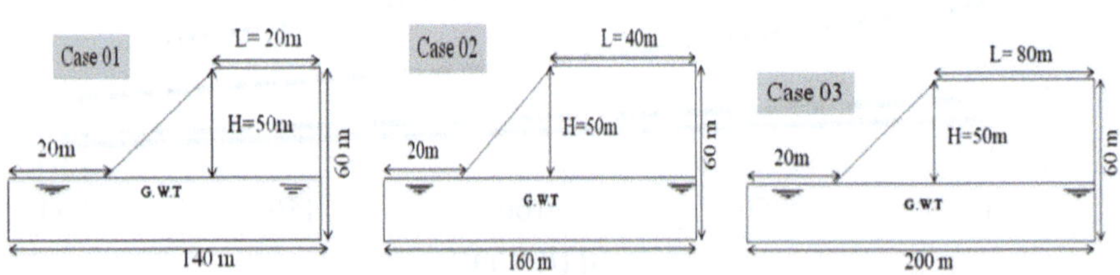

Fig. 15. different model cases used in this study

A surcharge with 20 m width was applied on the 3 profiles and it is observed that the factor of safety increases with increase in slope length. A profile with L = 80 m is more stable during the application of different surcharge values in comparison with the two previous cases. For the case 03, the factor of safety is still constant (2.12) for the surcharge range varying [100–200 kPa], and then it decreases gradually with increase of surcharge (Fig. 16) (Shukla et al. 2009).

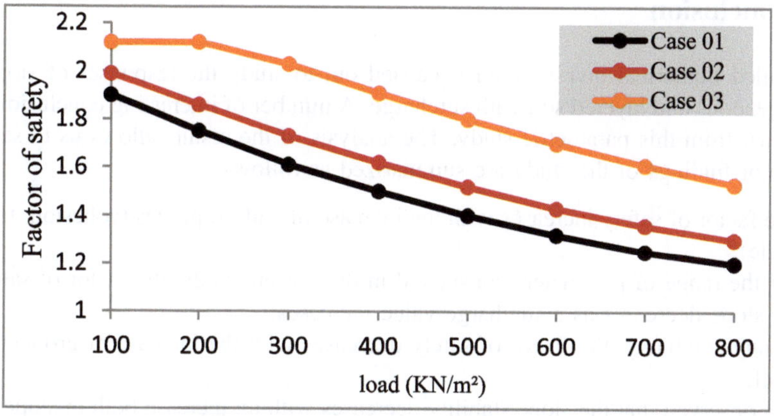

Fig. 16. variation of factor of safety for different slope length values

4.7 Effect of the Position of Surcharge

The case 03 which is mentioned in the previous section was used as a typical model for the analysis of the influence of surcharge position on the factor of safety. The stability increases if the load is applied far from the crest of the slope. The surcharge width B is fixed (B = 15 m). For a distance d = 15 m left from the crest, FS was found 1.47. However, if this distance is tripled to be 45 m, in this case, the stability increased (FS = 1.62).

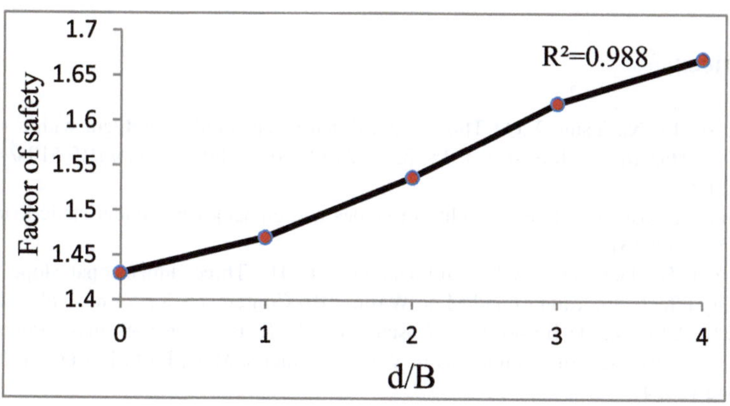

Fig. 17. Variation of factor of safety in function of distance to crest values

As shown in Fig. 17, factor of safety varies linearly with the position of surcharge (with $R^2 = 0.988$). i.e. that the safety factor is a function of the position of surcharge from the crest of slope.

5 Conclusion

A detailed numerical investigation is carried out to study the responses of slopes of homogeneous and layered soil with surcharge. A number of interesting conclusions can be drawn from this parametric study. The analysis of the results allows us to say that the major findings of the study are summarized as follows:

- The factor of safety increases with the increase of soil properties (cohesion, friction angle).
- For the range of parameters considered in the present study, the factor of safety of the slope decreases as a surcharge value increases;
- At any surcharge, the factor of safety decreases with the increase of ground water level;
- It is observed that the slope stability decreases with an increase both in slope angle and slope height. On the other hand, it increases with increase in the slope length;
- The factor of safety of slope of layered soil decreases as the weak layer thickness increases;
- Factor of safety increases with the increase of the distance of surcharge from the crest of the slope;
- As the load width increased, the failure zone extended and propagated towards the toe of the slope.
- All the parameters are correlated with the factor of safety during a loading event, some of them are linear (such as: cohesion, friction angle, surcharge width; surcharge position) while the ground water level, slope angle are polynomial in which the curve varies as a polynomial second order.

References

Cheng, Y.M., Li, N., Yang, X.Q.: Three-dimensional slope stability problem with a surcharge load. Nat. Hazards Earth Syst. Sci. **15**, 2227–2240 (2015). https://doi.org/10.5194/nhess-15-2227-2015

Chowdhury, I., Dasgupta, Sh.P.: Stability of slopes and embankments under static load. EJGE, 5351–5382 (2013)

Chaudhary, K.B., Domingos, V.H., Gitirana, G., Lu, H.: Three dimensional slope stability: geometry effects. In: Tailings and Mine Waste 2016 Conference, Keystone, Colorado (2016)

Crosta, G.B., Utili, S., De Blasio, F.V., Castellanza, R.: Reassessing rock mass properties and slope instability triggering conditions in Valles Marineris, Mars. Earth Planet Sci. Lett. **388**, 329–342 (2014)

Falak, M.E., Bonab, M.H.: Numerical studying the effects of gradient degree on slope stability analysis using limit equilibrium and finite element methods. Int. J. Acad. Res. **4**(4) (2012). https://doi.org/10.7813/2075-4124.2012

Griffiths, D.V., Lane, P.A.: Slope stability analysis by finite elements. Geotechnique **49**(3), 387–403 (1999). https://doi.org/10.1680/geot.51.7.653.51390

Han, J., Leshchinsky, D.: Limit equilibrium and continuum mechanics-based numerical methods for analyzing stability of MSE walls. In: 17th ASCE Engineering Mechanics Conference (EM) (2004)

Jiewen, T., Aiping, T., Yuejun, L., Ketong, L.: Effect of surcharge on the stability of rock slope under complex conditions. J. Eng. Sci. Technol. Rev. (2013)

Kainthola, A., Verma, D., Thareja, R., Singh, T.N.: A review on numerical slope stability analysis. Int. J. Sci. Eng. Technol. Res. (IJSETR) **2**(6), 1315–1320 (2013). ISSN: 2278 – 7798

Kanjanakul, Ch., Chub-uppakarn, T.: Comparison between numerical and limit equilibrium methods for slope stability analysis. In: 18th National Conference on Civil Engineering (NCCE), pp. 104–110 (2013)

Li, N., Cheng, Y.M.: Laboratory and 3-D distinct element analysis of the failure mechanism of a slope under external surcharge. Nat. Hazards Earth Syst. Sci. **15**, 35–43 (2015). https://doi.org/10.5194/nhess-15-35-2015

Moni, M.M.A., Sazzad, M.M.: Stability analysis of slopes with surcharge by LEM and FEM. Int. J. Adv. Struct. Geotech. Eng. **04**(04), 216–225 (2015). ISSN 2319-5347

Napoli, M.L., Barbero, M., Ravera, E., Scavia, C.: A stochastic approach to slope stability analysis in bimrocks). Int. J. Rock Mech. Min. Sci. **101**, 41–49 (2018)

Noroozi, A.Gh., Hajiannia, A. The effects of various factors on slope stability. Int. J. Sci. Eng. Invest. **4**(46), 44–48 (2015). ISSN: 2251–8843

Osiński, P., Rickson, R.J., Hann, M.J., Koda, E.: Assessment of slope stability influenced by vegetation cover and additional loads applied. In: SGGW, pp. 81–91 (2014)

Richards Kevin, S., Reddy Krishna, R.: Slope failure of embankment dam under extreme flooding conditions: comparison of limit equilibrium and continuum models. In: Proceedings of the Geo-Frontiers Conference, ASCE (2005)

Sarma, S.K.: Stability analysis of embankments and slopes. Geotechnique **23**(3), 423–433 (1973)

Sazzad, M.M., Habib, M.A., Karim, R.: Modeling the undrained behavior of granular materials from drained test using dem. In: 2nd International Conference on Advances in Civil Engineering, pp. 319–324 (2014a)

Sazzad, M.M., Rahman, F.I., Moni, M.A.A.: A numerical study on the effect of surcharge on the stability of slope. In: 2nd International Conference on Advances in Civil Engineering, pp. 275–280 (2014b)

Sazzad, M.M., Haque, M.F.: Effect of surcharge on the stability of slope in a homogeneous soil by FEM. In: 2nd International Conference on Advances in Civil Engineering, pp. 314–318 (2014)

Sazzad, M.M., Mamun, M.A.A., Rahman, F.I.: Effect of material model on the FEM based stability analysis of slope. In: International Conference on Mechanical, Industrial and Materials Engineering (ICMIME) (2015a)

Sazzad, M.M., Rahman, F.I., Mamun, M.A.A.: Mesh effect on the FEM based stability analysis of slope. In: International Conference on Recent Innovation in Civil Engineering for Sustainable Development (IICSD-2015), pp. 387–391 (2015b)

Sazzad, M.M., Rahman, F.I., Moni, M.A.A.: Effects of water-level variation on the stability of slope by LEM and FEM. In: Proceedings of the 3rd International Conference on Civil Engineering for Sustainable Development (ICCESD) (2016). ISBN: 978-984-34-0265-3

Singh, A.K., Kundu, J., Sarkar, K.: Stability analysis of a recurring soil slope failure along NH-5, Himachal Himalaya, India. Nat. Hazards (2017a). https://doi.org/10.1007/s11069-017-3076-z

Shukla, S.K., Khandelwal, S., Verma, V.N., Sivakugan, N.: Effect of surcharge on the stability of anchored rock slope with water filled tension crack under seismic loading condition. Geotech. Geol. Eng. **27**, 529–538 (2009). https://doi.org/10.1007/s10706-009-9254-3

Singh, D., Jha, J.N., Gill, K.S.: Effect of canal water level on stability of its embankment and side slopes. Int. J. Eng. Sci. Technol. (IJEST) **9**, 645–651 (2017b). ISSN: 0975-5462

Sun, D.M., Zang, Y.G., Semprich, S.: Effects of airflow induced by rainfall infiltration on unsaturated soil slope stability. Transp. Porous Med. **107**, 821–841 (2015)

Tang, G., Zhao, L., Liang, L., Zuo, S., Zhang, R.: Stability design charts for homogeneous slopes under typical conditions based on the double shear strength reduction technique. Arab. J. Geosci. **10**, 1–16: 280 (2017). https://doi.org/10.1007/s12517-017-3071-4

Toropov, S.Y., Toropov, E.S., Kozlov, A.V., Toropov, V.S.: Behavior of a double-layer pipeline under static load. IOP Conference Series: Materials Science and Engineering, vol. 154, p. 012033 (2016). https://doi.org/10.1088/1757-899x/154/1/012033

Ureel, S., Momayez, M.: An investigation of the limit equilibrium method and numerical modeling for rock slope stability analysis. In: Rock Mechanics and Its Applications in Civil, Mining, and Petroleum Engineering, GSP 237, pp. 218–227. ASCE (2014)

Usluogullar, O.F.: A novel algorithm for slope stability analysis. In: Proceedings of the Institution of Civil Engineers, pp. 1–9. ICE Publishing (2014)

Wei, L., Koutnik, Th., Woodward, M.: A slope stability case study by limit equilibrium and finite element methods. In: GeoFlorida 2010: Advances in Analysis, Modeling & Design (GSP 199), pp. 3090–3099 (2010)

Xie, Z., Xie, R., Lu, X.: Stability analysis on high and steep slope of open-pit based on limit equilibrium method. Applied Mechanics and Materials, vol. 777, pp. 106–111. Trans Tech Publications, Switzerland (2015). https://doi.org/10.4028/www.scientific.net/AMM.777.106

Zhang, Y., Chen, G., Zheng, L., Li, Y., Zhuang, X.: Effects of geometries on three-dimensional slope stability. Can. Geotech. J. **50,** 233–249 (2013). https://doi.org/10.1139/cgj-2012-0279

Zhu, H.H., Wang, E.Y., Shi, B., Wing Wong, J.K.: Feasibility study of strain based stability evaluation of locally loaded slopes: insights from physical and numerical modeling. Eng. Geol. **208**, 39–50 (2016)

Evaluation of Load-Settlement Behaviour of Square Model Footings Resting on Geogrid Reinforced Granular Soils

B. A. Mir(✉) and Saba Ashraf

Department of Civil Engineering, National Institute of Technology, Srinagar, Srinagar 190006, Jammu and Kashmir, India
p7mir@nitsri.net, bamiriitb@gmail.com,
bashiriisc@yahoo.com, sabaashraf09@gmail.com

Abstract. Reinforced soil foundation is a soil foundation containing reinforcing elements or inclusions in the form of strips, grids or ties which are placed horizontally. The purpose of this research was to study the load settlement behavior of square footings placed on geogrid reinforced sandy soil foundation using model tests. While the geometric parameters are specific to certain specifications and conditions such as gradation, relative density of soil beneath the foundation etc., the type, aperture size, stiffness and thickness etc. of geogrid may have some effect on the critical parameters of the reinforced soil foundation. Therefore, two types of geogrid have been used in this study to investigate the load settlement behavior for model footings reinforced with uniaxial (UNX100) and biaxial (BIX100) polyester geogrids resting on poorly graded medium sand. Parameters of this study include the depth to the first reinforcement layer, spacing between reinforcement layers, plan area of the reinforcement and the grade of reinforcement. Load tests were conducted on model footings made of mild steel plates (100 mm * 100 mm size) in four phases to evaluate the effects of single and multiple layers of geosynthetic reinforcement placed below shallow spread footings and to quantify the difference in performance between the uniaxial and biaxial Geogrid reinforcement. Based on the current study, values for optimum embedment depth, optimum spacing between geogrid layers, and optimum plan area of reinforcement for both types of geogrid have been proposed. It was observed that inclusion of reinforcement increased the bearing capacity as well as decreased the settlement of the foundation. The presence of geogrid in the soil makes the relationship between the settlement and applied pressure of the reinforced soil almost linear. Biaxial geogrid performs better than uniaxial geogrid, though the uniaxial reinforcement also leads to significant improvement in performance of footings compared to unreinforced footings.

Keywords: Square footing · Geogrid · Sand bed · Laboratory model tests
Uniaxial geogrid · Biaxial geogrid · Load tests · Bearing capacity ratio

H. Shehata and B. Das (Eds.): GeoMEast 2018, SUCI, pp. 103–126, 2019.
https://doi.org/10.1007/978-3-030-01923-5_9

1 Introduction

Reinforced earth is a material composed of soil fill strengthened by the reinforcement such as rods, bars, fibers and nets. Reinforcements interact with soil by means of frictional resistance. The concept of strengthening the soil by reinforcement is of very old origin. From the olden times attempts have been made to improve mechanical properties of soil by; adding straw or bamboo strips etc. Reinforced earth systems are used today in a variety of civil engineering applications owing to its particular advantages such as cost reduction, simplicity and ease of construction. The reinforced soil foundation system has been a technical and commercial success due to its numerous advantages and thus garnered the attention of the academic circles for research and development in this field. One of the applications of reinforced soil is that it can prove to be an effective alternative to the conventional ground improvement techniques in the improvement of bearing capacity of weak soils. Reinforced earth systems have nowadays found a wide range of civil engineering applications. The reinforced soil foundation derives its reinforcing function from the frictional bond between the reinforcement and the soil. This is the reason that granular cohesionless soils have usually been specified in literature. Vidal (1968) did some pioneering work to illustrate the use of reinforced soil foundation systems for the improvement of the bearing capacity of soft subsoil foundation. Reinforced foundation beds show a good improvement in the bearing capacity and in reducing foundation settlement. In the past, beneficial effect of incorporating tensile reinforcing strips in granular soil fills has been described. In this pursuit, soil reinforced with strips, fabrics, grids and cells are becoming very common. Research is being carried out to study the effect of inclusions below the foundations in increasing the bearing capacity of foundations, all around the globe. During the last three decades, several laboratory model tests and a few field tests have been performed to investigate the ultimate and allowable bearing capacity of shallow foundations supported by single or multiple layered geogrid reinforced sand or clay. Some of the investigators have evaluated the useful effect of using geogrid as a soil reinforcement under the footings by conducting laboratory model tests (Guido et al. 1986; Omar et al. 1993; Das et al. 1994; Das and Khing 1994; Das and Omar 1994; Yetimoglu et al. 1994; Huang and Menq 1997; Das et al. 1998; Shin et al. 2002; Kumar and Saran 2003; Kumar et al. 2007). Adams and Collin (1997) conducted large scale model footing tests to study the ultimate bearing capacity on Geogrid reinforced soil.

Binquet and Lee (1975a, b) have presented the results of model tests with strip footing on reinforced sand foundations for three conditions and showed that the load-settlement and ultimate bearing capacity of footing can be improved by a factor of about 2 to 4 times the bearing capacity of unreinforced soil at comparable settlements for otherwise identical conditions. Akinmasuru and Akinbolade (1981) presented results of laboratory model tests with square footing on a deep homogeneous sand bed reinforced with flat strips of rope fiber material and reported that different types of failure mode observed are in agreement with the modes of failure identified by Binquet and Lee (1975a, b). Similar studies have been reported by many other researchers (e.g. Fragaszy and Lawton 1984; Mahmoud and Abdrabbo 1989; Chadbourne 1994; Dash et al. 2003; Abu-Farsakh et al. 2013; Panda and Ray 2014; Prasad et al. 2016; Budania et al. 2017). Singh (1988) has conducted model plate load tests on reinforced sand beds using aluminum strips, and grid reinforcements and concluded that for a given reinforcement material, the thickness of the reinforcement will not influence mobilization of interfacial friction. Further, many researchers (e.g. Guido et al. 1987; Mandal and Sah 1992; Shin et al. 1993; Fahel et al. 2000; Chen 2007; Chen et al. 2007; Abu-Farsakh et al. 2007; Abu-Farsakh et al. 2008; Sharma et al. 2009; Ornek 2009; Laman et al. 2009; Abdelhadi 2013; Pathan et al. 2017) have reported extensive experimental and numerical study on clayey soil reinforced soil foundation. Therefore, in this paper, authors have made an attempt to investigate load-settlement behavior of square model footings resting on Geogrid reinforced granular soils.

In the present study, the load settlement behavior was studied for model footings reinforced with uniaxial geogrid (UNX100) and biaxial geogrid (BIX100), both polyester geogrids, resting on sand classified as poorly graded at a relative density equal to 50% and quantum of improvement was observed for each grade. Parameters of the testing program include the spacing between reinforcement layers, the depth to the first reinforcement layer, plan area of the reinforcement, the grade of reinforcement. Load tests were conducted on model footings made of mild steel plates (100 mm * 100 mm size) in four phases to evaluate the effects of single and multiple layers of geosynthetic reinforcement placed below shallow spread footings and to quantify the difference in performance between the uniaxial and biaxial geogrid reinforcement. Based on the current study, values for optimum embedment depth, optimum spacing between geogrid layers, and optimum plan area of reinforcement for both types of geogrid have been proposed in the results. It was observed that inclusion of reinforcement increased the bearing capacity as well as decreased the settlement of the foundation. The increase in bearing capacity was expressed in a non-dimensional form, called Bearing Capacity Ratio.

2 Objective of Current Investigation

Although the basic principles of reinforced soil foundation are simple and easy to translate in the field, the theoretical aspects involved in the reinforced soil concept are numerous. Thus warranting extensive research regarding various parameters involved. The purpose of this research was to study the load settlement behaviour of square footings placed on geogrid reinforced sandy soil foundation and to observe the various parameters instrumental to their performance using model tests. The main objective of this work is to study the behaviour of axially loaded square footing on sand bed with and without reinforcement, study the various parameters contributing to their performance and to find out improvement in bearing capacities using two types of geogrid reinforcement. Parameters of the testing program include the depth to the first reinforcement layer, the spacing between reinforcement layers, plan area of the reinforcement, the grade of reinforcement.

A review of the existing literature revealed that most of the experimental studies on geogrid reinforced sand foundation were carried out using a single type of geogrid. Although the geometric parameters of the reinforced soil foundation are governed by certain specifications and conditions such as gradation, Relative density of soil beneath the foundation among others; the type, aperture size, stiffness and thickness etc. of geogrid may also have some effect on the critical parameters of the reinforced soil foundation, therefore two types of geogrid have been used in this study which will help to shed some light on the matter.

3 Experimental Study and Sample Preparations

The experimental program has been designed to study the bearing capacity of axially loaded square footing on single and multi-layered geogrid reinforced sand bed. For this purpose, the laboratory model tests were conducted on square footings in one density (50%). Load tests were performed to evaluate the effects of single and multiple layers of geosynthetic reinforcement placed below shallow spread footings. Two types of geogrid are evaluated, a biaxial geogrid and a uniaxial geogrid. Parameters of the laboratory model tests consist of the embedment depth of reinforcement, spacing between individual reinforcement layers, plan area of the reinforcement, the grade of reinforcement. All tests were performed in a test pit constituted by a steel tank of dimensions 0.5 m × 0.5 m × 0.6 m. Figure 1 shows the typical diagram depicting various geometric parameters of reinforced soil foundation.

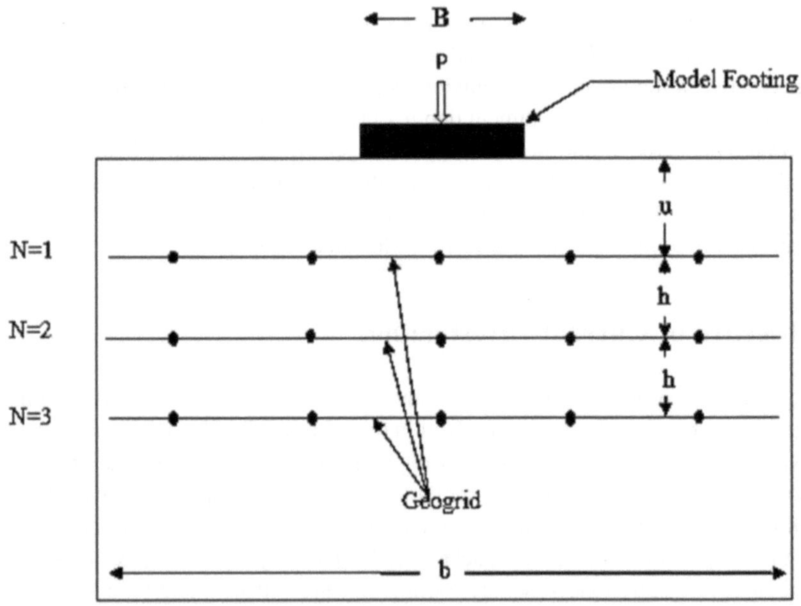

Fig. 1. Typical diagram depicting various geometric parameters of reinforced soil foundation

3.1 Materials

1. *Sand*

   Locally available river sand was used as granular bed during the testing program. The relevant properties of the materials were determined following Indian standard test procedures. The sand used in the experimental program has been mined from the river bed of River Sindh. It was cleaned and washed properly to get rid of any tree roots, pebbles and other unwanted matter. The sand is classified as poorly graded sand (SP) by the Indian Standard Soil Classification System. The grain-size distribution for the sand is shown in Fig. 2. Table 1 lists the physical properties of the sand used in the study. All the tests have been conducted as per relevant Codal procedures (IS: 1498; IS: 2720).

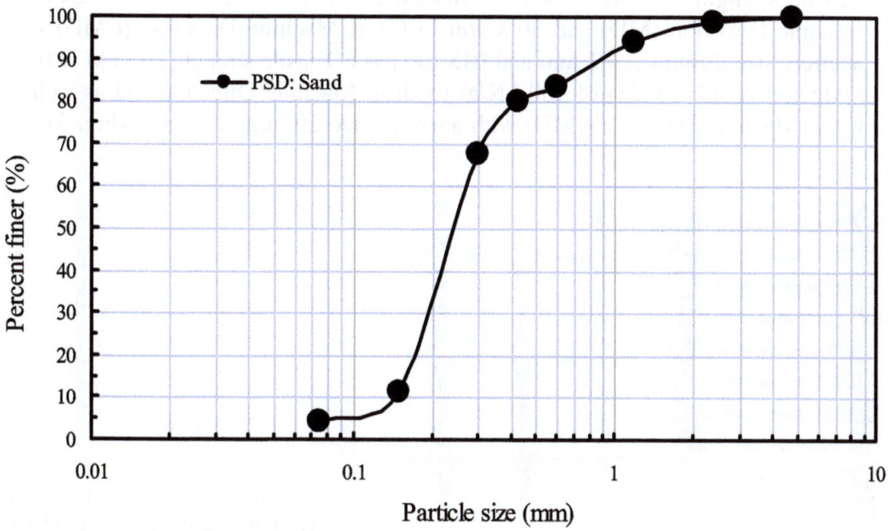

Fig. 2. Particle size distribution curve of sand

Table 1. Physical properties of sand used in the study

Property	Value
D_{10} (mm)	0.145
D_{30} (mm)	0.19
D_{60} (mm)	0.26
D_{50} (mm)	0.24
Coefficient of curvature, C_C	0.96
Coefficient of uniformity, C_U	1.79
Specific gravity, G	2.67
Maximum dry unit weight (kN/m^3)	17.2
Minimum dry unit weight (kN/m^3)	15.1
Friction angle, φ (at D_R = 50%)	35.92°
e_{max}	0.77
e_{min}	0.55

2. *Geogrid*
   Based on the stress transfer, two types of geogrid- uniaxial (UNX100) and biaxial (BIX100) were used as reinforcement. These are polyester (PET) geogrid reinforcement for soil. These geogrids are constructed of high molecular weight and high tenacity knitted polyester yarns with a proprietary coating. UNX100 has a Tensile strength (ASTM D 6637 – Method A; ASTM D4354) of 100 kN/m in machine Direction (MD) and 30 kN/m in Cross Machine Direction (CMD) with aperture size 29 mm × 62 mm and BIX100 has a Tensile strength (ASTM D 6637 - Method A; IS: 13325) of 100 kN/m (both in Machine Direction-CD as well as Cross Machine Direction-CMD) with aperture size 29 mm × 29 mm (Fig. 3).

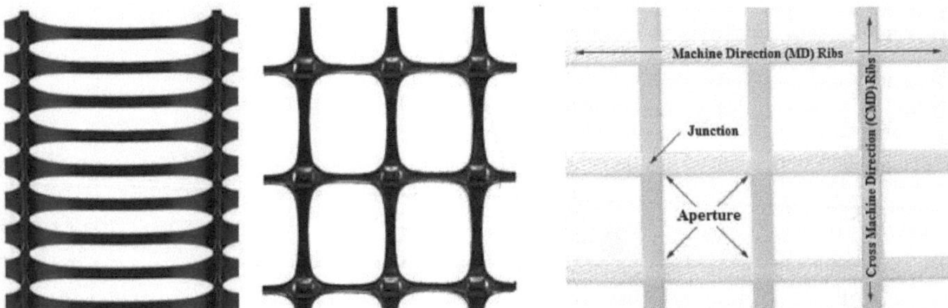

Uniaxial geogrid | Biaxial geogrid | Geogrid's rib formation in machine and cross machine directions during manufacturing process

Fig. 3. Types of geogrid used in the study

3.2 Equipment Used

Test Tank

The load tests were conducted in a tank measuring 0.5 m × 0.5 m × 0.6 m and made up of mild steel of 12 mm thickness. Scales were fitted on the internal walls of the tank so that it will be helpful in maintaining the required density accurately. The sides of the tank are heavily braced to avoid lateral yielding. As per the provisions of IS 1888-1962, the width of the test pit should not be less than 5 times the width of the test plate, so that the failure zones are freely developed without any interference from sides. Also, Chummar (1972) suggested that in case of cohesionless soil, the failure zone extends upto 5B to the sides and 3B below the footing. In order to ensure the development of failure zones fully and freely without any interference from the sides and bottom of the tank, a model footing of size 100 mm*100 mm was adopted for this tank size. Additionally the sides and bottom of the tank were lined with polythene sheets for the same purpose. Figures 4 and 5 show the schematic section and plan respectively of the test tank used.

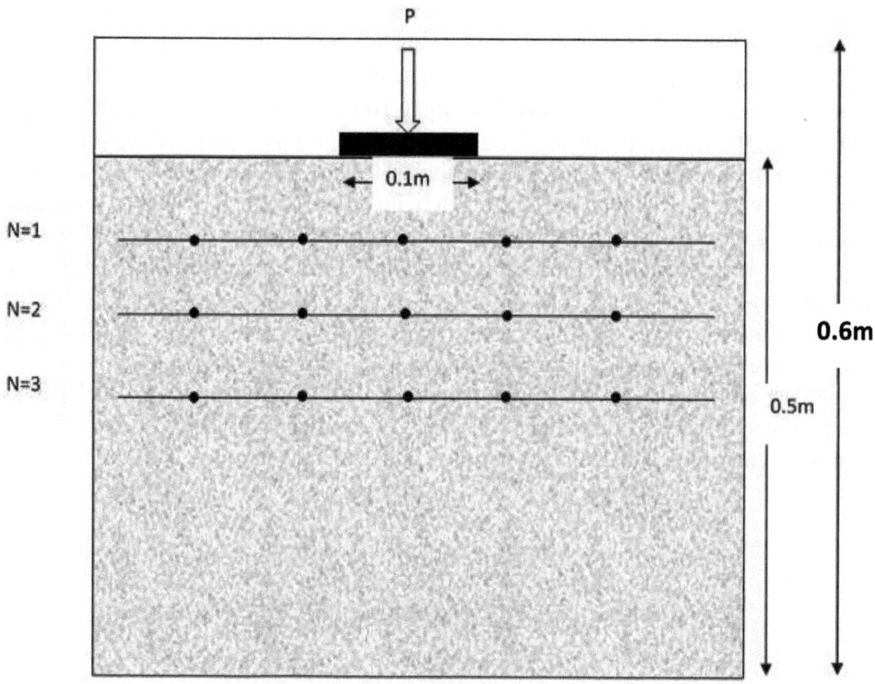

Fig. 4. Schematic section of the test tank used

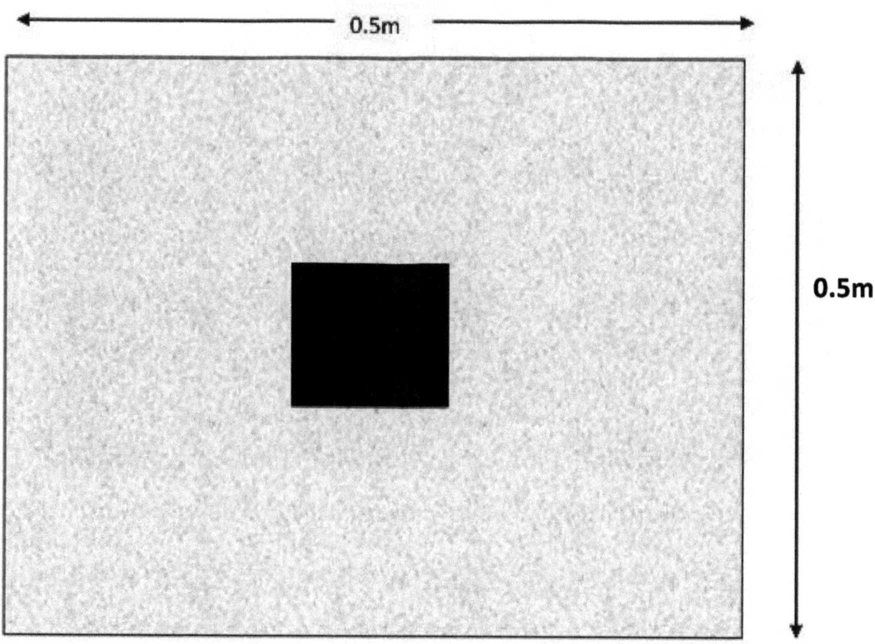

Fig. 5. Schematic plan of the test tank used

The applied loads were resisted against a reaction frame and the frame consisted of four columns supporting two horizontal beams on which the Load transferring shaft rests. The footings were loaded with a manually operated screw jack. The model foundation used for this study was 100 mm * 100 mm in size. It was made out of a mild steel plate with a thickness of 25 mm. Square footing was selected to minimize the dimensional effects (strip and rectangular footings). The bottom of the model foundation was made rough by coating it with glue and then rolling it over sand. Proving ring of 5 kN capacity was used which had a least count of 6.017 N. Two Dial gauges with least count 0.01 mm and range 25 mm were used in the test program. The dial gauges were kept on the rims of the tank as the entire tank had steel strips welded wide enough to accommodate the magnetic base of the dial gauge. The dial gauge needles were placed over the footing loaded with the load transferring column. Settlement occurring as a result of load application was recorded by two dial gauges. Required settlement was taken as the average of the two dial gauge readings in mm. Figures 6a and 6b depicts the photographic image of the equipment used in this study.

Fig. 6a. Photographic image of equipment used in the study

Fig. 6b. Photographic image of sand sample at the start of experiment

3.3 Sample Preparation

The test foundation was prepared to a depth of 4B beneath the model footing to be tested in accordance with the Boussinesq stress-distribution theory. According to this theory, the stress below a footing reduces to a negligible value at a depth of about 3B from the base of footing. Also, Chummar (1972) suggested that in case of cohesionless soil, the failure zone extends upto 5B to the sides and 3B below the footing. Therefore, to be on the safer side, using an excavation depth of 4B should assure that the results were not influenced by previous tests.

The internal dimensions of the test tank were measured precisely with a steel scale and volume of sand to be filled in each layer was calculated. After fixing a density, at which all the tests are to be done by we can calculate the weight of sand needed for that particular thickness of sand layer. Here the density to be maintained was 1.61 g/cc and the sand was placed in 100 mm lifts. All tests were performed at 50% relative density. The box was filled by sand using sand raining technique. Sand was poured into the test tank in 100 mm layers from a fixed height by raining technique to achieve the desired average unit weight of compaction. The height of fall was fixed by making several trials in the test tank prior to the model test to achieve the desired unit weight. Once the backfilling process was completed, the sand was carefully leveled in the areas directly below the footing. This was to ensure that the surface footings had full contact with the sand and that the load applied to the footing was normal. For the unreinforced test or the reference test, the model footing was placed on the surface. For the reinforced case, the first layer of geogrid was placed at the proposed depth below the base of the footing while the other consequent layers of geogrid were placed at particular depths to be tested. After placing the geogrid, small weights were placed all around the geogrid layer to keep the geogrid in position while the calculated weight of sand was being poured over it using sand raining technique.

3.4 Testing Procedure

A series of laboratory model footing tests were conducted on geosynthetic reinforced sand foundation (ASTM 1196, IS 1888). Two types of geogrid types UNX100 (uniaxial) and BIX100 (biaxial), both polyester geogrids were used as reinforcement inclusions in these tests. The footing was placed in position in the test tank. Load was applied through the load gauge and was maintained manually. Load was measured with the proving ring. A rocker plate and ball joint were fitted to the proving ring. Two dial gauges were placed on diagonally opposite sides to record settlements on the corners of the footings. The load was applied in increments. Load was increased by the next increment only when there was no substantial change (>2 mm/min) in settlement between any two time intervals. Each load increment was maintained manually. The settlement was recorded at 1, 2.25, 4, 6.25, 9, 16, 25, and 30 min intervals from the start of each load increment. After each series was finalized, the sand layer was carefully removed and the reinforcement layer was visually examined for distortions and/or damages.

3.5 Testing Program

Experimental studies were carried out to obtain the load-settlement response of a model square footing resting on unreinforced and reinforced granular beds. The parameters studied in this research were the effect of top layer spacing, vertical spacing between layers, grade of reinforcement, and embedment depth and plan area of reinforcement. From the laboratory model tests, settlement ratio versus bearing pressure response of footings is brought out for all the cases. The research was conducted in four phases (A, B, C & D) as outlined below:

1. *Phase-A*
   Phase-A consisted of the test for evaluating the response of model footing placed on unreinforced soil.
2. *Phase-B*
   This phase consisted of series B-1 and B-2. The primary purpose of Phase B was to evaluate the effects of different reinforcement depth with respect to footing size (u/B) and to quantify the difference in performance between the uniaxial and biaxial Geogrid reinforcement.
3. *Phase-C*
   This phase consisted of series C-1 and C-2. The primary purpose of Phase C was to evaluate the effects of different reinforcement widths with respect to footing size

(b/B) and to quantify the difference in performance between the uniaxial and biaxial Geogrid reinforcement.

4. *Phase-D*

   This phase consisted of series D-1 and D-2. The primary purpose of Phase D was to evaluate the effects of different reinforcement spacing with respect to footing size (h/B) and to quantify the difference in performance between the uniaxial and biaxial Geogrid reinforcement.

The testing parameters of the laboratory model tests are given in Table 2.

Table 2. Laboratory testing parameters

Test series	Model foundation	Type of geogrid	Constant parameters	Variable parameter	Objective
A-1	Square	–	D_R = 50%	–	To determine q_u
B-1	Square	UNX 100	N = 1, D_R = 50%, b/B = 5	u/B = 0.2,0.4,0.6	To determine $(u/B)_{opt}$
B-2	Square	BIX100	N = 1, D_R = 50%, b/B = 5	u/B = 0.2,0.4,0.6	To determine $(u/B)_{opt}$
C-1	Square	UNX 100	N = 1, D_R = 50%, u/B = 0.4	b/B = 4,4.5,5	To determine $(b/B)_{opt}$
C-2	Square	BIX100	N = 1, D_R = 50%, u/B = 0.5	b/B = 4,4.5,5	To determine $(b/B)_{opt}$
D-1	Square	UNX 100	N = 3, D_R = 50%, u/B = 0.4	h/B = 0.2,0.3,0.4	To determine $(h/B)_{opt}$
D-2	Square	BIX100	N = 3, D_R = 50%, u/B = 0.5	h/B = 0.2,0.3,0.4	To determine $(h/B)_{opt}$

Where: UNX - Uniaxial geogrid type based on strength or stress transfer
BIX- Biaxial geogrid type based on strength or stress transfer
B = Width of foundation (m)
u = Location of the top layer of reinforcement from the bottom the foundation (m)
N = Number of reinforcement layers
h = Vertical distance between two consecutive Geogrid layers (m)
b = width of reinforcement layer (m)
D_R = Relative Density of Sand (%)

4 Results and Discussions

Laboratory bearing capacity tests were conducted on model footings made of mild steel plates (100 mm * 100 mm size). It was observed that inclusion of reinforcement increased the ultimate bearing capacity as well as decreased the settlement of the foundation. The increase in ultimate bearing capacity is generally expressed in a non-dimensional form, called Bearing Capacity Ratio, BCR_u (Binquet and Lee 1975a, b), which is expressed as:

$$BCR_u = q_{u(R)}/q_u \quad (1)$$

Where, $Q_{u(R)}$= ultimate bearing capacity of reinforced soil.
q_u= ultimate bearing capacity of reinforced soil.
q_u = ultimate bearing capacity of unreinforced soil.

In most cases, shallow foundations in cohesion less soils are designed for limited settlement levels as design is often governed by settlement and not by ultimate bearing capacity and, hence the calculation of magnitude of BCR becomes meaningless. Hence it is necessary to determine the BCR at various levels of settlement to aid in the design process of foundation.

$$BCR_s = q_R/q_u \quad (2)$$

Where, q_R = load/unit area of reinforced soil at settlement level "*s*".
q_u= load/unit area of unreinforced soil at settlement level "*s*".

4.1 Test Series B-1 and B-2: To Find the Optimum Depth of Embedment "*u*"

Previous studies have shown that in order to obtain maximum advantage from the geogrid reinforcement layer, it is desirable that "u" be less than 0.67*B*. For larger *u/B*, the failure surface will be fully located above the top layer of reinforcement. For Square model footing on sand, Guido et al. (1985) reported u/B_{opt} as 0.5, Yetimoglu et al. (1994) reported u/B_{opt} as 0.25. Chung and Cascante (2007) reported u/B_{opt} as 0.3–0.5. To verify these results *u/B* was varied in this test series.

To find out the optimum depth of reinforcement, '*u*' was varied by adopting values of 0.2, 0.4, 0.6 times the width of the footing, '*B*' for uniaxial geogrid UNX100 and 0.2,0.4,0.6 times the width of footing B for biaxial geogrid BIX100, (Test series-B1& B2). For both the series B-1 & B-2, other constant parameters were: relative density (D_R) = 50%, number of reinforcement layers (*N*) = 1, and plan area of reinforcement (*b/B*) = 5. Figures 7 and 8 show the plot of *q* vs. *s/B* ratios for various u/B ratios, presenting the variation of bearing pressure with settlement ratio for various reinforcement depth ratios. It can be seen from Figs. 6a and 6b that for geogrid grade

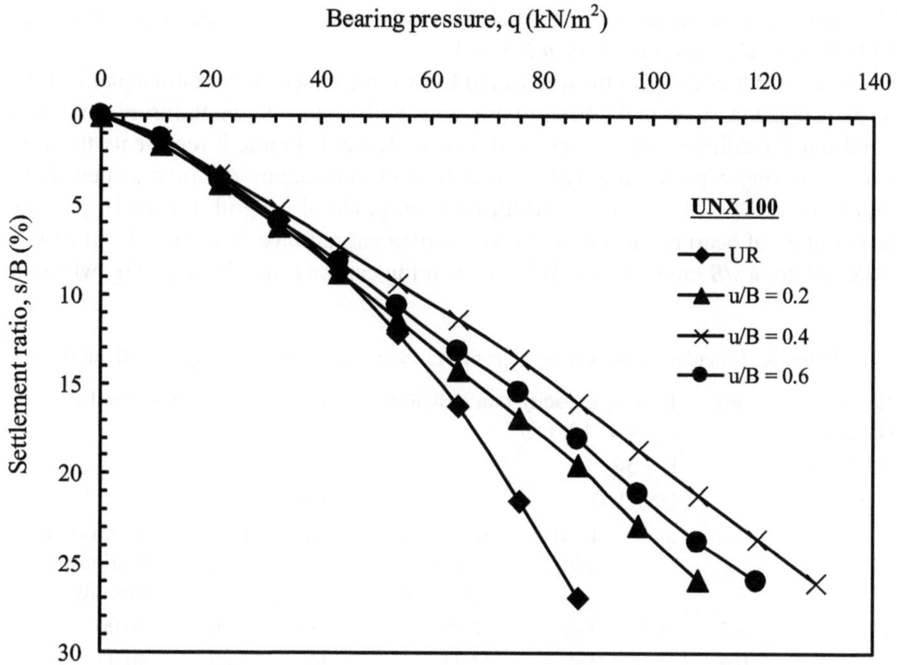

Fig. 7. Plot of q VS. s/B ratio for N = 1, b/B = 5 (series B-1)

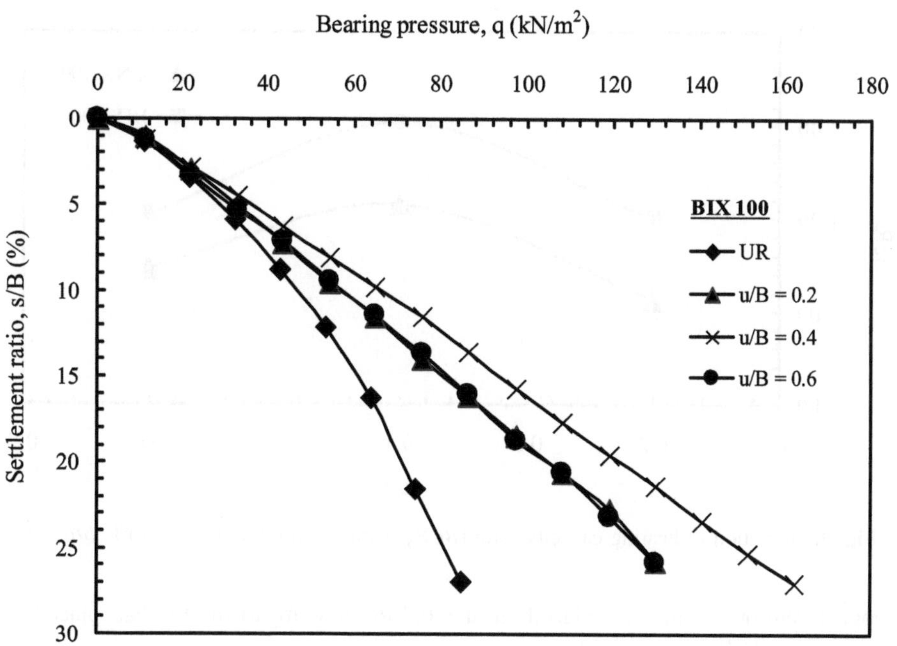

Fig. 8. Plot of q VS. s/B ratio for N = 1, b/B = 5 (series B-2)

UNX100, q_R is maximum for u/B = 0.4. Similarly as seen in Fig. 8, for geogrid grade BX100, q_R is also maximum for u/B = 0.4.

As the depth of the geogrid was varied for a value of settlement ratio equal to 10%, the percentage increase in the bearing pressures with respect to unreinforced sand were found out for different u/B values as shown in Table 3. Figure 9 represents the variation of bearing capacity ratio (BCR) with respect to changing u/B ratio achieved at a settlement of 10% of the footing width for each grade of geogrid. Figure 10 presents the variation of bearing capacity ratio with settlement ratio for Test series B. In case of UNX100, for a s/B ratio equal to 10%, the bearing capacity ratio increased by when the

Table 3. Calculation of bearing capacity ratio at s/B = 10% for varying u/B ratio

No. of Geogrid layers, N	u/B	Bearing capacity ratio (BCRs) = q_R/q_u obtained at s/B(%) = 10% (q_u = 47 kN/m^2)					
		b = 5B					
		UNX100			BIX100		
		q_R	BCRs q_R/q_u	Increase in bearing capacity (%)	q_R	BCRs q_R/q_u	Increase in bearing capacity (%)
1	0.2	48.2	1.03	2.55	56.4	1.20	20.00
1	0.4	57.8	1.23	22.98	66	1.40	40.43
1	0.6	51.4	1.09	9.36	56.8	1.21	20.85

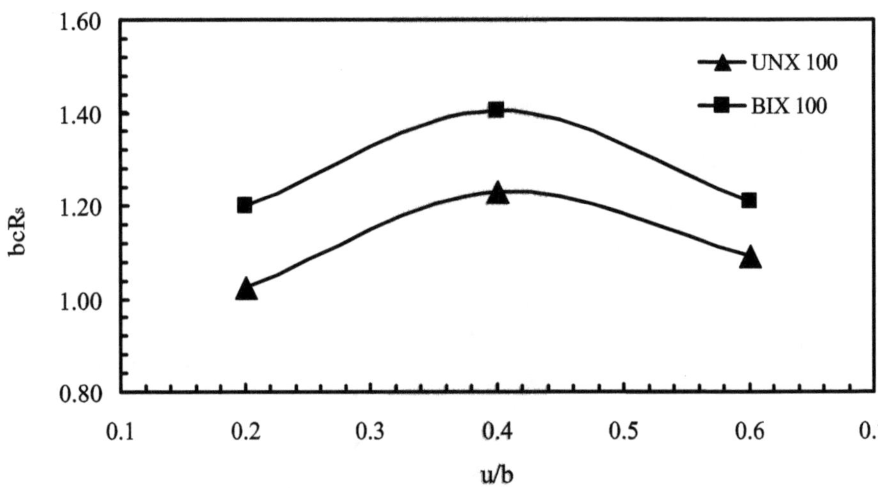

Fig. 9. Variation of bearing capacity ratio (BCR_S) with u/B for N = 1, D_R= 50%, b/B = 5

geogrid reinforcement was placed at u = 0.40B in comparison to that placed at u = 0.2B or 0.6B.

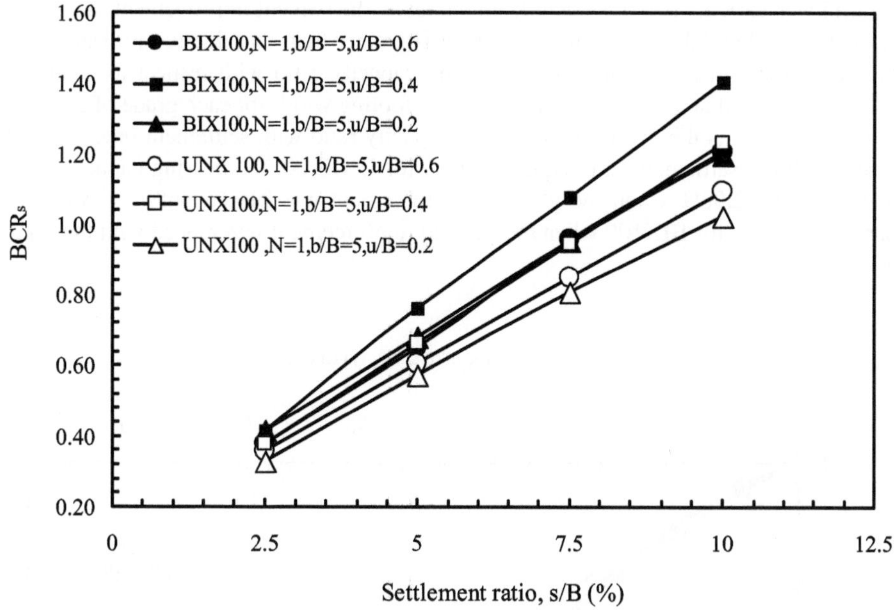

Fig. 10. Variation of bearing capacity ratio (BCR_S) with *s/B* for N = 1, D_R= 50%, *b/B* = 5

4.2 Test Series C-1 & C-2: To Find the Optimum Width of Reinforcement in Terms of b/B

For the purpose of studying the effect of width of the reinforcement on load settlement behavior of the model foundation, '*b*' was varied by adopting the values equal to 4*B*, 4.5*B*, and 5*B* in the tests, while the depth of reinforcement was maintained at the optimum depth (0.4B) and the other parameters were kept at a constant value viz. N = 1 & R_D= 50%. Figures 11 and 12 present the variation of bearing pressure with settlement ratio for different widths of geogrid placed in sand for UNX100 and BIX100 respectively. Omar et al. (1993) reported a value of 4.5*B* for square footing placed Geogrid reinforced sand. Yetimoglu et al. (1994) concluded an optimum width of 4.5*B* for a square footing. Ghazavi and Mirzaeifar (2010) reported an optimum value of 3*B*. Latha and Somwanshi (2009) determined the optimum width of the reinforcement to be about 4*B* for a square footing placed on geogrid reinforced foundation.

The load-settlement response of footing for each variation of width of reinforcement was drawn, and it was observed that ultimate bearing capacity increases with increase in width of reinforcement for both types. Although first portions of the curves show minimal improvement, the latter portions show significant improvement in bearing capacity. This improvement in ultimate bearing capacity with width of reinforcement is often attributed to the extension of reinforcement into the adjacent soil mass beyond the active zone resulting in wide slab effect. Thus concluding the fact that the bearing capacity of reinforced sand increases considerably with increase in reinforcement size when placed within a certain effective zone. Table 4 shows the percentage increase in bearing capacity at a settlement ratio of 10%, as the plan area of the geogrid was varied. As it is evident from the table, the percent improvement in bearing capacity with biaxial grade as reinforcement is greater than that due to uniaxial grade. Figure 13 represents the variation of bearing capacity ratio with respect to changing *b/B* ratio obtained at a settlement of 10% of the footing width for each grade of geogrid. Figure 14 presents the variation of bearing capacity ratio with settlement ratio for Test series C. For a settlement ratio equal to 10%, for UNX100, the bearing capacity ratio increased by 22.98% when the geogrid reinforcement was placed with $b = 5B$ compared to 40.43% for BIX100 when the geogrid reinforcement was placed with the same width.

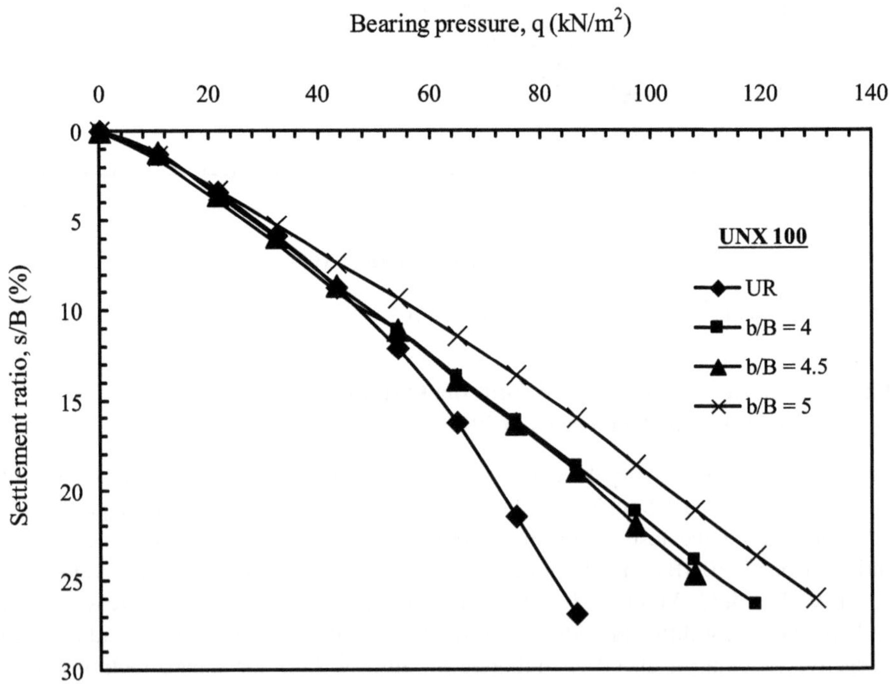

Fig. 11. Plot of q VS. s/B ratio for $N = 1$, $u/B = 0.4$, $D_R = 50\%$ (series C-1)

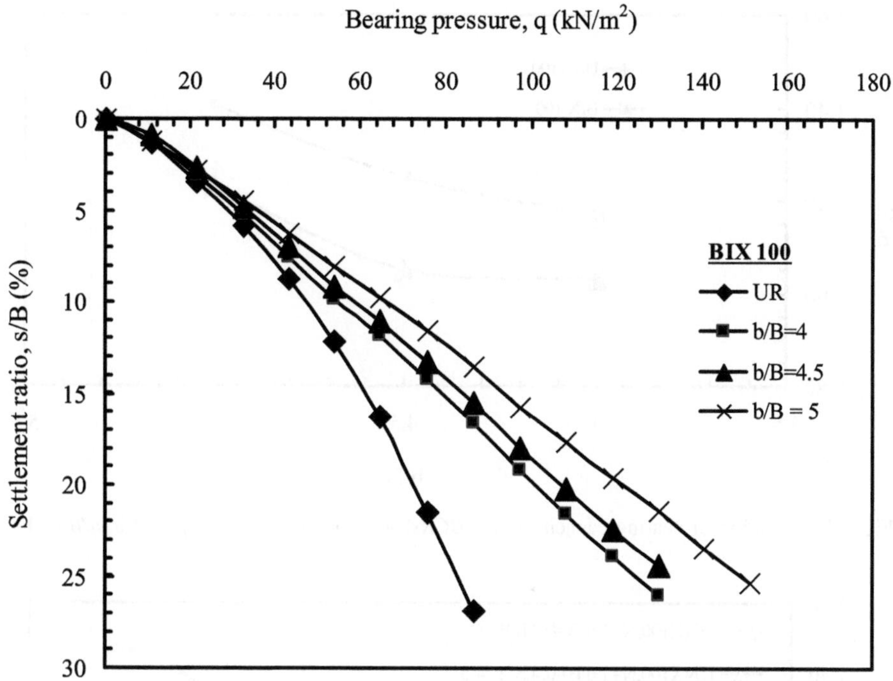

Fig. 12. Plot of q VS. s/B ratio for N = 1, u/B = 0.4, D_R= 50% (series C-2)

Table 4. Calculation of bearing capacity ratio at s/B = 10% for varying b/B ratio

No. of Geogrid layers N	b/B	Bearing capacity ratio (BCR_s) = q_R/qu obtained at s/B(%) = 10% (q_u = 47 kN/m^2)					
		u/B = 0.4					
		UNX100			BIX100		
		q_R	BCRs q_R/qu	Increase in bearing capacity (%)	q_R	BCRs q_R/qu	Increase in bearing capacity (%)
1	4	48.5	1.03	3.19	54.8	1.17	16.60
1	4.5	49.4	1.05	5.11	58.8	1.25	25.11
1	5	57.8	1.23	22.98	66	1.40	40.43

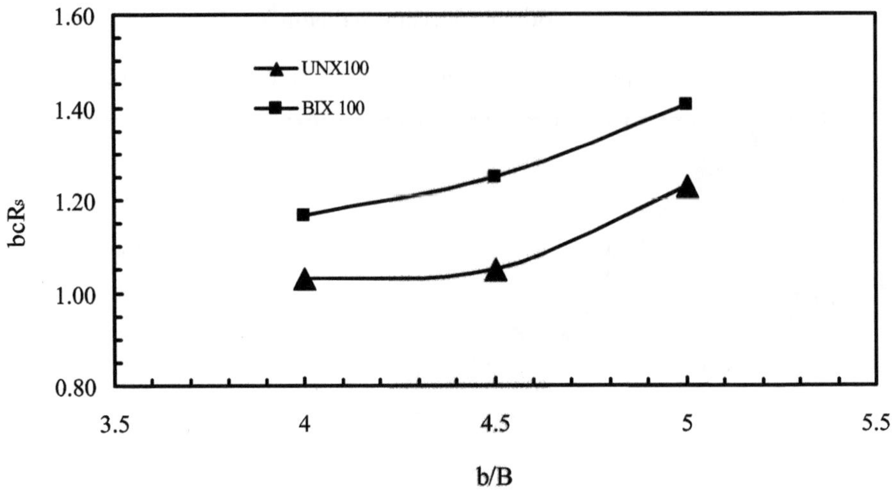

Fig. 13. Variation of bearing capacity ratio (*BCRs*) with *b/B* for $N = 1$, $D_R = 50\%$, $u/B = 0.4$

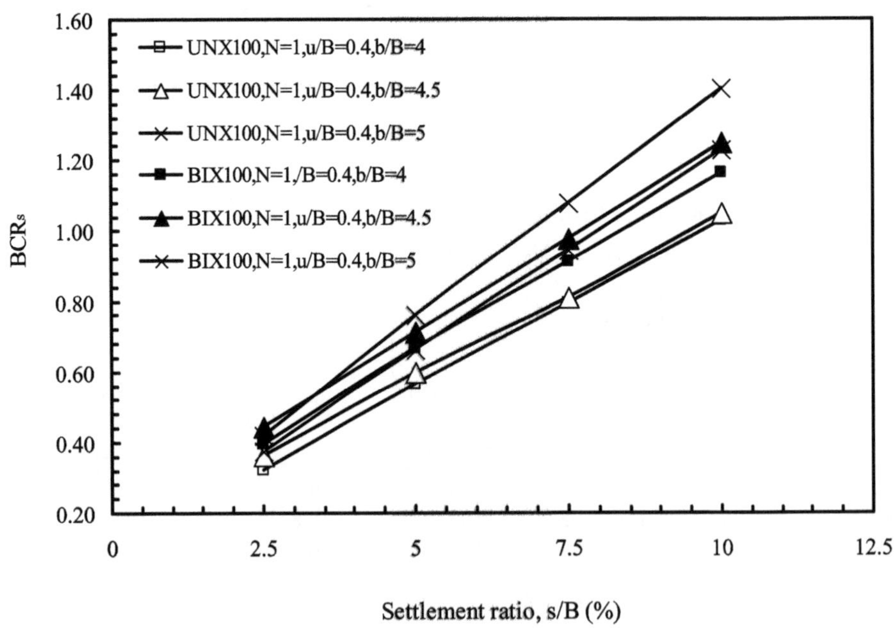

Fig. 14. Variation of bearing capacity ratio (*BCRs*) with *s/B* for $N = 1$, $D_R = 50\%$, $u/b = 0.4$

4.3 Test Series D-1 & D-2: To Find the Optimum Spacing Between Reinforcement Layers "h"

The effect of vertical spacing of reinforcement layers (*h*) on settlement ratio, BCR_S was investigated for surface footings by using three layers of geogrid in each test of series D-1 and D-2. The vertical spacing was varied from 0.2*B*, 0.3*B* to 0.4*B*. Figures 15 and 16 present the variation of bearing pressure with settlement ratio for different spacing of geogrid layers placed in sand for UNX100 and BIX100 respectively. As it is evident from the graphs, the q_R calculated at s/b ratio of 10% shows maximum value in case of h/b = 0.3 in case of UNX100 whereas in case of biaxial geogrid BIX100, optimum h/b is obtained at a value of *h/b* = 0.4. Omar et al (1993) reported a value of 0.33*B* for square footing placed geogrid reinforced sand. Yetimoglu et al. (1994) reported an optimum value between 0.2 and 0.4 times width of foundation. Similarly Adams and Collin (1997) reported a value of 0.25–1.25*B* for various sizes of square footings. Ghazavi and Mirzaeifar (2010) reported an optimum value of 0.3*B*.

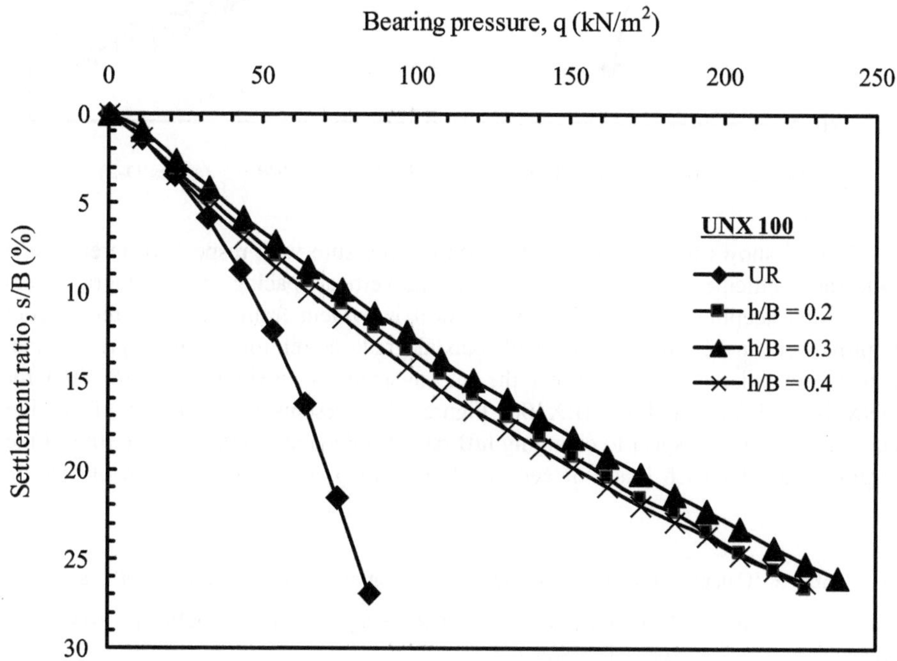

Fig. 15. Plot of q VS. s/B ratio for $N = 3$, $u/B = 0.4$, $D_R = 50\%$, (series D-1)

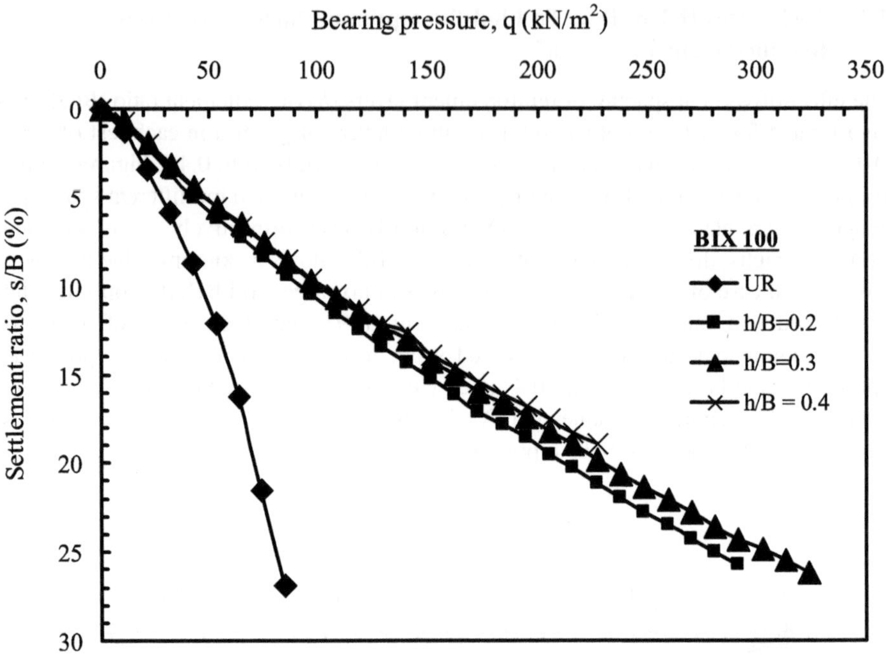

Fig. 16. Plot of q VS. s/B ratio for N = 3, D_R= 50%, b/B = 5 (series D-2)

Table 5 shows the improvement in bearing pressure with respect to unreinforced sand for a settlement ratio equal to 10%, as the vertical spacing of the geogrid layers was varied keeping N = 3 & D_R= 50%. As it is evident from the table, the percent improvement in bearing capacity with biaxial grade as reinforcement is greater than that due to uniaxial grade. Further the improvement is maximum for h/B = 0.3 for UNX100 and h/B = 0.4 for BIX100. Figure 17 represents the variation of bearing capacity ratio with respect to changing h/B ratio obtained at a settlement of 10% of the footing width for each grade of geogrid. Figure 18 presents the variation of bearing

Table 5. Calculation of Bearing Capacity Ratio at s/B = 10% for varying h/B ratio

No. of Geogrid layers N	h/B	Bearing capacity ratio(BCR_s) = q_R/qu obtained at s/B(%) = 10% (q_u = 47 kN/m^2)					
		u/B = 0.4					
		UNX100			BIX100		
		qR	BCRs q_R/qu	Increase in bearing capacity (%)	qR	BCRs q_R/qu	Increase in bearing capacity (%)
3	0.2	70.2	1.49	49.36	92.8	1.97	97.45
3	0.3	77.5	1.65	64.89	101	2.15	114.89
3	0.4	65	1.38	38.30	102	2.17	117.02

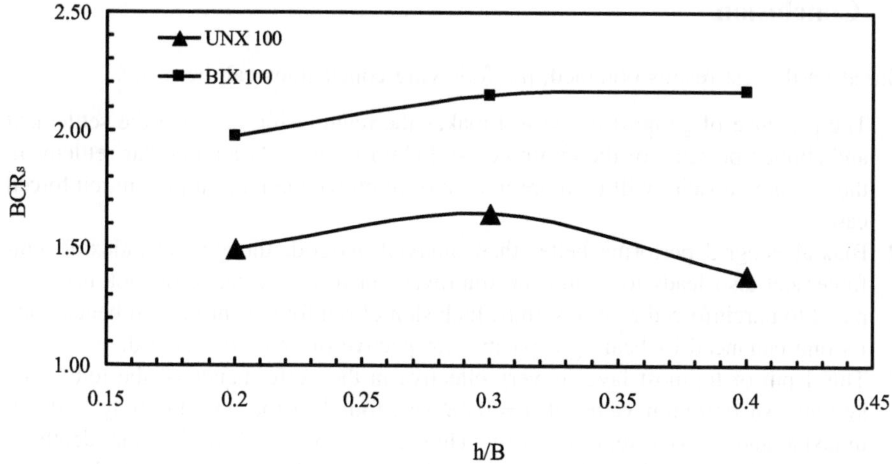

Fig. 17. Variation of bearing capacity ratio (*BCRs*) with *b/B* for $N = 3$, $D_R = 50\%$, $u/B = 0.4$

capacity ratio with settlement ratio for Test series D. For a settlement ratio equal to 10%, for UNX100, the bearing capacity ratio increased by 64.89% when the geogrid reinforcement was placed with $h = 0.3B$ compared to 117.02% for BIX100 when the geogrid reinforcement was placed with $h = 0.4B$.

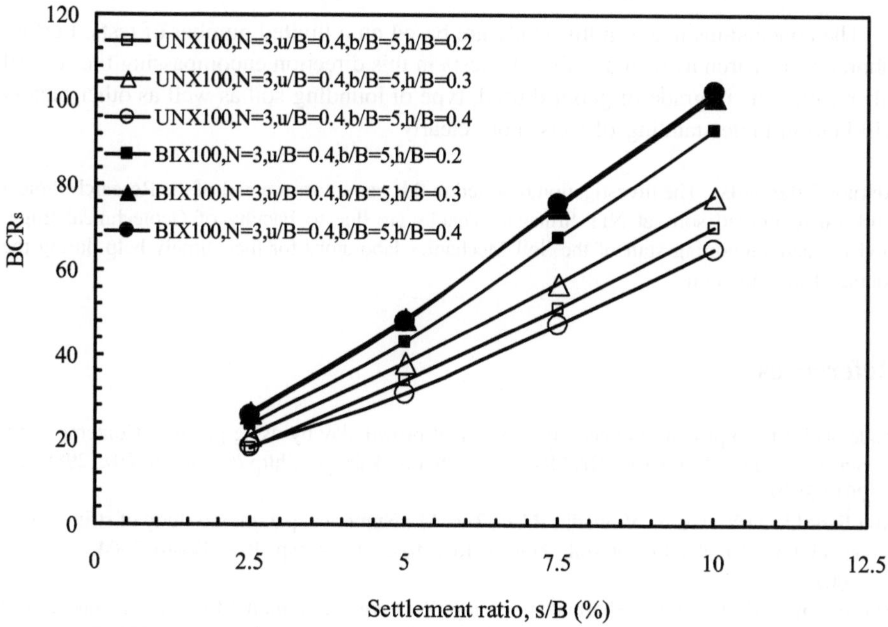

Fig. 18. Variation of bearing capacity ratio (BCR_s) with *s/B* for $N = 3$, $D_R = 50\%$, $u/b = 0.4$

5 Conclusion

Based on the test results obtained, the following conclusions can be made:

1. The presence of geogrid in the soil makes the relationship between the settlement and applied pressure of the reinforced soil almost linear. At a particular settlement, the bearing capacity will be more in case of reinforced condition than unreinforced case.
2. Biaxial geogrid performs better than uniaxial geogrid, though the uniaxial reinforcement also leads to significant improvement in performance of footings compared to unreinforced footings, thus, Inclusion of reinforcement in sand beneath the footing enhanced its bearing capacity, irrespective of grade of geogrid.
3. The depth of topmost layer is very effective in the performance of the reinforced system. An optimum value of $u = 0.4B$ was found in the present study both for uniaxial and biaxial reinforcement. Thus it can be inferred that the depth of embedment 'u' is independent of the grade of geogrid in the current study.
4. For the same soil and geogrid configuration, the ultimate bearing capacity increases with the increase in the width of geogrid, b and no. of layer of geogrid N. Maximum improvement was found at $5b$ in both cases. Hence concluding that improvement in bearing capacity by virtue of increase in width is not affected by grade of geogrid. No optimum no. of geogrid layers is tested in this study.
5. The spacing between the individual reinforcement layers (h) affects the improvement in bearing capacity. The improvement is maximum for $h/B = 0.3$ for UNX100 and $h/B = 0.4$ for BIX100. Thus the spacing between individual layers h is governed by the grade of geogrid reinforcement used.

The conclusions made in this study are based on a limited number of tests. Further laboratory research as well as full scale tests in this direction encompassing the several other variations in grade of geogrid used, type of founding soil as well as other factors will help in understanding of facts more clearly.

Acknowledgments. The investigation reported in this paper forms a part of the Research Project work on reinforced soils at NIT Srinagar. Thanks are due to Faculty of Geotechnical Engg. Division and supporting staff of the Soil Mechanics laboratory for their timely help during the course of investigation.

References

Abdelhadi, M.: Improving the bearing capacity of brown clay by using geogrid. Contemp. Eng. Sci. **6**(5), 213–223 (2013) HIKARI Ltd. www.m-hikari.com, http://dx.doi.org/10.12988/ces.2013.3529

Abu-Farsakh, M.Y., Gu, J., Voyiadjis, G.Z., Tao, M.: Numerical parametric study of strip footing on reinforced embankment soil. Transp. Res. Rec.: J. Transp. Res. Board **2004**, 132–140 (2007)

Abu-Farsakh, M., Chen, Q., Sharma, R., Zhang, X.: Large-scale model footing tests on geogrid reinforced marginal embankment soil. Geotech. Test. J. ASTM **31**(5), 413–423 (2008)

Abu-Farsakh, M., Chen, Q., Sharma, R.: An experimental evaluation of the behavior of footings on geosynthetic-reinforced sand. Soils Found. **53**(2), 335–348 (2013)

Adams, M.T., Collin, J.G.: Large model spread footing load tests on geosynthetic reinforced soil foundations. J. Geotech. Geoenviron. Eng. ASCE **123**(1), 66–72 (1997)

Akinmasuru, J.O., Akinbolade, J.A.: Stability of loaded footing on reinforced soil. J. Geotech. Eng. **107**(GT6), 819–827 (1981)

ASTM D6637: Standard test method for determining tensile properties of geogrids by the single or multi-rib tensile method. ASTM International, West Conshohocken, PA, 2015 (2015) www.astm.org

ASTM D4354: Standard practice for sampling of geosynthetics and rolled erosion control products(RECPs) for testing. ASTM International, West Conshohocken, PA, 2012 (2012). www.astm.org

ASTM D1196: Standard test method for nonrepetitive static plate load tests of soils and flexible pavement components, for use in evaluation and design of airport and highway pavements. Reapproved 1997, pp. 112–113 (1993)

Binquet, J., Lee, K.L.: Bearing capacity analysis of reinforced earth slabs. J. Geotech. Eng. Div. ASCE **101**(GT12), 1257–1276 (1975a)

Binquet, J., Lee, K.J.: Bearing capacity tests on reinforced earth slabs. J. Geotech. Eng. Div. ASCE **101**(GT12), 1241–1255 (1975b)

Budania, R., Arora, R.P., Singhvi, B.S., Kumar, H.V.: Experimental study of rectangular footing resting over geo-grid reinforced sand. Int. J. Adv. Eng. Res. Dev. **4**(1), 292–299 (2017)

Chadbourne, W.: An investigation into the performance of shallow spread footing in reinforced cohesionless soil. M.S. Thesis, Department of Civil and Environmental Engineering, Tufts University, Boston (1994)

Chen, Q.: An experimental study on characteristics and behavior of reinforced soil foundation. Ph.D. dissertation, Louisiana State University, Baton Rouge, USA (2007)

Chen, Q., Abu-Farsakh, M., Sharma, R., Zhang, X.: Laboratory investigation of behavior of foundations on geosynthetic-reinforced clayey soil. Transp. Res. Rec.: J. Transp. Res. Board **2004**, 28–38 (2007)

Chummar, A.V.: Bearing capacity theory from experimental results. Proc. ASCE J. Soil Mech. Found. Div. **99**(sm7), 575–577 (1972)

Chung, W., Cascante, G.: Experimental and numerical study of soil reinforcement effects on the low-strain stiffness and bearing capacity of shallow foundations. J. Geotech. Geol. Eng. **25**(3), 265–281 (2007)

Das, B.M., Omar, M.T.: The effects of foundation width on model tests for the bearing capacity of sand with geogrid reinforcement. Geotech. Geol. Eng. **12**(2), 133–141 (1994)

Dash, S.K., Sireesh, S., Sitharam, T.G.: Behavior of geocell-reinforced sand beds under circular footing. Ground Improv. **7**(3), 111–115 (2003)

Fahel, A.R.S., Palmeria, E.M., Ortigao, J.A.R.: Behaviour of geogrid reinforced abutments on soft soil in the BR 101-sc highway Brazil. Proc. ASCE Spec. Geotech. Publ. **103**, 257–270 (2000)

Fragaszy, R.J., Lawton, E.: Bearing capacity of reinforced sand subtypes. J. Geotech. Eng. ASCE **110**(10), 1500–1507 (1984) https://doi.org/10.1061/(ASCE)0733-9410(1984)110:10(1500)

Ghazavi, M., Mirzaeifar, H.: Bearing capacity of multi-edge shallow foundations on geogrid-reinforced sand. In: Proceedings of the 4th International Conference on Geotechnical Engineering and Soil Mechanics, November 2–3, 2010, Tehran, Iran, Paper No. 600, pp. 1–9 (2010)

Guido, V.A., Chang, D.K., Sweeney, M.A.: Comparison of geogrid and geotextile reinforced earth slabs. Can. Geotech. J. **23**(4), 435–440 (1986)

Guido, V.A., Knueppel, J.K., Sweeny, M.A.: Plate load tests on geogrid—reinforced earth slabs. Proc. Geosynth. **87**, 216–225 (1987)

Huang, C.C., Menq, F.Y.: Deep-footing and wide-slab effects in reinforced sandy ground. J. Geotech. Geoenviron. Eng. ASCE **123**(1), 30–36 (1997)

IS: 1498: Method of test on soils: classification and identification of soils for General Engineering purposes. Bureau of Indian Standards, New Delhi (1970)

IS: 2720-Part. 4: Method of test on soils: grain size analysis. Bureau of Indian Standards, New Delhi (1985)

IS: 2720-Part. III: Method of test on soils: determination of specific gravity of fine, medium & coarse grained soil. Bureau of Indian Standards, New Delhi (1980)

IS: 2720-Part. 14: Method of test on soils: determination of density index (R.D) of cohesion less soil. Bureau of Indian Standards, New Delhi (1983)

IS 1888: Method of load test on soils. Bureau of Indian Standards, New Delhi (1982)

IS: 13325: Methods of test for soils: determination of tensile properties of extruded polymer geogrids using the wide strip - test method. Bureau of Indian Standards, New Delhi (1992)

Laman, M., Yildiz, A., Ornek, M., Demir, A.: Geogrid reinforcement on soft clay deposits. TUBITAK Scientific Research Project (No: 106M496), Ankara, Turkey, 528 p. (2009)

Latha, G.M., Somwanshi, A.: Bearing capacity of square footings on geosynthetic reinforced sand. Geotext. Geomembr. **27**, 281–294 (2009)

Mahmoud, M.A., Abdrabbo, F.M.: Bearing capacity tests on strip footing resting on reinforced sand subtypes. Can. Geotech. J. **26**(1–4), 1989 (1989)

Mandal, J.N., Sah, H.S.: Bearing capacity tests on geogrid-reinforced clay. Geotext. Geomembr. **11**(3), 327–333 (1992)

Omar, M.T., Das, B.M., Puri, V.K., Yen, S.C.: Ultimate bearing capacity of shallow foundations on sand with geogrid reinforcement. Can. Geotech. J. **30**, 545–549 (1993)

Ornek, M. Geogrid reinforcement on soft clay deposits. Ph.D. thesis, University of Cukurova, Turkey (2009)

Panda, S., Ray, N.H.S.: An investigation on behavior of centrally loaded shallow foundation on sand bed reinforced with geogrid. Int. J. Eng. Res. Gen. Sci. **2**(4), 128–146 (2014)

Pathan, S.M.K., Nadeemallah, L.R., Shwetha, V.C., Ali, M.H., Shivakumar, R.H.: Role of geogrid in enhancing the bearing capacity of soil. Int. J. Eng. Sci. Comput. **7**(5), 11295–11298 (2017)

Prasad, B.D., Hariprasad, C., Umashankar, B.: Load-settlement response of square footing on geogrid reinforced layered granular beds. Int. J. Geosynth. Ground Eng. **2**(36) (2016). https://doi.org/10.1007/s40891-016-0070-6

Shin, E.C., Das, B.M., Puri, V.K., Yen, S.C., Cook, E.E.: Bearing capacity of strip foundation on geogrid-reinforced clay. Geotech. Test. J. ASTM **16**, 534–541 (1993)

Singh, H.R.: Bearing capacity of reinforced soil beds. Ph.D. thesis, Indian Institute of Science, Bangalore, India (1988)

Yetimoglu, T., Wu, J.T.H., Saglamer, A.: Bearing capacity of rectangular footings on geogrid reinforced sand. J. Geotech. Eng. Div. ASCE **120**(12), 2083–2099 (1994)

Ferro-Cement Injection Grout for Foundation Strengthening

Maria Avdusheva(✉) and Alexander Nevzorov

Northern (Arctic), Federal University (NArFU), Arkhangelsk, Russia
m.avdusheva@yandex.ru

Abstract. Stone or masonry foundations of historic buildings and objects of cultural heritage used to be built using the lime-based mortar and such practice was quite common mainly due to the availability of the local materials. The main disadvantage of such foundations is the missing or crumbling of the mortar between stones under the impact of freezing-thawing annual cycles and the groundwater aggressivity affects. The recovery of their bearing capacity usually requires special operations such as high pressure cement grout injections. However the quality control of the injections is a challenging task mainly because the degree of the cracks and the cavities filling may be estimated only by inspection of foundation surface after the excavation of pits.

In order to increase the reliability of injections the authors developed a novel method. They add fine electrical conductive substance directly to the grout and place electrodes on the foundation surface prior the reconstruction. That allows to control the injection process by the measurement of the electrical conductivity of the foundation substance. Moreover, the application of ferro-materials as additive allows to increase the degree of filling of the cracks and the cavities due to the magnetic field effect.

The main aim of the current study is to analyze the properties of cement grout with ferro-additives in various concentrations. The electrical conductivity, rheological performance and strength of the grout were determined by different laboratory experiments.

The results of the experiments clearly showed the high perspective of this method and confirmed that the magnetic field effect may considerably simplify the strengthening of old stone foundations.

1 Introduction

Usually the foundations constructed before the wide use of concrete are required underpinning and strengthening works due to the unsatisfactory condition of the stones, the bricks or mortar and increasing loads after reconstruction [1, 2]. For such foundations the local materials were normally used, as limestone mined from a nearby quarries and a lime-based grout. Nowadays the main problem of the stone and masonry foundations is the unacceptable condition of the grout. The long-term groundwater aggressivity affects and repeated cycles of freezing and thawing in conditions of high wetness significantly decrease the foundation strength, lead to the missing and crumbling of the mortar and to the development of cracks and cavities (Fig. 1).

H. Shehata and B. Das (Eds.): GeoMEast 2018, SUCI, pp. 127–135, 2019.
https://doi.org/10.1007/978-3-030-01923-5_10

Fig. 1. Strength loss of the stone foundations

The most common way to strength the foundations is the injections of cement grout into the cracks and cavities [3]. Some plasticizers, modifiers and other substances may be introduced into the mixture to obtain grout with certain properties. For example, then a nano- and fine powders of titanium dioxide TiO_2, aluminum oxide Al_2O_3, ferric oxides Fe_2O_3 and Fe_3O_4, clay and other materials are normally used as additives [7–9].

However, the quality control is challenging because the excavation of the ground around the old stone foundations is dangerous and work consuming, the control is usually performed by drilling the boreholes at several points through the foundations after the cement grout has set.

In the current study the authors suggested the novel method to monitor the quality of injection process [4]. According to this method the electrodes are placed on the foundation surfaces prior the renovation and fine electrical conductive substance is added to the grout before injection. The control of penetration is performed by measuring the electrical conductivity. Moreover, the application of ferro-materials as additive allows to increase the degree of filling of the cracks and the cavities due to the magnetic field effect.

The research works of J.J. Soto-Bernal, R. Gonzalez-Mota (2015), R.V. Lesovik, M.S. Ageeva (2011) on the effect of magnetic field on physical, mechanical and microstructural properties of cement grout are well known [5, 6]. The experiments demonstrated the increase of the cement hydration rate and improvement of the grout microstructure by reducing its porosity.

The purpose of the current research is to study the properties of cement grout with magnetite and ferrosilicium additives. During the laboratory experiments the electrical conductivity, rheological performance and the strength properties of grout have been investigated.

2 Materials

For experiments the authors applied the magnetite from Kiryabinskoye field in the Republic of Bashkortostan and ferrosilicium IMEXSAR 270D normally used in mining industry for minerals processing in the form of fine powder. According to the manufacturer data 96.5% of the magnetite particles are less than 0.14 mm and the material consists at least of 77% of Fe_2O_3 and 22% of FeO. The ferrosilicium powder consists of 80% of ferrum and the total amount of silicium is equal 14–16%. From these materials four different types of additives with various dispersion were prepared: M-1 – magnetite powder obtained by sieving and separating particles smaller than 0.1 mm; M-2 – fine powder of magnetite; FS-1 – ferrosilicium powder with the particles less than 0.1 mm and FS-2 – fine powder of ferrosilicium.

The additives M-2 and FS-2 were milled to nano sizes by dry grinding down with planetary ball mill Retsch PM-100 at 420 rpm for 60 min. The distribution of the particle sizes after grinding was determined in the glycerin-based suspension by Photon Correlation Spectroscopy with Delsa^{TM}Nano (Fig. 2, Table 1). Portland cement of grade CEM I 42,5 N (ASTM 150 C) from Peterburgcement was used.

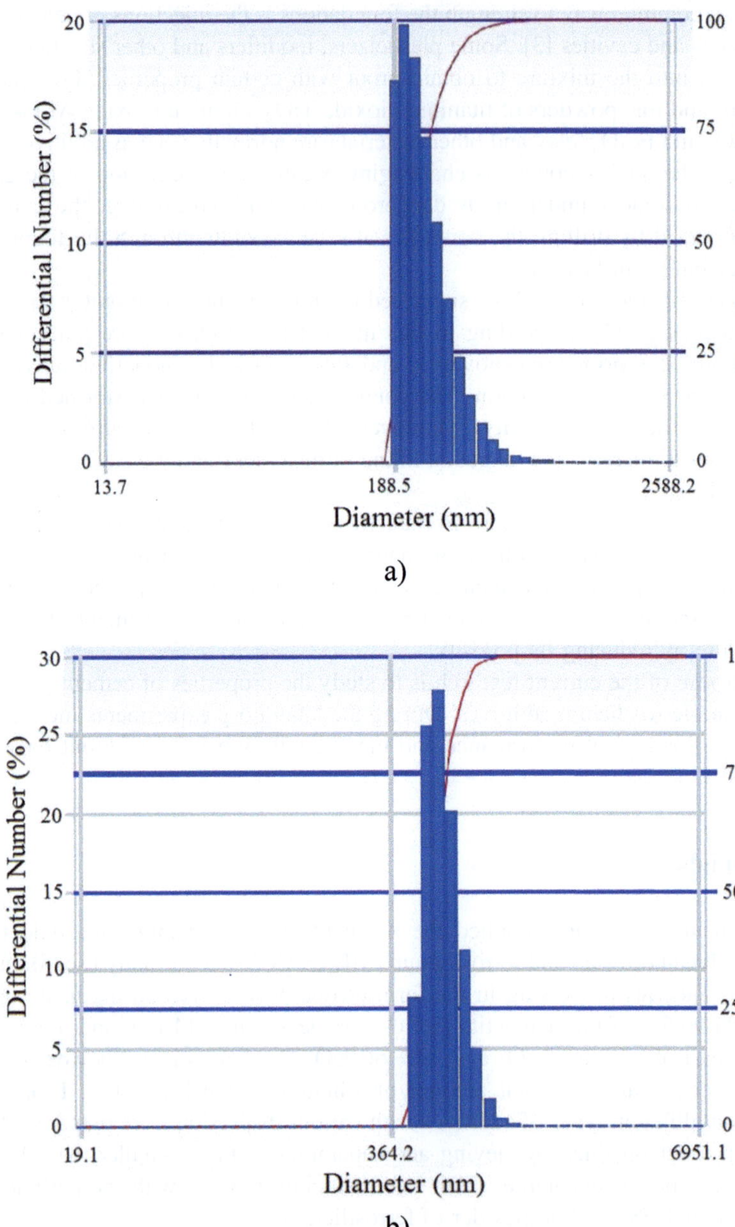

Fig. 2. Distribution of particle size of the additives: (a) M-2; (b) FS-2

Table 1. Description of the additives grain size

Material	Magnetite		Ferrosilicium	
Additive code	M-1	M-2	FS-1	FS-2
Particle size	<0.1 mm	258 ± 74 nm	<0.1 mm	610 ± 108 nm

The mixtures of various compositions were prepared for evaluation of the effect of the different additives on the properties of cement grout. The grout with 1–10% additives was used to study the flow properties and depth of grout penetration into the material with large cracks and cavities. To study flow properties the water/cement (w/c) ratio was taken as: 0.4, 0.45 and 0.5, and to measure the penetration depth w/c ratio 0.4 and 0.5 was used.

To evaluate the strength and electrical conductivity of the grout three component mixtures of sand, cement and water with additives in the amount from 1 to 10% were prepared. The grain size distribution curve of sand is shown on Fig. 3.

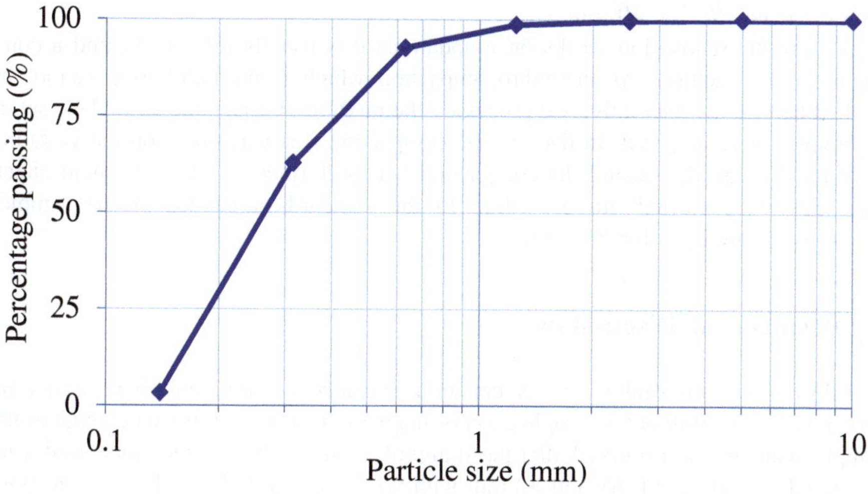

Fig. 3. Grain size distribution curve of sand

3 Methods

The flow properties of the cement mixture was studied by a truncated metal cone opened at both ends with 120 cm^3 volume, 20 mm top diameter and 40 mm bottom diameter. The cement and additive mixture was mixed with water for three minutes and placed into the cone placed on glass sheet. After removing the cone the mixture was exposed by a magnetic field generated by neodymium magnet F200, which was located and moved below the glass sheet. The flowability was evaluated according to the average diameter of the sample spread.

The capacity of the grout mixture to penetrate into the cracks and cavities of stone foundation was studied in transparent cylinders 75 mm high filled by dry or moisturized crushed stone with particle sizes 5–10 mm or 10–20 mm. The experiments were performed with and without magnetic field effect. The parameter controlled during the experiments was the relative penetration depth of the cement grout into the sample calculated by the following formulae:

$$\varepsilon = \frac{h_p}{h_0} \qquad (1)$$

where h_p - depth of grout penetration;

h_0 - height of crushed stone sample.

The electrical conductivity of the mixtures was measured by E7-20 RLC-meter at electrical current frequencies from 25 Hz to 1 kHz.

The strength of grout with various compositions was measured in the samples with dimensions 4 × 4 × 16 cm, aged for 28 days. The flexural strength was defined in a three-point bending test with the length of the support span equaled to 10 cm. The halves of the samples were further tested for compression with the load applied through the steel plates 62.5 × 40 mm.

During the renovation works on foundations it is usually difficult to find a compacting grout method. As the vibrocompaction which is normally used cannot be applied here, as an alternative we processed the mixture by magnetic field. Two series of samples were prepared. In the first the compaction was performed on the vibrating table for 40 s. In the second the compaction was performed by magnetic field effect. The tests were carried out according to the standard procedure in the samples 4 × 4 × 16 cm, aged for 28 days.

4 Results and Discussion

The flowability was studied on cement grout prepared without sand. The results are shown in Fig. 4. Without the application of magnetic field no significant change of the sample diameter was observed after the slump of mixture. The diameter increased from the initial 40 mm to 54, 68 and 82 mm with w/c ratio 0.4, 0.45, and 0.5, correspondently. The experiments demonstrated that the effect of the magnetic field resulted in a remarkable increase of the flow properties. Thus, only 1% of the additive relative to the cement weight increased the sample diameter from 18% to 97% depending on the w/c ratio while adding 10% of the additive resulted in 76–190% increase of diameter. Taking together, the additives FS-1 and FS-2 showed the highest magnetic susceptibility.

Next, we evaluated the change of the penetration depth of the grout in crushed stone under magnetic field effect. Adding 1% to 5% of the additives did not produce any effect and the grout remained in the upper part of the sample. While adding of 10% FS-1 combined the magnetic field resulted in significant increase of penetration depth of the grout (Table 2). The experiments demonstrated that the magnetic field effect increases the depth of cement mixture penetration into the voids by 2 to 3 times.

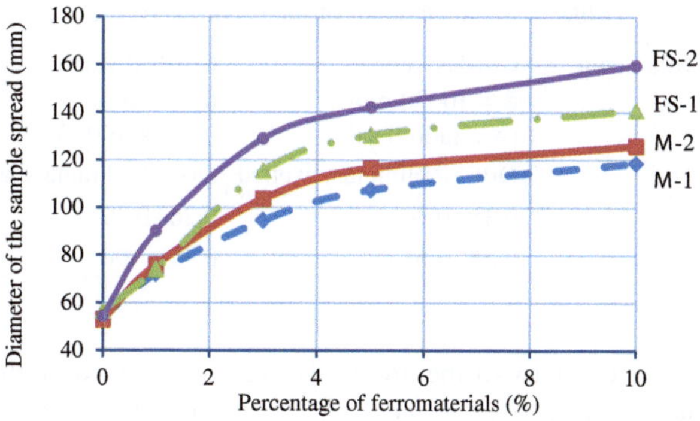

a)

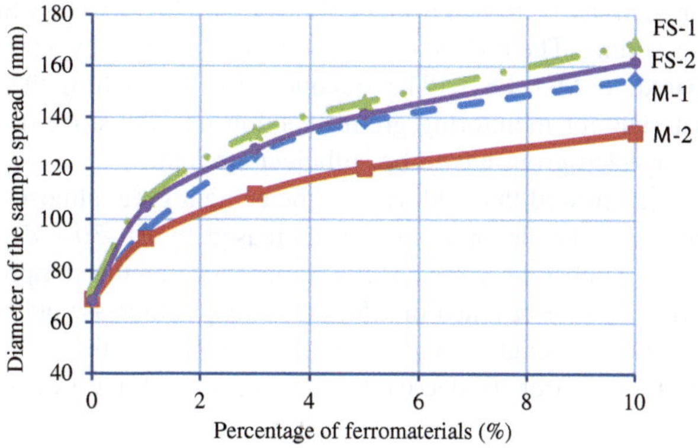

b)

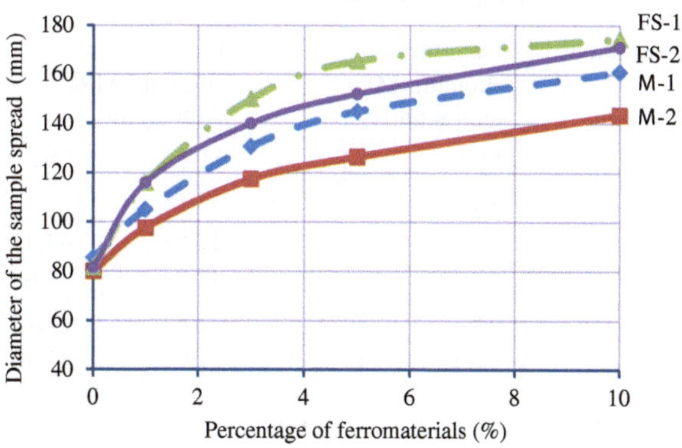

c)

Fig. 4. Flowability of the grout under magnetic field effect: w/c ratio: (a) 0.4, (b) 0.45, (c) 0.5

Table 2. Relative depth of the grout penetration

Additive amount, %	Crushed stone d = 10–20 mm (w/c ratio 0.4)		Crushed stone d = 5–10 mm (w/c ratio 0.5)	
	Dry	Preliminary wetted	Dry	Preliminary wetted
0	0.35	0.36	0.26	0.29
10	0.58	0.63	0.47	1.00

The electrical conductivity of the grout mixture was assessed on the samples with the additive content of 1%, 3%, 5% and 10%. The results showed that ferro-materials significantly increase the electrical conductivity. Thus, at the 50 Hz electrical current the electrical conductivity increased from 0.022 Sm/m to 0.028–0.038 Sm/m as a result of ferro-materials adding. The highest conductivity was reached by adding M-2 or FS-2 additives compared to M-1 FS-1. The obtained results confirm the advantages of adding ferro-materials for monitoring grout injection into the stone or masonry foundations. But the injected grout has to be sufficient strength.

The experiments showed that additives in the cement grout almost always increase its strength (Table 3). The flexural strength increased by 18–30% depending on the amount of additive. Almost the same effect was produced on the compression strength. Adding 3% of M-1 additive resulted in 1.5 times strength increase, adding 1% and 5% increased compression strength only by 12–15% while adding 10% of dispersed magnetite showed a 5% strength decrease compared to additive-free grout.

Table 3. Strength of the grout

Additive amount, %	Flexural strength, MPa, with additive		Compression strength, MPa, with additive	
	M-1	M-2	M-1	M-2
0	5.13		37.15	
1	5.13	7.02	41.57	44.41
3	6.08	6.41	48.74	42.68
5	6.62	6.28	42.85	41.75
10	6.80	6.36	41.99	35.19

Finally the possibility to substitute vibrocompaction with the magnetic field effect was assessed. After placing cement mixture with ferro additives into the moulds some of the samples were vibrocompacted while others were processed by magnetic field. The tests demonstrated that exposing the grout by the magnetic field resulted in the same grout strength as vibrocompaction.

5 Conclusion

The tests performed on cement grout showed that adding ferro additives produces electro conductive material with a number of new properties:

1. Under the magnetic field effect cement grout became more fluid. Its flowability increased by 20% to 190% depending on the additive used and the water/cement ratio of the grout.
2. The magnetic field facilitated the increase of the grout penetration depth into the stone voids by 2–3 times.
3. The electrical conductivity of the mixture may serve as an indicator for evaluating the quality of filling cracks and cavities stone or masonry foundations.
4. The additives in the cement grout changed its strength. For instance, the compression strength increased up to 30% depending on the type and amount of the additive.
5. The magnetic processing may replace the vibration effect for grout compaction.

References

1. Shuliatiev, O.A., Mozgacheva, O.A., Pospekhov, V.S.: Development of the underground space of cities: scientific publication. Publishing House ACB, Moscow (2017)
2. Mangushev, R.A.: Process settlement of buildings and structures in the area affected by underground construction. Publishing House ACB, Moscow (2017)
3. Il'ichev, V.A., Mangushev, R.A.: Geotechnical reference book. Basements, foundations and underground structures. Publishing House ACB, Moscow (2014)
4. Avdusheva, M.A., Nevzorov, A.L., Aizenshtadt, A.M.: Patent of RF No.2642762. Method for foundation underpinning. Bulletin No. 3 (2018)
5. Soto-Bernal, J.J., Gonzalez-Mota, R., Rosales-Candelas, I., Ortiz-Lozano, J.A.: Effects of static magnetic fields on the physical, mechanical, and microstructural properties of cement pastes. Adv. Mater. Sci. Eng. (2015). https://doi.org/10.1155/2015/934195
6. Lesovik, R.V., Ageeva, M.S., Chernysheva, N.V.: Activation of fine-grained concrete on ferrous technogenic sands by magnetic field. Bulletin of V.G. Shukhov BSTU 2011. No. 1 (2011)
7. Silvestre, J., Silvestre, N., de Brito, J.: Review on concrete nanotechnology. Eur. J. Environ. Civ. Eng. (2015). https://doi.org/10.1080/19648189.2015.1042070
8. Amin, M.S., El-Gamal, S.M.A., Hashem, F.S.: Effect of addition of nano-magnetite on the hydration characteristics of hardened Portland cement and high slag cement pastes. J. Therm. Anal. Calorim. (2012). https://doi.org/10.1007/s10973-012-2663-1
9. Lee, H.-S., Kwon, S.-J.: Effects of magnetite aggregate and steel powder on thermal conductivity and porosity in concrete for nuclear power plant. Adv. Mater. Sci. Eng. (2016). https://doi.org/10.1155/2016/9526251

Improvement of Fine Soils Through Microbial-Induced Calcite Precipitation

Fuchen Teng[1(✉)] and Shao-Chi Chien[2]

[1] Department of Civil and Construction Engineering, National Taiwan University of Science and Technology, Taipei, Taiwan
FTeng@mail.ntust.edu.tw

[2] Department of Information Tourism, Hsuan Chuang University, Hsinchu City, Taiwan

Abstract. A bacterial solution that generated microbial-induced calcite precipitation (MICP) was injected into kaolinite specimens through the electroosmotic technique in this study. Suitable electroosmotic injection times were selected to improve kaolinite specimens with the bacterial solution. After being treated for 360 h with electroosmotic injection of the bacterial solution, the cone resistance of the kaolinite reached a maximum of 2,000 kPa in the area close to the anode (approximately 10 times the original CPT cone resistance (200 kPa)) and 1,000 kPa in the area close to the cathode (approximately 5 times of the original CPT cone resistance). Based on the results, the electroosmotic injection method can overcome the difficulty of injecting a bacterial solution into low-permeability soft clay and can improve its strength.

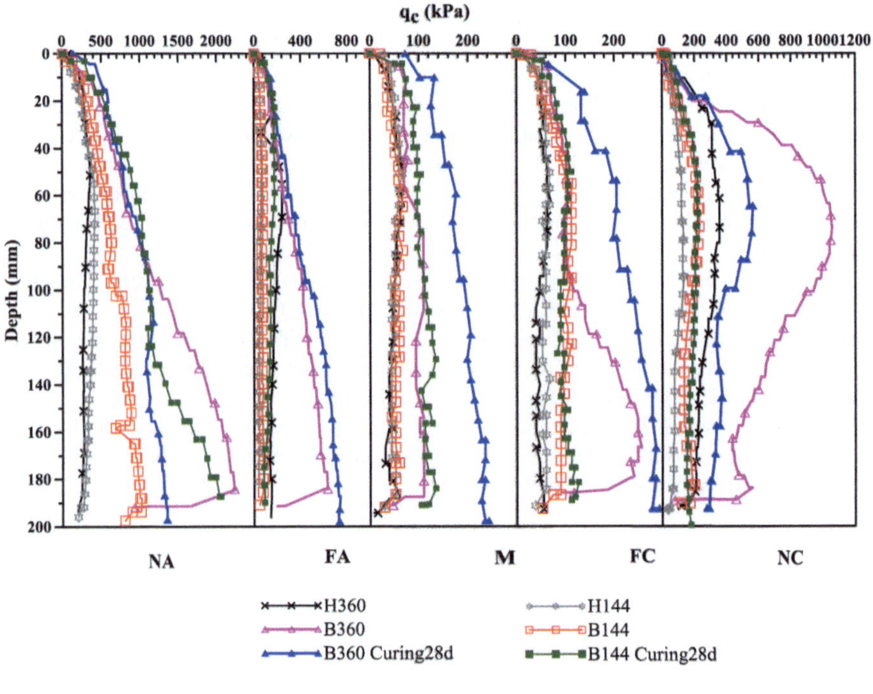

H. Shehata and B. Das (Eds.): GeoMEast 2018, SUCI, pp. 136–150, 2019.
https://doi.org/10.1007/978-3-030-01923-5_11

1 Introduction

Soil improvement techniques integrate artificial materials with soil particles using energy (pressure, electric energy, etc.) to improve the engineering properties of the soil. However, green energy and ecological and environmental protection have gradually garnered attention in recent years. Karol (2003) noted that most of the chemical grouts used for ground improvement (except for sodium silicate) are toxic and dangerous. Therefore, sustainable, low-pollution improvement methods are needed to reduce the use of macromolecular chemical grouts and high-energy consumption equipment.

In recent years, the microbial-induced soil improvement technique has attracted attention. This technique generates a series of chemical reactions by controlling the activities of the microorganisms in the soil. The by-products of this type of biochemical reaction may alter the soil microstructure, which affects the macroscopic engineering properties of the soil. In the microbial-induced soil improvement technique, the keys to generating chemical reactions are the pH of the soil and the concentration of chemical substances in the soil. Chemical reactions can be classified into inorganic matter precipitation (biomineralization), organic matter precipitation (biofilm) and gas generation (biogas). The key to controlling the chemical reactions lies in the timing of the injection of nutrients (culture medium) as well as in the distribution and activity of microorganisms when using microorganisms to induce reactions. It is worth noting that similar biochemical soil processes have already been used to solidify metallic substances in soil (Fujita et al. 2008), and biomineralization has been used to repair stone structures (DeMuynck et al. 2008; Mastromei et al. 2008; Shen and Cheng 2008). Recently, a comprehensive review on the state of research on bio-mediated soil has been made by DeJong et al. (2010). Based on the aforementioned facts, the use of microorganisms to improve soil is a new technique that meets the requirements of low energy consumption, low carbon emission and low chemical pollution. Relevant studies conducted worldwide have demonstrated that the microbial-induced soil improvement technique is effective in improving sandy soil and may become a new choice for soil improvement in the future.

In sandy soil layers, the focus is placed on reducing the coefficient of permeability to seal the layers. However, in engineering practice, soft clay layers are often the soil layers in need of improvement to increase their strength. Therefore, the use of the new microbial-induced soil improvement technique in practical applications depends on whether the technique is applicable to soft clay layers and then effectively improve the soil.

The electroosmotic chemical grouting method that drained the pore water in the soft clay through the electroosmotic technique and simultaneously injected a chemical grout is thus applied to conquer the difficulty of injection microorganisms into soft clays. The effectiveness of this method has been verified through laboratory and field experiments (Chien et al. 2009, 2010, 2012; Teng et al. 2013). The technique is sometimes called electroosmotic chemical treatment (ECT).

In this study, a bacterial solution was injected into relatively large kaolinite specimens via electroosmosis, and the effectiveness of the bacterial solution in improving the properties of kaolinite under various electroosmotic injection times was observed.

The aim was to bond soil particles or fill pores with by-products from microorganisms that was injected into kaolinite by electroosmotic technique.

2 Experiments Configurations

2.1 The Experimental Cell

As shown in Fig. 1, the test container for the electroosmotic test was made of a plexiglass plate and was 230 mm wide, 440 mm long and 320 mm deep. Two platinum (Pt) coated titanium (Ti) meshed plates, 350 mm wide, connected to a power supply device served as the anode and cathode. A chamber connected to the anode, 20 mm wide and 150 mm long, was used to store part of the injected chemical solutions, which allowed the chemical solutions to be uniformly injected into the soil under the electric field. Another chamber with the same dimensions was connected to the cathode and was used to store drained water, which was then discharged through a drain tube. The top plate also served as a bearing plate, and the air pressure acted to provide the designated consolidation pressure. Platinum (Pt)-coated titanium (Ti) mesh was used as the anode to prohibit metal ions from entering the soil due to oxidization of the anode during electroosmosis.

2.2 Soil Sample Preparation

A kaolinite with a liquid limit of 46 and a plastic limit of 25 was used in this study. The kaolinite powder was thoroughly mixed with a sufficient amount of distilled deionized water and resulted in a water content of 60%. The soil was placed in the test container in layers. After putting the soil into the experimental cell, it was covered with filter paper, and the electrodes (platinum-coated titanium mesh) were attached. A 30 kPa of vertical pressure was then applied to the soil sample in increments to produce the normally consolidated condition. The water content after consolidation was equal to 51%.

2.3 Experimental Bacterial Species

Bacillus smithii was selected in this study to improve the kaolinite. Table 1 lists the experimental information for Bacillus smithii in detail. Tables 2 and 3 list the chemical composition and basic physical properties of the experimental kaolinite, respectively.

Bacillus smithii (strain number: 14179) was cultured in a solid-state culture medium that consisted of 3.0 g/L of beef extract, 5.0 g/L of peptone and 15.0 g/L of agar and water and had a pH of 7. Bacillus smithii was inoculated into culture vessels that had been sterilized at 121 °C for 30 min. Afterward, Bacillus smithii was cultured in a 45 °C environment. Each month, Bacillus smithii was subcultured. Subsequently, the bacterial colony was inoculated into 50 mL of a liquid culture medium, which consisted of 3.0 g/L of beef extract, 5.0 g/L of peptone and water and had a pH of 7, for culture. The bacterial colony was inoculated into culture vessels that had been sterilized at 121 °C for 30 min. The bacterial colony was then cultured in a shake flask at 100 rpm and 45 °C for 1 d and then inoculated in a 5 L fermentation tank. The tank

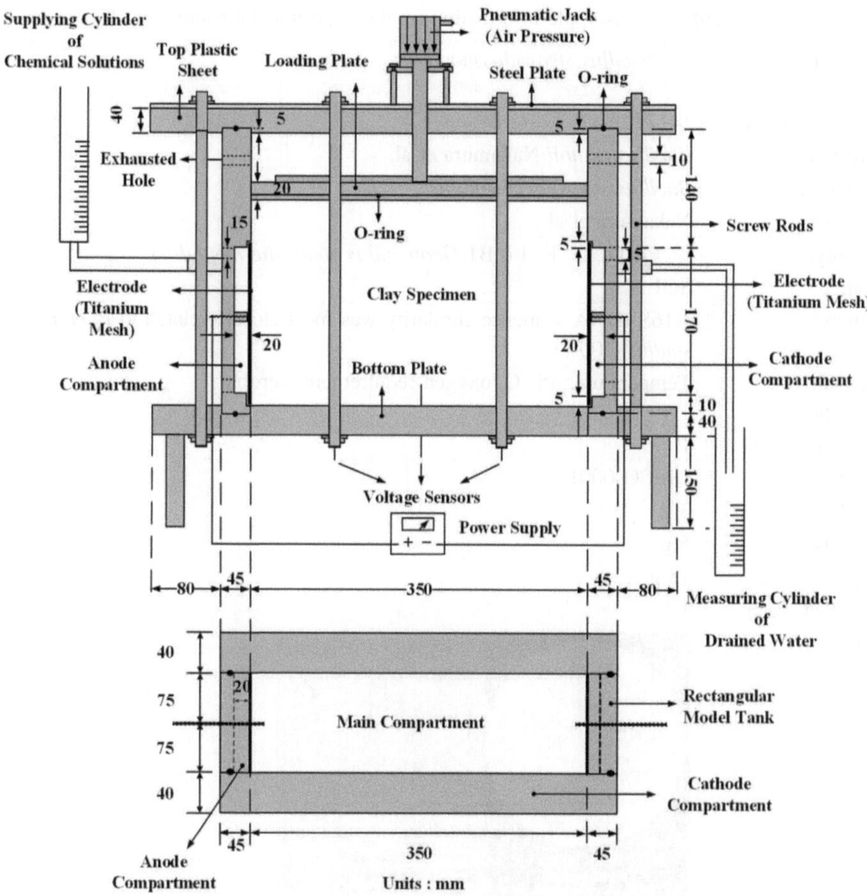

Fig. 1. Schematic diagram of the experimental apparatus for electroosmotic chemical grouting

was sterilized at 121 °C for 30 min and contained a medium consisting of 3.0 g/L of beef extract, 5.0 g/L of peptone and water at a pH of 7 for propagation and culture of large numbers for 2 d at 45 °C. Afterward, the bacteria were harvested.

2.4 Test Plan

Eight ECT tests were performed. A voltage gradient of 50 V/m was adopted for all tests, as suggested by Mitchell (1993). After setting up the specimen, a series of ECT tests were conducted by injecting Bacillus smithii over 144, 360, 480, and 600 h, and the tests were denoted as ECTB144, ECTB360, ECTB480 and ECTB600, respectively. Another series of ECT tests were performed by injecting distilled deionized water over 144 and 360 h, and these tests were denoted as ECTH144 and ECTH360, respectively. For the ECT tests, the Bacillus smithii was first injected through the anode for 144 and 360 h during electroosmosis, followed by curing for 28 d, and these samples were

Table 1. Example information on the experimental bacteria

Scientific Name:	*Geobacillus stearothermophilus*
BCRC number:	14179
Organism:	*Bacillus smithii* Nakamura et al.
Synonymy:	*Bacillus stearothermophilus*
Author:	Nakamura et al.
History:	≪ FIRDI, Y. K. Lo B1(*Geobacillus stearothermophilus*)
Source:	Soil
Others:	1.16S rDNA sequence similarity was most closely related to *Bacillus smithii* (100%).
Growth conditions:	Temperature: 45 °C, oxygen requirement: aerobic
Biosafety level:	1
Nutrient agar	DIFCO 0001
Beef extract	3 g
Peptone	5 g
Agar	15 g
Distilled water	1 L
Image	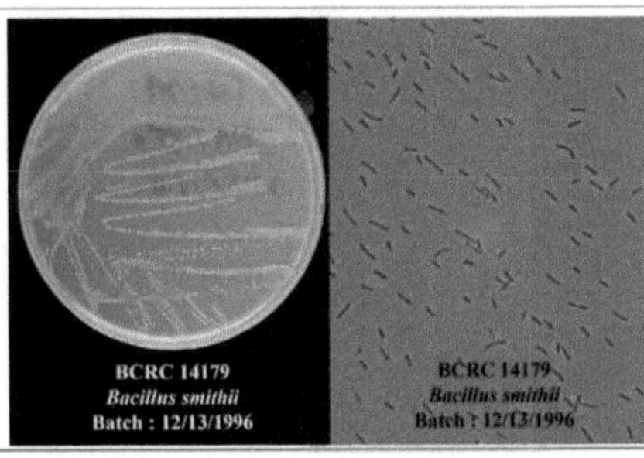

Table 2. Analysis of the chemical composition of the experimental kaolinite

Items of analysis	Mean percent by weight
Silicon Dioxide (SiO_2)	45.5%
Aluminum Oxide (Al_2O_3)	38.00%
Titanium Dioxide (TiO_2)	1.65%
Iron Oxide (Fe_2O_3)	0.90%
Calcium Oxide (CaO)	0.25%
Magnesium Oxide (MgO)	0.03%
Potassium Oxide (K_2O)	0.23%
Sodium Oxide (Na_2O)	0.04%
Loss on Ignition (L.O.I.)	13.40%

represented by ECT B144 Curing 28d and ECT B360 Curing 28d, respectively. Figure 1 shows the schematic diagram of the experimental apparatus. Figure 2 shows the flow chart of the electroosmosis experimental procedure. Figure 3 shows the sampling locations and CPT after the electroosmosis experiment was complete. Table 4 lists the conditions for each of the eight tests.

Table 3. Basic physical properties of the experimental kaolinite

Liquid limit (%)	46
Plastic limit (%)	25
Plastic index (%)	21
Specific gravity	2.61
Particle size (μm)	1.0–2.0
Unified soil classification result	CL
Whiteness (GBE)	80
pH	6–8
Surface area (m^2/g)	24
Permeability (cm/s)	Approximate 10^{-7}

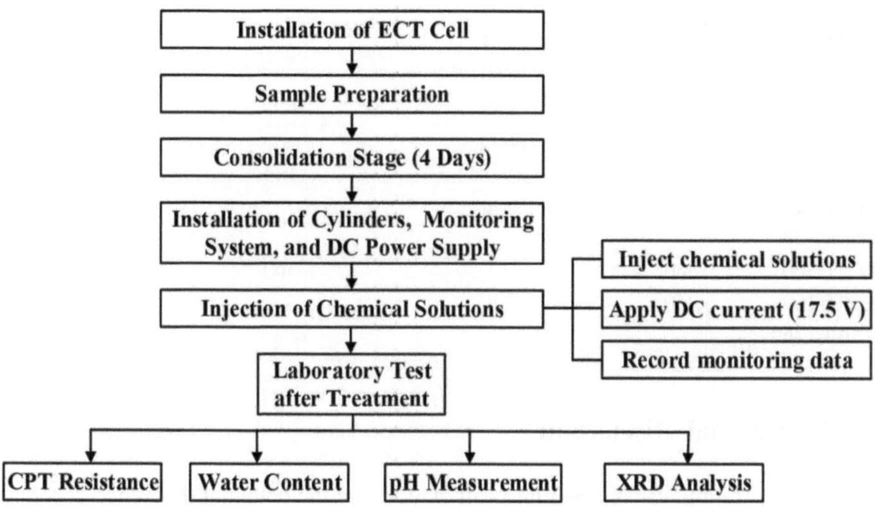

Fig. 2. Flow chart of the electroosmosis experiment

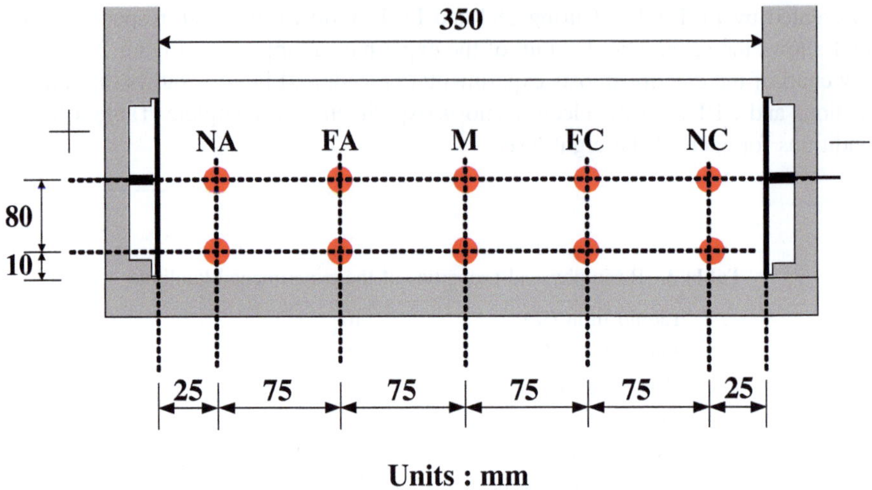

Fig. 3. Sampling locations

Table 4. Symbols and conditions of the eight tests

Test conditions test symbols	Electroosmotic time (h)	Bacterial solution injection time (h)	Deionized water injection time (h)	Curing time (day)
ECTB144	144	144	–	–
ECTB360	360	360	–	–
ECTB480	480	480	–	–
ECTB600	600	600	–	–
ECTH144	144	–	144	–
ECTB144 Curing 28d	144	144	–	28
ECTH360	360	–	360	–
ECTB360 Curing 28d	360	360	–	28

3 Results and Discussion

A bacterial solution was injected into soft clays using the potential difference generated via electroosmosis in this study. An electrolytic phenomenon occurred during the experiment. An acidic environment, a basic environment and a neutral environment were formed at the anode, cathode and middle areas, respectively. The bacterial solution was injected through the anode reservoir and replaced each day. The bacterial solution was injected via electroosmosis for 144, 360, 480 and 600 h. After the experiment was completed, the changes in the physical and chemical properties of the kaolinite were analyzed.

3.1 Effect of the Microorganism Injection Time

3.1.1 CPT Cone Resistance

As shown in Fig. 4, the CPT cone resistance of the kaolinite in the NA area increased with increasing injection time and reached a maximum (over 2,000 kPa), which was approximately 10 times the original CPT cone resistance (approximately 200 kPa), after being treated with 360 h of bacterial solution injection (process ECTB360). However, in the test with 480 h of bacterial solution injection (process ECTB480), the CPT cone resistance of the kaolinite in the NA area decreased significantly, and the maximum CPT cone resistance of the kaolinite was less than 500 kPa.

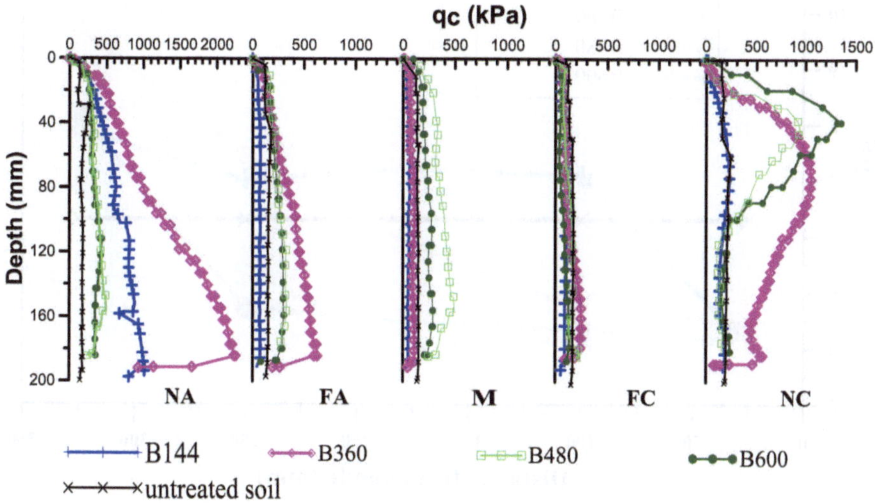

Fig. 4. CTP cone resistance of the kaolinite after being treated with electroosmotic injection of microorganisms

A relatively high CPT cone resistance of the kaolinite in the FA area was achieved after being treated with process ECTB360; however, the resistance of the kaolinite in this area after being treated with process ECTB480 was less than that after being treated with process ECTB360. Only the CPT cone resistance of the kaolinite treated with process ECTB360 reached 500 kPa. The maximum CPT cone resistance of the kaolinite in the M and FC areas was less than 500 kPa, and the resistance of the kaolinite in these areas decreased significantly after being treated with 600 h of bacterial solution injection (process ECTB600).

The CPT cone resistance of the kaolinite in the NC area increased significantly after being treated with the bacterial solution injection for more than 360 h, and the mean maximum CPT cone resistance in this area exceeded 1,000 kPa.

3.1.2 pH Measurements

Figure 5 shows the pH measurements. After being treated with 144 h of bacterial solution injection (process ECTB144), the overall pH of the clay did not change significantly, except in the NA area, where the inherent neutral environment of the bacterial solution and the clay was maintained. However, after being treated with process ECTB360, the pH of the kaolinite in the FA and M areas decreased significantly, and the environments in these two areas became acidic.

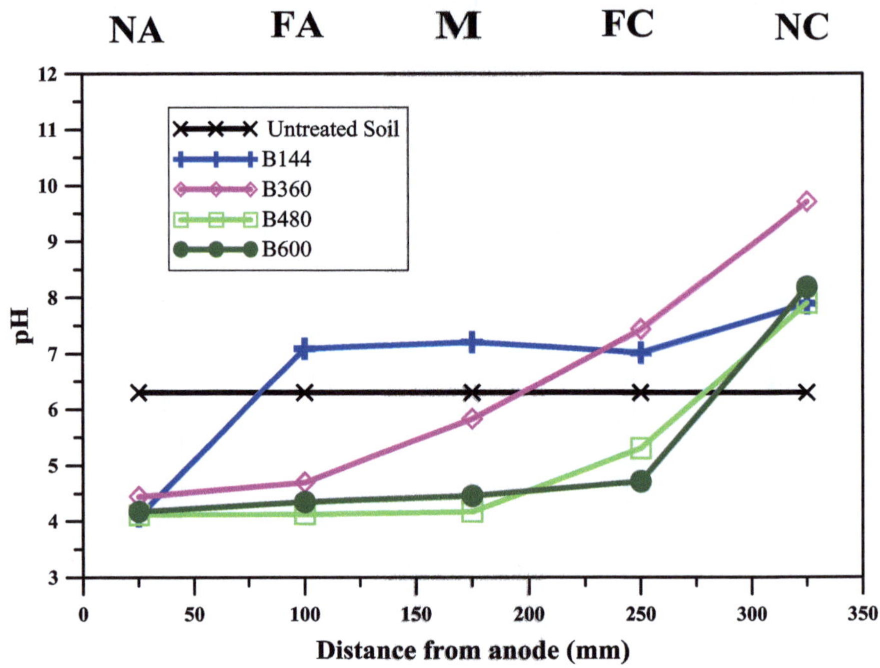

Fig. 5. pH distribution in the kaolinite after electroosmotic injection of microorganisms

The pH of the kaolinite in the NC area increased significantly. It can be inferred that this phenomenon was a result of electrolysis. In addition, after being treated with process ECTB480, the kaolinite had an acidic pH in the NA, FA, M and FC areas, whereas the pH of the kaolinite in the NC area was close to that of the kaolinite treated with process ECTB144 in the same area.

Based on these results, it can be concluded that after being treated with process ECTB360, the overall pH of the kaolinite decreased to an acidic level, except in the NC area, where the pH was maintained at a neutral level. There are two possible causes of this phenomenon. One possible cause is that the living bacteria in the kaolinite consumed the oxygen in the water and released hydrogen ions (H^+ ions). Another possible cause is that after the bacteria died, their proteins decomposed, and H^+ ions were released from the amino acids, which resulted in a decrease in the overall pH of the kaolinite. It can also be inferred from the results that H^+ ions were released at relatively high rates after the bacterial solution was injected for more than 360 h because of the oxygen consumption by the bacteria and the decomposition of the amino acids. Consequently, the elevated pH of the kaolinite treated with processes ECTB480 and

ECTB600 in the FC and NC areas decreased again, and the pH in these areas was approximately 8.

3.2 Effect of Curing After Injection of Microorganisms

Kaolinite samples treated with processes ECTB144 and ECTB360 were cured for 28 days and the properties of the treated samples were analyzed. These respective tests are denoted by ECTB144-curing28d and ECTB360-curing28d. In addition to these two extended tests, two other tests, denoted as ECTH144 and ECTH360, were also conducted as controls. In these control tests, deionized water was injected into the kaolinite instead of a solution containing microorganisms, and the changes in the properties of the kaolinite were analyzed. Furthermore, when curing the kaolinite, the anode and cathode reservoirs were filled with water, and each valve of the anode and cathode reservoirs was closed until the curing process was complete.

3.2.1 CPT Cone Resistance

Figure 6 shows that the CPT cone resistance of the kaolinite. Cone resistance in the NA area in ECTB144 significantly increased after curing and was significantly higher than that of the kaolinite treated with a 144 h injection of deionized water (process ECTH144). In ECTB360, the CPT cone resistance of the kaolinite in the NC area increased significantly and had a maximum value of over 1,000 kPa. However, after being treated with process ECTB360-curing28d, the CPT cone resistance of the kaolinite in the NC area decreased to approximately 500 kPa. It is worth noting that in the FA area, the CPT cone resistance of the kaolinite specimens treated with processes ECTB360 and ECTB360-curing28d both reached a maximum of 500 kPa.

3.2.2 pH Measurements

As shown in Fig. 7, the pH in the NA area was almost the same among all the kaolinite specimens treated with various processes. The pH of the kaolinite did not change significantly in the areas after being treated with process ECTB144 and process ECTB144-curing28d, except in the FA area, where the pH decreased. Based on the pH of the kaolinite treated with process ECTB360 and process ECTB360-curing28d, the pH of the kaolinite remained the same after curing in all areas except the M area, where the pH increased. The pH of the kaolinite in the NC area increased to 9 to 10 after process ECTB360 and process ECTB360-curing28d. This basic environment was not achieved with process ECTB144 or process ECTH360.

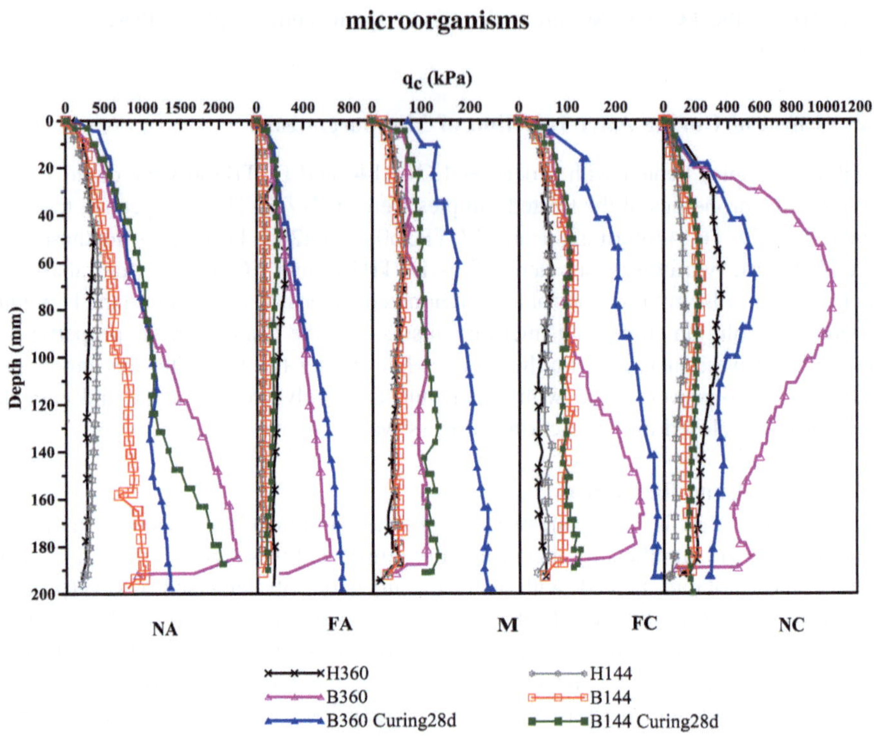

Fig. 6. CPT cone resistance of the kaolinite after electroosmotic injection of microorganisms

3.2.3 X-Ray Powder Diffraction (XRD) Results

The XRD analysis of the kaolinite specimens treated with process ECTB144 and process ECTB144-curing28d was shown in Fig. 8. The XRD analysis results show that a new peak did not appear after process ECTB144; in addition, a new peak did not appear after curing. Therefore, it was concluded that no new crystalline compounds were formed after injection of the bacterial solution and curing.

3.2.4 Discussion

The results in Figs. 6 and 7 showed that the CPT cone resistance of the kaolinite treated with process ECTB144 increased in the NA area after 28 d of curing, whereas no significant difference in the CPT cone resistance occurred in the NA area after 28 d of curing for the kaolinite treated with process ECTB360. However, the pH values of the kaolinite specimens treated with process ECTB144, process ECTB360, process ECTB144-curing28d and process ECTB360-curing28d were the same in the NA area (pH = 4.2).

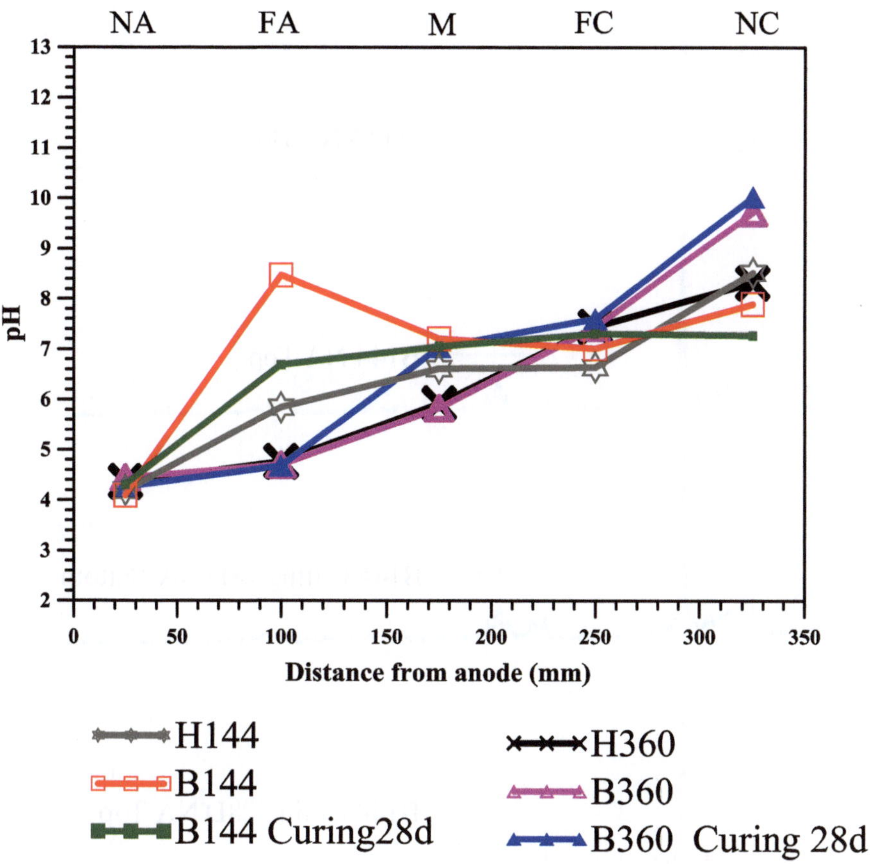

Fig. 7. pH distribution in the kaolinite after electroosmotic injection of microorganisms

Electroosmotic injection of microorganisms can improve the soil strength in the NA area to a sufficient level, although in general electroosmotic injection of microorganisms is not as efficient as electroosmotic chemical grouting in improving soil strength. In addition, because environmental protection awareness has been gaining ground in recent years, an environmentally friendly method will become an important consideration in the selection of a method for soil improvement. Based on this, electroosmotic injection of microorganisms may be favored over electroosmotic chemical grouting. However, in the present study, soil improvement was only found in the NA and NC areas. Therefore, the mechanism by which electroosmotic injection of microorganisms improves soil requires further investigation.

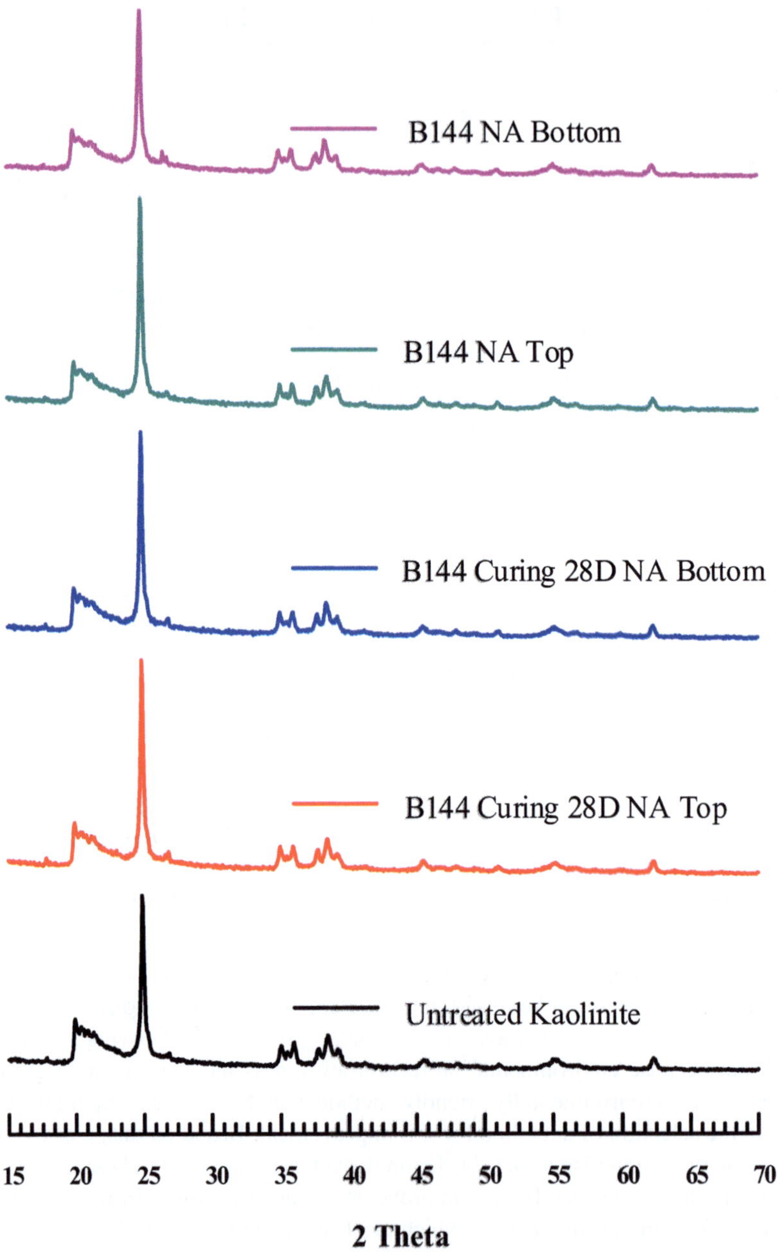

Fig. 8. XRD analysis results

4 Conclusions

Recent studies on the electroosmotic chemical grouting method have shown that while this method can satisfactorily improve all the areas of a soil mass, the highly basic metallic solution used in this method has an impact on the environment. Therefore, the present study investigated electroosmotic injection of microorganisms to develop a method that has a relatively low negative impact on the environment. The results showed that electroosmotic injection of microorganisms was effective in improving kaolinite in the NA and NC areas (process ECTB144 and process ECTB360); in addition, this method can result in a relatively neutral pH, compared to electroosmotic chemical grouting, and has a relatively low impact on the environment. The results of the present study are discussed and relevant suggestions are provided in the following:

1. Injection of microorganisms via electroosmosis could achieve good soil improvement results with 360 h of injection of the bacterial solution; however, this method was only effective in improving the NA area (CPT cone resistance greater than 2 MPa) and the NC area (CPT cone resistance greater than 1 MPa) of the kaolinite.
2. Compared to electroosmotic chemical grouting, electroosmotic injection of microorganisms resulted in a relatively neutral pH; in addition, no significant change in the pH of the kaolinite treated with electroosmotic injection of microorganisms occurred after curing.
3. In the NC area, the CPT cone resistance of the kaolinite treated with process ECTB360 was almost the same as that of the kaolinite treated with electroosmotic chemical grouting. However, the CPT cone resistance of the kaolinite treated with process ECTB360 did not increase after 28 d of curing. This is because the water content of the kaolinite in the cathode area increased significantly after curing, which resulted in a decrease in the CPT cone resistance. The loss of deionized water from the cathode reservoir during the curing process indicated that the increase in the water content was a result of the absorption of water by the kaolinite.

References

Chien, S.C., Ou, C.Y., Wang, M.K.: Injection of saline solutions to improve the electro-osmotic pressure and consolidate foundation soil. Appl. Clay Sci. **44**(3–4), 218–224 (2009)

Chien, S.C., Ou, C.Y., Lee, Y.C.: A novel electro-osmotic chemical treatment for soil improvement. Appl. Clay Sci. **50**(4), 481–492 (2010)

Chien, S.C., Ou, C.Y., Wang, Y. H.: Feasibility Study on clay with interbedded sand using E.C. T. Geotech. Eng. (SCI) (2012). https://doi.org/10.1680/geng.11.00076

DeMuynck, W., Debrouwer, D., DeBelie, N., Verstraete, W.: Bacterial carbonate precipitation improves the durability of cementitious materials. Cem. Concr. Res. **38**, 1005–1014 (2008)

DeJong, J.T., Mortensen, B.M., Martinez, B.C., Nelson, D.C.: Bio-mediated soil improvement. Ecol. Eng. (2010). https://doi.org/10.1016/j.ecoleng.2008.12.029

Fujita, Y., Taylor, J.L., Gresham, T.T., Delwiche, M., Colwell, F., McLing, T.L., Petzke, L.M., Smith, R.W.: Stimulation of microbial urea hydrolysis in groundwater to enhance calcite precipitation. Environ. Sci. Technol. **42**, 3025–3032 (2008)

Karol, R.H.: Chemical Grouting and Soil Stabilization, p. 558. Marcel Dekker, New York (2003)

Mastromei, G., Marvasi, M., Perito, B.: Studies on bacterial calcium carbonate precipitation for stone conservation. In: 1st International Conference on Bio-Geo-Civil Engineering, The Netherlands, pp. 103–106 (2008)
Shen, J., Cheng, X.: Laboratory investigation on restoration of Chinese ancient masonry buildings using microbial carbonate precipitation. In: 1st International Conference on Bio-Geo-Civil Engineering, The Netherlands, pp. 28–34 (2008)
Teng, F.-C., Chien, S.-C., Ou, C.-Y.: Electro-osmotic chemical treatment of clay on shear modulus. Geotech. Test. J. **36**(2), 248–255 (2013)

Effect of "Matching" on the Prediction of Excess Pore Water Pressure in Plane Strain Analyses

Md Rajibul Karim(✉) and Md Mizanur Rahman

School of Natural and Built Environments, University of South Australia, Mawson Lakes, SA, Australia
Rajibul.Karim@unisa.edu.au

Abstract. Consolidation analyses of embankments are often conducted on the assumption of the plane strain condition. This poses a problem if the effects of vertical drains (used for accelerating consolidation) are to be modelled. The conditions for the consolidation around vertical drains are close to axial symmetry. "Matching" procedure (Hird et al. 1992) is often used for estimating the plane strain equivalent hydraulic conductivity and/or the drain spacing for modelling simplicity. "Matching" is achieved by comparing the average degree of consolidation between axisymmetric and plane strain unitcells (a single drain surrounded by area of influence). As the process is based on average degree of consolidation, the excess pore water pressure at corresponding points (plane strain and axisymmetric) in an unitcell can be different. The discrepancy can be maximum at the boundary of the unitcell even though the settlement may match with good accuracy. In this paper, a relationship is developed to correct the calculated excess pwp in a plane strain analysis. A correction factor is proposed and the variation of the correction factor with different variables (hydraulic conductivity, level of imposed stress, mechanical properties of the soil) are discussed. A series of finite element consolidation analyses using elastic, elastoplastic and elastic viscoplastic constitutive models have been used for this purpose.

1 Introduction

For embankments or similar structures, when the foundation soil is improved with vertical drains, the modelling of consolidation becomes problematic. The flow condition around each drain are more close to axial symmetry. To make modelling of consolidation in a plane strain condition possible, a workaround is the "matching" procedure (Hird et al. 1992, 1995). By comparing average degree of consolidation between an axisymmetric unitcell and an equivalent plane strain unitcell, the matching procedure allows calculation of equivalent plane strain hydraulic conductivity (K_{pl}) of a soil. When K_{pl} is used in a plane strain simulation, it often produces good estimation for settlement. Figures 1 and 2 shows a comparison of flow characteristics between an axisymmetric and plane strain unitcell and their idealization. While "matching", a user can choose from a number of different options i.e. hydraulic conductivity matching (spacing between drains are kept the same in the plane strain analysis and hydraulic

H. Shehata and B. Das (Eds.): GeoMEast 2018, SUCI, pp. 151–161, 2019.
https://doi.org/10.1007/978-3-030-01923-5_12

conductivity is adjusted); space matching (hydraulic conductivity is kept the same but the spacing between drains are adjusted) and combined matching (both spacing and hydraulic conductivity are adjusted).

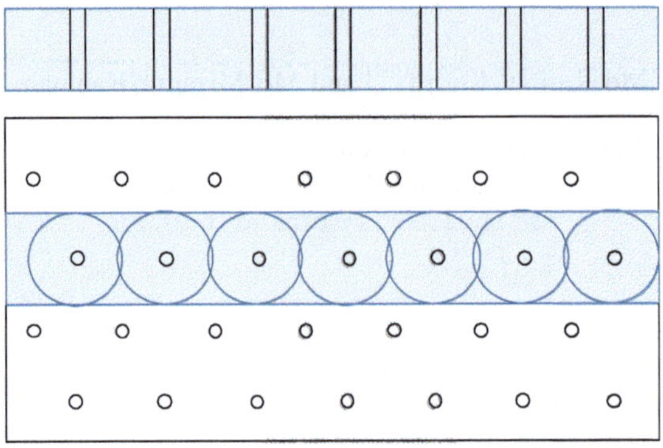

Fig. 1. Plan view of a vertical drain improved ground (bottom) and equivalent plane strain section (top).

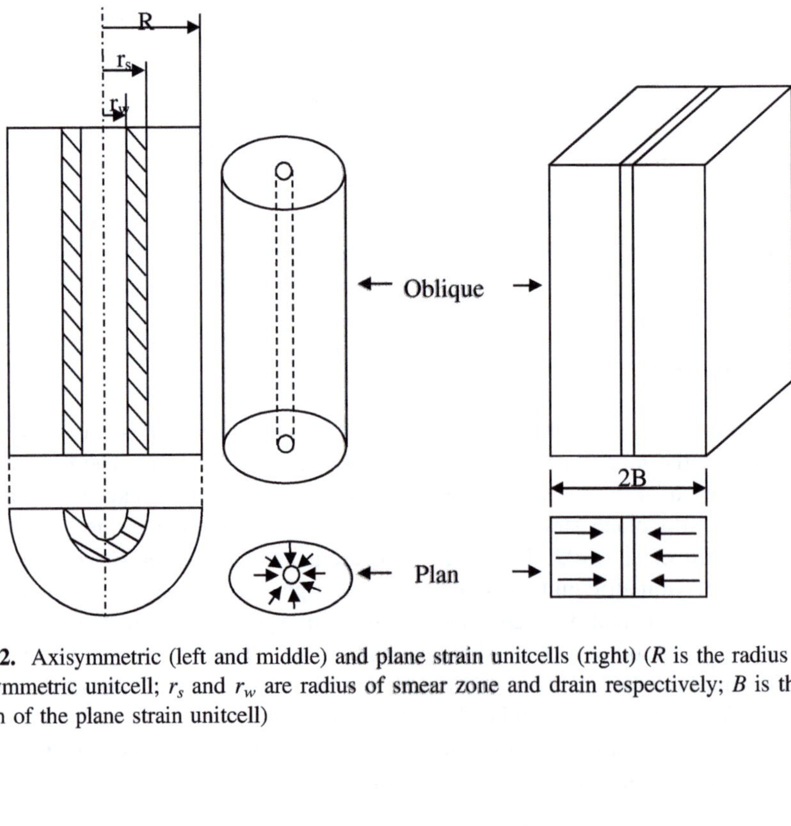

Fig. 2. Axisymmetric (left and middle) and plane strain unitcells (right) (*R* is the radius of the axisymmetric unitcell; r_s and r_w are radius of smear zone and drain respectively; *B* is the half width of the plane strain unitcell)

Following Eqs. (1, 2 and 3) are reproduced after Hird et al. (1992) for space matching, hydraulic conductivity matching and combined matching respectively,

$$\frac{B}{R} = \left\{ \frac{3}{2} \left[\ln\left(\frac{R}{r_s}\right) + \frac{K_{ax}}{K_s} \ln\left(\frac{r_w}{r_s}\right) - \frac{3}{4} \right] \right\}^{\frac{1}{2}} \tag{1}$$

$$\frac{K_{pl}}{K_{ax}} = \frac{2}{3\left[\ln\left(\frac{R}{r_s}\right) + \frac{K_{ax}}{K_s} \ln\left(\frac{r_w}{r_s}\right) - \frac{3}{4} \right]} \tag{2}$$

$$\frac{K_{pl}}{K_{ax}} = \frac{2B^2}{3R^2\left[\ln\left(\frac{R}{r_s}\right) + \frac{K_{ax}}{K_s} \ln\left(\frac{r_s}{r_w}\right) - \frac{3}{4} \right]} \tag{3}$$

Here, K_{ax} and K_s are hydraulic conductivity in horizontal direction of soil in the undisturbed zone and smear zone respectively. Other symbols are defined in Fig. 2.

Matching technique has been used by many researchers in their plane strain analyses (Chai et al. 2013; Indraratna et al. 2005; Karim 2013; Karim et al. 2012, 2014, 2011; Manivannan et al. 2011) capturing the settlement behaviour with good accuracy. While a good match can be obtained for the average degree of consolidation (settlement), in plane strain and axisymmetric analyses pore water pressure (pwp) at corresponding points in the geometry may not be the same, which is often overlooked. The calculated excess pwp in the two cases can be different throughout the radius of the unitcell. The maximum value of the excess pwp, which occurs on the periphery of the unitcell, may be significantly different in plane strain and axisymmetric analyses (Hird et al. 1992, 1995). A better approach is needed to model the consolidation in a plane strain analysis for vertical drain improved soil.

The first step in the process can be understanding the factors that affect the variations and identify the sources of discrepancy. Also a correction factor is proposed that can quantify the differences in calculated pwp. Later the effect of different parameters on the correction factor is investigated to better understand the problem. The correction factor is described below.

Assuming the excess pwp calculated in the axisymmetric unitcell is more realistic and reliable because of the drainage and other boundary conditions, a relationship can be developed to adjust the calculated excess pwp in plane strain analyses. By comparing the excess pwp at the boundary of the unitcells, the following equation, can be proposed,

$$pwp_{pl}^{eq} = pwp_{pl} \times \frac{pwp_{ax}}{pwp_{pl}} \tag{4}$$

Here, pwp_{ps}^{eq} is the corrected excess pwp in the plane strain analysis, pwp_{ax} is the excess pwp calculated in an axisymmetric analysis and pwp_{ps} is the calculated excess pwp in the equivalent plane strain analysis. The last term of the equation (pwp_{ax}/pwp_{pl}) can be treated as a correction factor (c_f). It can easily be estimated by running two unitcell analyses, i.e., one axisymmetric and one equivalent plane strain and then comparing the calculated excess pwp at the boundaries.

It is important to realize that the correction factor may vary along the radius of a unitcell as well as with progression of consolidation. In this paper, the investigation is only limited to the excess pwp calculations at the boundary of the unitcells as the field piezometers (when used) are usually considered to be located there. This paper investigates the variation of c_f with hydraulic conductivity, soil elastic modulus, B/R ratio, stresses and linear and nonlinear material behaviour using numerical simulations. However, the same procedure can also be applied to any other location in the unitcell and is beyond the scope of this paper.

2 Numerical Simulations

The simulations have been conducted using a Finite Element (FE) program - AFENA (Carter and Ballam 1995). The axisymmetric and plane strain FE meshes used in the analyses are presented in Fig. 3. The axisymmetric unitcell has the drain at the centre with radius equal to the radius of the influence zone (assumed R = 0.7894 m) of the drain. The plane strain analysis on the other hand, has a chosen half width B. A soil thickness of 4 m is selected. The hydraulic conductivity (K_{ax}) for different axisymmetric analysis are presented in Table 1. K_{pl} is estimated using "combined matching" (Hird et al. 1992) technique in this case.

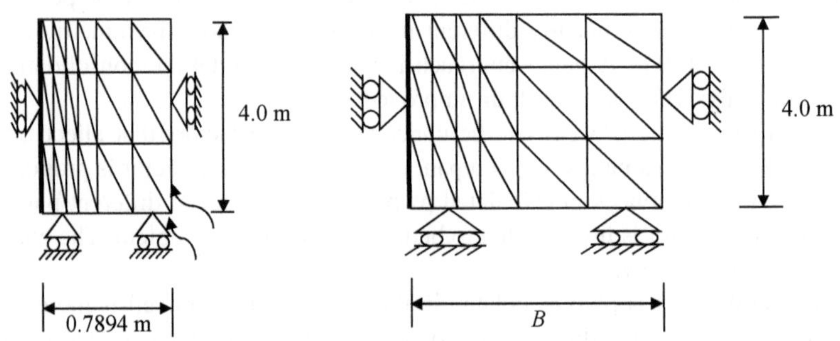

Fig. 3. FE axisymmetric (left) and plane strain (right) mesh used for the analyses

The bottom and the outside boundaries are modelled as impermeable (no flow). The top boundary is also modelled as impermeable to confirm uniform radial flow along the depth of the unitcell. The inside boundary, representing the periphery of the PVD is modelled as zero excess pwp boundary. The displacement boundary conditions are shown in Fig. 3. The results from the analyses using different group of models with increasing levels of complexity are discussed below.

Table 1. Details of analyses A01 to A10

Analyses	E (kPa)	K_{ax}	R (m)	B (m)	K_{pl}/K_{ax}	Imposed stress (kPa)
A01	3000	2.5E−5	0.7895	2.00	1.35	50
A02	3000	1.0E−5	0.7895	2.00	1.35	50
A03	3000	5.0E−6	0.7895	2.00	1.35	50
A04	3000	5.0E−6	0.7895	2.00	1.35	150
A05	3000	2.5E−6	0.7895	2.00	1.35	150
A06	6000	5.0E−6	0.7895	2.00	1.35	150
A07	6000	5.0E−6	0.7895	1.50	0.764	150
A08	6000	2.5E−6	0.7895	1.50	0.764	150
A09	6000	5.0E−6	0.7895	2.50	2.12	150
A10	6000	2.50E−6	0.7895	2.50	2.12	150

2.1 Relationships for Linear Elastic Material Behaviour

First set of simulations are conducted with the soil modelled as a linear elastic material. Table 1 presents the details of different analyses: elastic modulus, B, K_{ax} along with the K_{pl}/K_{ax} values (obtained by "matching").

Effect of Hydraulic Conductivity on c_f

Analyses A01 to A03 were conducted to examine the effect of hydraulic conductivity values on c_f and its variation with progression of consolidation. The elastic modulus and the applied stresses were kept the same. Different K_{ax} values were used for the axisymmetric analyses and the corresponding K_{pl} were calculated using the "combined matching" (Hird et al. 1992) technique. Figure 4 presents the variation of c_f for analyses A01 to A03 with time (along with others to be discussed later). It can be seen from the figure that, c_f at the outer impermeable boundary varies with time differently if hydraulic conductivity is changed. c_f decreased (non linearly) with the progression of consolidation and eventually asymptote at zero. For the soil with higher hydraulic conductivity, c_f reached zero quicker and opposite was the case for soil with lower hydraulic conductivity.

Effect of Magnitude of Imposed Stress on c_f

To check the effect of the magnitude of imposed stresses on the variation of c_f, analyses A04 was conducted. All the analyses parameters for A04 and A03 are same except for the imposed stress on the top boundary. Instead of 50 kPa, a 150 kPa of stress was applied. It can be seen from Fig. 4 that, the plots from analyses A03 and A04 are identical. This indicates that, in a linear elastic analysis, the value of c_f is independent of the magnitude of imposed stresses.

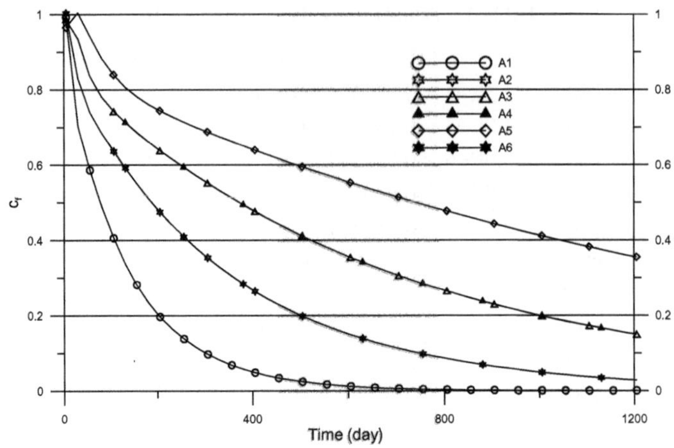

Fig. 4. c_f plotted against time for linear elastic analyses A01 to A06

Effect of Material Parameters on c_f

To examine the effect of material strength parameters on the evaluation of c_f, analysis A06 was conducted. The elastic modulus, in the analysis, was increased to 6000 kPa (3000 kPa in A04) and all other parameters were kept the same. It is interesting to note that, the plot of analysis A06 coincides with plot traced by A02, which, used half the hydraulic conductivity as of A06.

A Unique Relationship

For the 6 analyses discussed so far, different c_f plots against time were obtained. In Fig. 5, data from Fig. 4 are plotted against normalized excess pwp which is expressed as,

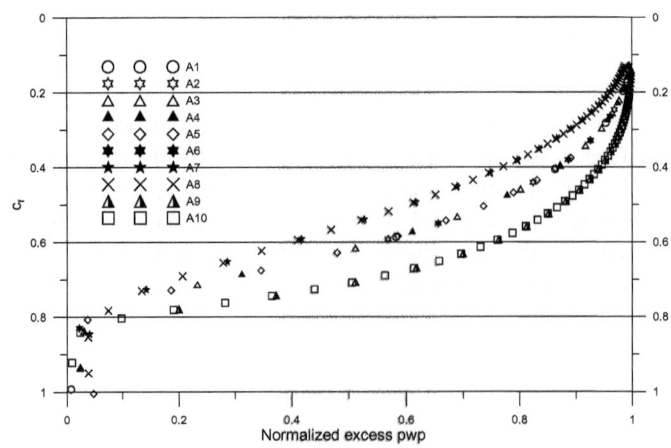

Fig. 5. Evaluation of c_f against normalized excess pwp for linear elastic analyses (A01 to A10)

$$\ddot{U} = \frac{\sigma_a - \Delta U}{\sigma_a} \tag{5}$$

where, Ü is the normalized excess pwp, σ_a is the applied effective stress, and ΔU is the current excess pwp at the equivalent plane strain unitcell boundary.

In Fig. 5 a unique c_f relationship is achieved. It is independent of hydraulic conductivity, magnitude of imposed stress and the elastic modulus in this case. Careful observation of eq. [5] will reveal that, its definition is similar to degree of consolidation for an instantaneous loading scenario. However, the definition of degree of consolidation does not hold for the case of continuous and stagged loading situations. The term "normalized excess pwp" will be used in this paper to avoid confusion.

Effect of Different Equivalent Unitcell Width on c_f

To check the effect of width of equivalent plane strain unitcell (2 B) on c_f, analyses A07 to A10 were conducted. Different values for B were used. Corresponding K_{pl}/K_{ax} values were calculated using "combined matching" (Hird et al. 1992) technique. The results are added in Fig. 5. Analyses A07 and A08 had B = 1.5 m whereas the analyses A09 and A10 had B = 2.5 m (analyses A01 to A06 used B = 2.0 m). It can be seen from Fig. 5 that, the c_f plots with the same values of B follows identical trend. This indicates, if the geometry is kept the same, in a linear elastic analysis, the strength parameters, imposed stresses or hydraulic conductivity have no effect on the variation of c_f against Ü.

2.2 Relationships for Elastoplastic Material Behaviour

Over-Consolidated Soil

More realistic material behaviour is taken into account in this section. Table 2 summarizes the analyses details. Analyses B01 and B02 were conducted to check the variation of c_f with hydraulic conductivity for Modified Cam Clay (MCC) (Roscoe and Burland 1968) material model. The value of preconsolidation stress (p_{c0}) was set very high to ensure nonlinear elastic behaviour. The other MCC material parameters used for the analyses were, recompression index, κ = 0.066, compression index, λ = 0.33, slope of the critical state line, M = 1.113, Poisson's ratio, v = 0.3 and the critical state void ratio, e_{CS} = 2.70. The initial stresses for all the analyses were kept σ_v = 10 kPa and σ_h = 5.305. The hydraulic conductivity values were varied to check its effect on c_f. Figure 6 plots c_f against Ü for analyses B01 and B02. From the figure, the value of hydraulic conductivity has insignificant influence on c_f when plotted against Ü.

The effect of stresses on c_f is investigated in analyses B03 and B04 (also presented in Fig. 6). It can be seen from the figure that, if the applied stress changes, the c_f plots become slightly different, however, it closely follows the line traced by of analyses B01 and B02. That is, in a nonlinear elastic analysis, the evaluation of c_f can be affected by the differences in the applied stresses and hydraulic conductivity of the soil has no or insignificant role to play.

Table 2. Details of analyses B01 to D02

Analyses	p_{c0} (kPa)	K_{ax} (m/d)	R (m)	B (m)	K_{pl}/K_{ax}	Imposed stress (kPa)
B01	400	2.0E−6	0.7895	2.0	1.35	100
B02	400	1.0E−6	0.7895	2.0	1.35	100
B03	400	2.0E−6	0.7895	2.0	1.35	50
B04	400	1.0E−6	0.7895	2.0	1.35	50
C01	10	5.0E−7	0.7895	2.0	1.35	100
C02	10	2.0E−5	0.7895	2.0	1.35	100
C03	50	5.0E−7	0.7895	2.0	1.35	100
C04	50	2.0E−5	0.7895	2.0	1.35	100
D01	50	5.0E−7	0.7895	2.0	1.35	100
D02	50	1.0E−6	0.7895	2.0	1.35	100

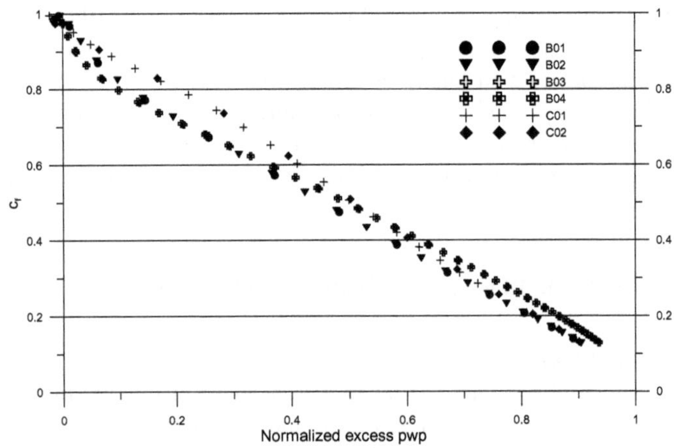

Fig. 6. c_f plotted against normalized excess pwp for analyses B01 to C02

Normally Consolidated Soil

To check the effect of hydraulic conductivity on the variation of c_f for fully elasto-plastic material behaviour, analyses C01 and C02 were carried out. The input p_{c0} values were calculated from the initial field stresses to make sure the soil was in normally consolidated state from the start of loading. All other MCC material parameters were kept the same as in the previous section. In this case also, (see Fig. 6) there was no significant effect of hydraulic conductivity on c_f.

Plots of B01 to C02 in Fig. 6 forms almost a straight line in c_f-normalized excess pwp space.

State of Soil Changing from Over to Normally Consolidated

To investigate the variation in c_f when soil behaviour changes from nonlinear elastic to elastoplastic, analyses C03 and C04 were conducted. The transformation was introduced in the analyses by setting up the appropriate p_{c0} values. The analyses were conducted with 100 kPa of applied stress on the top boundary. All other variables were kept the same as before. It can be seen from the figure that, the change in the state of soil is reflected by a change in the shape of the plot. The shapes of the curves are different from the ones established by elastoplastic analyses of normally and over-consolidated soils. The difference in hydraulic conductivity, in this case also, did not create any significant difference in the plots of the two analyses.

2.3 Relationships for EVP Material Behaviour

In this section, the variation of c_f is tested for elastic viscoplastic (EVP) material models. Analyses D01 and D02 were conducted with different hydraulic conductivity values and with the EVP model proposed by Karim et al. (2010). The same set of MCC material parameter as in analyses C03 and C04 were used. Additional parameters needed for the analyses where, $C_{\alpha max} = 0.085$ and $N = 0.052$. The p_{c0} was set to appropriate values to confirm elastic to EVP transformation during consolidation. The hydraulic conductivity values and other details are presented in Table 2. It can be seen from the Fig. 7, that, in this case also, there was insignificant effect of the hydraulic conductivity on the c_f plots. It can be concluded from the above discussion that, in case of more realistic material behaviour (MCC or EVP), the evaluation of c_f can be affected by the material parameters (e.g. p_{c0}) and the insitu stresses. However, the hydraulic conductivity values in all the cases had no or insignificant effect on their evaluation against Ü.

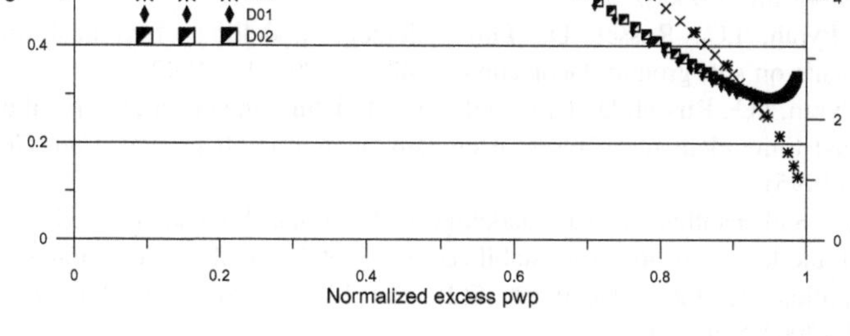

Fig. 7. c_f plotted against normalized excess pwp for analyses C03 to D02

3 Conclusion

The effect of using "matching" technique to estimate hydraulic conductivity of a soil in a plane strain analysis of vertical drain improved soil is discussed here. A correction factor that can be used to adjust the calculated pwp from a plane strain analysis is also discussed. The variation of the correction factor with different parameters in the analyses such as imposed stress, hydraulic conductivity and material (mechanical) behaviour have been investigated. For analyses with linear elastic material behaviour, there was a unique relationship between normalized excess pwp and the correction factor - independent of imposed stresses, hydraulic conductivity and even material behaviour. For the case of elastoplastic (normally or over consolidated) material behaviour, the correction factor was independent of hydraulic conductivity and imposed stresses and could be explained by a linear relationship between c_f and normalized excess pwp. The linear relationship did not hold for scenarios where soil state changes from over to normally consolidated. Also excess pwp generated from soil creep can affect the relationship. It should be noted that all the discussion in this paper are based on 1 step loading. Further investigation is warranted to better understand variation of pwp under stagged loading or ramp loading scenarios.

Acknowledgements. Part of this work was conducted when the first author's PhD study in University of New South Wales, Canberra. The guidance from Dr Gnanendran and Associate Professor S-C R Lo (supervisor and co-supervisor) towards the project are gratefully acknowledged.

References

Carter, J.P., Ballam, N.P.: AFENA users' manual version 5.0 [computer program]. Center for Geotechnical Research, University of Sydney, Sydney-2006, Australia (1995)

Chai, J., Bergado, D.T., Shen, S.-L.: Modelling prefabricated vertical drain improved ground in plane strain analysis. Proc. Inst. Civ. Eng.-Ground Improv. **166**(2), 65–77 (2013) https://doi.org/10.1680/grim.11.00007

Hird, C.C., Pyrah, I.C., Russel, D.: Finite element modelling of vertical drains beneath embankments on soft ground. Geotechnique **42**(3), 499–511 (1992)

Hird, C.C., Pyrah, I.C., Russel, D., Cinicioglu, F.: Modelling the effect of vertical drains in two-dimensional finite element analyses of embankments on soft ground. Can. Geotech. J. **32**, 795–807 (1995)

Indraratna, B., Sathananthan, I., Rujikiatkamjorn, C., Balasubramaniam, A.S.: Analytical and numerical modeling of soft soil stabilized by prefabricated vertical drains incorporating vacuum preloading. Int. J. Geomech. **5**(2), 114–124 (2005). https://doi.org/10.1061/(asce)1532-3641(2005)5:2(114)

Karim, M.R.: Behaviour of piles subjected to passive subsoil movement due to embankment construction - a simplified 3D analysis. Comput. Geotech. **53**, 1–8 (2013)

Karim, M.R., Gnanendran, C.T., Lo, S.-C.R.: Effect of geosynthetic reinforcement creep on the long term performance of an embankment. In: State of the Art and Practice in Geotechnical Engineering Geo-Congress 2012, Oakland, CA, pp. 1330–1339 (2012)

Karim, M.R., Gnanendran, C.T., Lo, S.-C.R., Mak, J.: Predicting the long-term performance of a wide embankment on soft soil using an elastic-visco-plastic model. Can. Geotech. J. **47**, 244–257 (2010)

Karim, M.R., Lo, S.-C.R., Gnanendran, C.T.: Behaviour of piles subjected to passive loading due to embankment construction. Can. Geotech. J. **51**(3), 303–310 (2014)

Karim, M.R., Manivannan, G., Gnanendran, C.T., Lo, S.-C.R.: Predicting the long-term performance of a geogrid-reinforced embankment on soft soil using two-dimensional finite element analysis. Can. Geotech. J. **48**(5), 741–753 (2011)

Manivannan, G., Karim, M.R., Gnanendran, C.T., Lo, S.-C.R.: Calculated and observed long term performance of Leneghans embankment. Geomech. Geoengin. **6**(3), 195–207 (2011)

Roscoe, K.H., Burland, J.B.: On the Generalized Stress-Strain Behavior of Wet Clay. Cambridge University Press, Cambridge (1968)

Variable Modulus of Subgrade Reaction - Settlement Profile Convergence Method

Apurba Tribedi(✉)

Tower Products, Bentley Systems Inc., Anaheim, CA, USA
apurba.tribedi@bentley.com

Abstract. The modulus of subgrade reaction (Ks), is a commonly used parameter for a mat foundation analysis. But, it is not one of the basic soil properties and depends on many parameters including soil stiffness, foundation stiffness, foundation shape, size etc. Researchers and engineers agree that the general practice of using a constant modulus of subgrade reaction, is over simplified and may produce erroneous results. In fact, assumed contact pressure distribution may vary significantly from the actual one and shows a considerable spatial variation [8]. However, due to the non-availability of any standard method, engineers tend to avoid a proper distribution of the modulus of subgrade reaction. A proper distribution cannot be assumed or pre-calculated, rather it must be a result of a soil-structure interaction analysis.

This paper proposes an automated method to accurately distribute subgrade modulus. It is an iterative solution, and relies on the convergence of displacement profiles of two successive analysis. Any of the standard equations can be used to calculate soil stress and the corresponding settlement at any point of the mat. First iteration of Ks at a given point is then estimated as the ratio of soil stress to the settlement. The solution will then iterate and calculate new sets of Ks after each successful analysis by dividing the base pressure with the displacement. A few case studies have been presented to verify the effectiveness of the proposed method. As expected, resulting distribution of the subgrade modulus shows a spatial variation. Also, the vertical displacement diagram of a mat under uniformly distributed load, takes the shape of a bowl. Another case study is done to understand the effect of foundation stiffness. It shows changes in contact pressure due to the change in foundation thickness. A flow-chart has been provided to help create a computer program.

1 Introduction

Soil-Structure-Interaction (SSI) is a broad and interesting topic among engineers and researchers. Because of its complexity and unpredictability, engineers often require using sophisticated analysis techniques, including specialized software.

Modulus of subgrade reaction (Ks) is one of the key parameters used in mat (or raft) foundation design. It approximately represents soil response at any given point underneath a mat foundation. This key parameter is very useful for FEM based mat foundation design. Spring constants can be calculated from Ks values and corresponding spring supports can be applied underneath a foundation. A detailed discussion on this topic can be found on a related topic titled Correlation between Soil Bearing Capacity and Modulus of Subgrade Reaction [9].

H. Shehata and B. Das (Eds.): GeoMEast 2018, SUCI, pp. 162–184, 2019.
https://doi.org/10.1007/978-3-030-01923-5_13

A mat foundation is often analyzed using a constant value of Ks. Many argue this approach is over-simplified and sometimes may yield erroneous output for the sake of convenience.

This paper tries to address this complex issue by first addressing the theory briefly and then providing a solution for FEM based mat foundation analysis. A new method (Settlement Profile Convergence Method) has been developed by the author to calculate and distribute Ks and apply spring supports at all supporting points.

1.1 Subgrade Reaction and Contact Pressure

As the term suggests *contact pressure* is the base pressure between the bottom of the foundation slab and the supporting subgrade medium. In other words, the contact pressure is the *subgrade reaction* on the footing base due to the application of load on top of the foundation.

From observations, one can easily conclude that a loaded foundation will deflect and, as a result, the supporting subgrade medium will settle. The magnitude and the profile of the settlement will depend on soil characteristics, foundation properties, and loading. Determination of the contact pressure is very important for foundation design, yet calculations can be very complex [1, 2]. For larger foundation like mats or rafts, determination of contact pressure can be even more complex. However, because of its importance, a simplified theory has been developed to estimate subgrade reaction and the theory is known as the *theory of subgrade reaction.*

The theory of subgrade reaction is primarily based on following two assumptions [5].

(1) The soil medium is a perfectly elastic material. So, it obeys Hooke's law and stress is linearly proportional to strain (or settlement).
(2) The ratio between base pressure q on the foundation and the corresponding soil settlement s will remain constant.

$$Ks = \frac{q}{s}$$

Where, q = contact pressure or subgrade reaction
s = soil settlement

Ks is known as the coefficient of subgrade reaction or modulus of subgrade reaction.

It is important to note, that the above-mentioned assumptions are considered erroneous as the real contact pressure significantly differs from those simplified assumptions.

1.2 Borowicka Stiffness Factor

Contact pressure distribution depends on several factors including foundation stiffness, size, shape, and depth. Borowicka (1938) showed that the distribution of the contact pressure depends on a factor, called K_r. For a perfectly flexible foundation, the factor will be calculated as Kr = 0, whereas Kr = ∞ indicates an infinitely rigid foundation.

$$K_r = \frac{1 - \mu_s^2}{6(1 - \mu_f)^2} \frac{E_f}{E_s} \left(\frac{h}{b}\right)^2$$

Where,
K_r = Borowicka stiffness factor
μ_s = Poisson's ratio of soil
μ_f = Poisson's ratio of foundation
E_s = Elastic modulus of soil, E_f = elastic modulus of foundation
h = thickness of the foundation
b = radius of a circular foundation or half of the width of a strip foundation

However, there is no foundation which is infinitely rigid or completely flexible. The stiffness is somewhere in-between, and the actual nature of the distribution will depend on the soil properties as well as foundation properties.

As shown in Fig. 1 (Das), for a uniformly loaded rigid circular foundation, q_c/q = 0.5, where qc = contact pressure at the center of the footing and q is the uniform pressure load applied on top of the footing. However, for the same condition, q_c/q = 0.67 for a rigid strip footing. It is interesting to note, that the contact pressure reaches towards infinity at the edge of the rigid foundation.

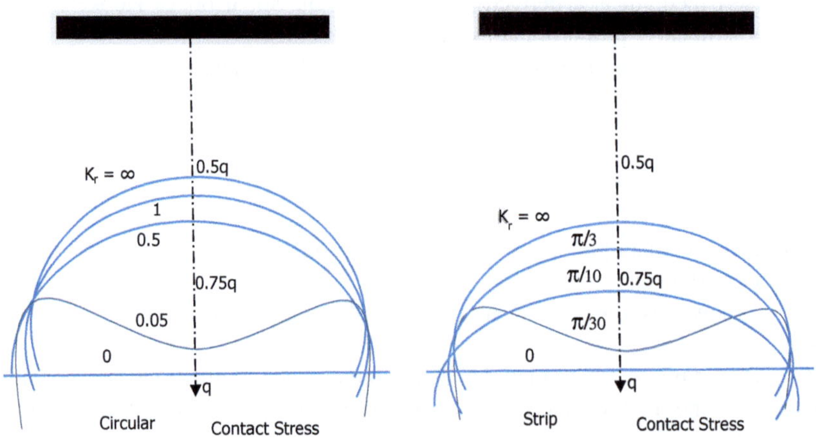

Fig. 1. Borowicka stiffness - contact stress for rigid circular and strip foundation

1.3 Contact Pressure and Settlement on a Flexible Foundation

A uniformly loaded flexible foundation on a perfectly elastic material of clay will have uniform base pressure and will take the shape of a bowl or trough. Evidently, a foundation will suffer the maximum deflection at the center [7] and will gradually reduce towards the edge (Fig. 2).

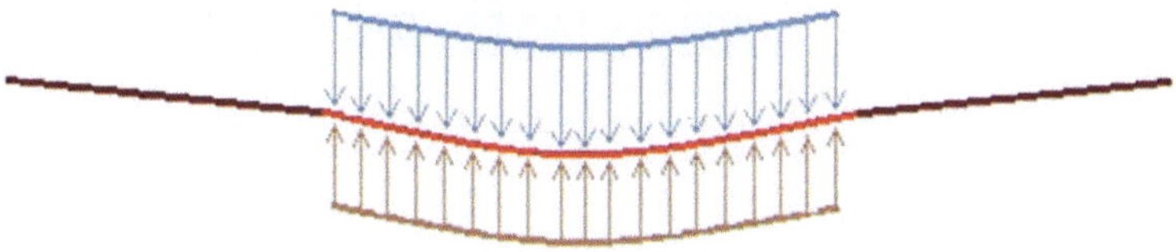

Fig. 2. Contact pressure and settlement on a flexible foundation

On the contrary, a uniformly loaded rigid foundation will settle uniformly (Fig. 3).

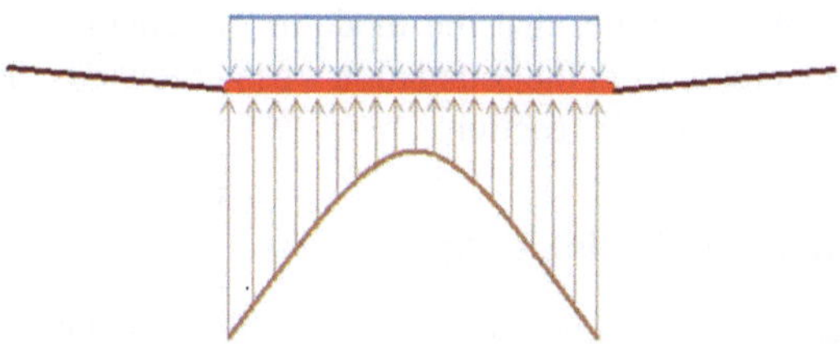

Fig. 3. Contact pressure and settlement on a rigid foundation

As discussed in earlier section (subgrade reaction and contact pressure), modulus of subgrade reaction is the ratio between the contact pressure (q) and its corresponding settlement (s). For a uniformly loaded flexible foundation, contact pressure remains uniform throughout the foundation but settlement decreases from the center towards the edge, so q/s or modulus of subgrade reaction increases from the center towards the edge (Fig. 4).

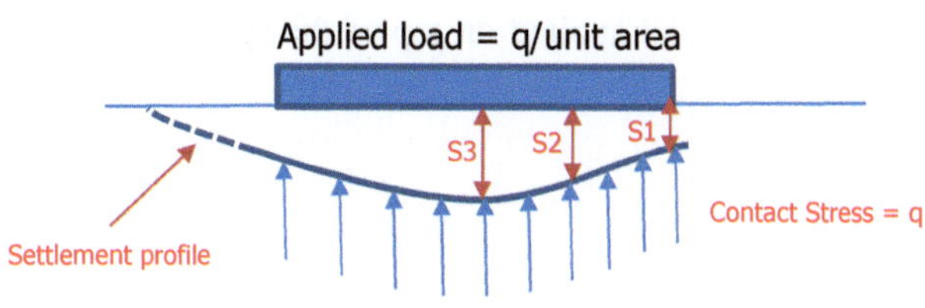

$$\frac{q}{s1} > \frac{q}{s2} > \frac{q}{s3}$$

Fig. 4. Subgrade modulus of a uniformly loaded flexible foundation

Conversely, for a uniformly loaded rigid foundation, contact pressure increases from the center towards the edge of the foundation while settlement remains uniform. It is interesting to note, like flexible foundation, the ratio between contact pressure and corresponding soil settlement (q/s) increases from the center of the foundation towards the edge (Fig. 5).

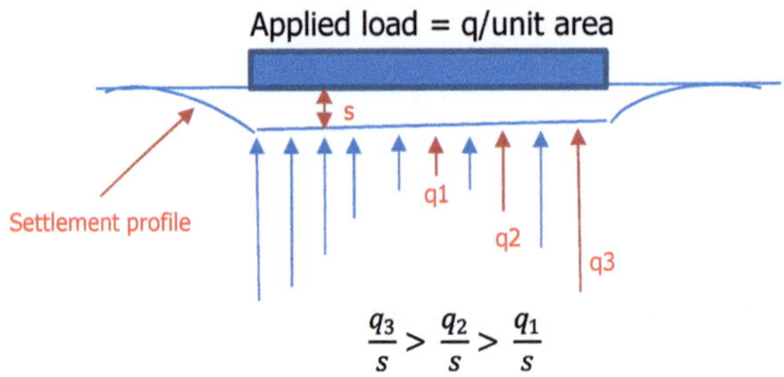

Fig. 5. Subgrade modulus of a uniformly loaded rigid foundation

So, it is reasonable to predict that the modulus of subgrade reaction will not be uniform, and it will increase from the foundation's center towards the edge. It is known as variable modulus of subgrade reaction.

1.4 Computer Model and Winkler Spring

The current practice is to use a commercially available FEM program to analyze mat foundations. The foundation is normally modeled as bending plates and the supporting soil is modeled as discrete *compression only* springs connected to plate nodes. Those springs are known as Winkler springs. It is a popular modeling paradigm because of its simplicity.

Winkler idealized a soil model consisting of closely spaced identical but mutually independent linear springs (Fig. 6). For such an idealized model, when the foundation is subjected to partial loading, foundation deformation is confined to the loaded region only. In other words, springs directly below the loaded area will be subjected to deflection and any other springs beyond the loaded area will remain unaffected. Figure 7 demonstrates different loading conditions and corresponding expected Winkler model behavior. For an arbitrary loaded area (Fig. 7a), spring deformation is directly proportional to the load applied on top of the spring.

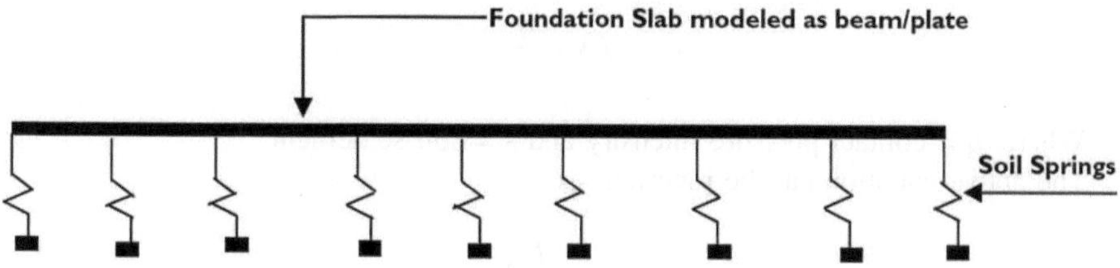

Fig. 6. Idealized Winkler spring model

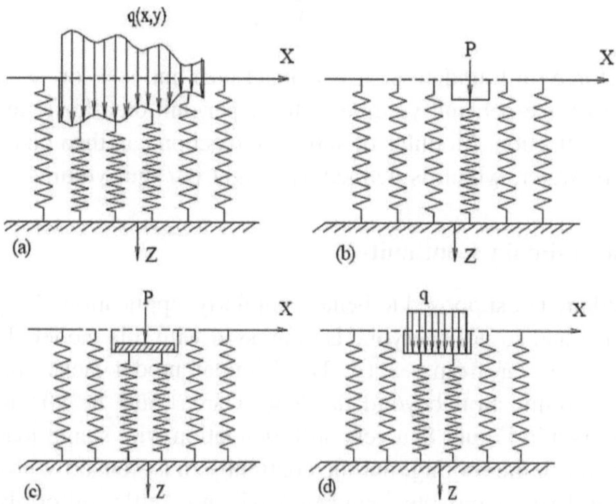

Fig. 7. Model behavior - deformation is confined to the loaded region only

1.5 Liquid Analogy

Winkler soil model resonates well with the liquid analogy. If an elastic plate floating on the liquid surface is uniformly loaded, the plate will deflect but liquid beyond the loaded area will not be affected. The plate deflection and the pressure under the plate will be uniform. This is similar to the Winkler model.

The pressure underneath the plate will be equivalent to the buoyancy force which can be calculated as the upward pressure acting on the plate multiplied by the area of the plate. The plate deflection (immersion) into the water can also be compared to foundation deflection. So, an elastic plate having weight W and surface area A, when placed on water, is immersed by a distance h and generates upward pressure q. Applying the same concept to foundation analysis, a footing subjected to load W and area A will undergo h deflection and generate base pressure q.

Terzaghi (1955) introduced modulus of subgrade reaction by the following mathematical expression.

$$Ks = \frac{q}{s}$$

Where, q = contact pressure intensity and s = soil settlement
The above equation can be rewritten as

$$Ks = \frac{P/A}{s}$$

$$Ks = \frac{P}{A \times s}$$

Where, P = force on foundation, A = contact area and s = soil settlement
It is similar to force per unit volume or force per unit area for a unit deflection or pressure for unit deflection. Modulus of subgrade reaction can then be compared to the specific weight of water, which is defined as weight per unit volume.

1.6 Soil Must Maintain Continuity

As soil and liquid are not supposed to behave similarly, application of liquid analogy to the soil is a stretch and so not everyone is impressed with this model. The behavior of an idealized model is far from reality. The Winkler model lacks continuity within supporting soil medium. Soil beyond the loaded area must be affected to maintain continuity. As shown in Figure 8, a very stiff foundation with stamp loading will settle uniformly just beneath the footing, but the continuity must remain with the supporting soil medium. Similarly, a flexible foundation will not settle uniformly under a uniformly distributed load and will maintain continuity with the supporting soil medium. Also, the springs cannot be completely independent, there must be couplings or linkage among springs.

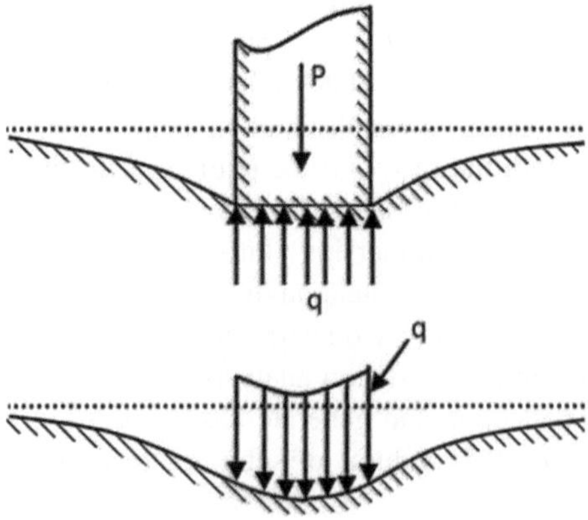

Fig. 8. Actual soil deformation to maintain continuity

1.7 Effect on a Flexible Mat Foundation Design

As we look at theories of subgrade reaction and study different proposed methods, it is important to study and understand the effect of these parameters on foundation design, and particularly the effects on flexible mat foundation design.

We would like to study three relevant parameters for both soil and structure, to understand the effect on mat foundation design. Those parameters are:

(1) Uniform Ks and the numerical value of Ks
(2) Distribution of Ks
(3) Structural rigidity

1.8 Uniform Ks and the Numerical Value of Ks

As discussed earlier, current practice is to use FEM based software packages to do mat foundation analysis. Also, most software packages accept input of a uniform Ks and automatically generate compression-only Winkler springs to model supporting soil medium underneath the mat slab.

1.9 Case Study

Let's take the example of a very simple, square mat foundation supporting four columns and analyze the foundation with two different values of Ks. First with Ks = 40 lb/in^2/in and then Ks = 80 lb/in^2/in.

Modeling -
Mat geometry: Square Mat 12 ft × 12 ft × 0.5 ft
Soil Modulus of Subgrade Reaction: 40 lb/in^2/in, 80 lb/in^2/in
Column Loading: 40 kips each (compressive)

Solution

A commercially available software STAAD Foundation has been chosen for the analysis. It uses compression-only springs to model supporting soil medium. Nodal springs are automatically calculated as a product of modulus of subgrade reaction and corresponding nodal tributary area.

We would like to study two types of output as followings:

(1) Effect on resulting base pressure distribution
(2) Effect on design bending moment

A summary of results is tabulated as below, which shows considerable differences in results for different uniform Ks values. So, it can be evidently concluded that engineers should select the value of Ks very carefully and should always seek help from qualified geotechnical engineers to provide the best-estimated value of modulus of subgrade reaction (Ks).

It should be noted that the above-mentioned model is a very simple mat foundation scenario with arbitrarily assumed mat dimensions as well as loading. In the next section, we will look at more realistic examples by referring to two published case studies (Fig. 9 and Table 1).

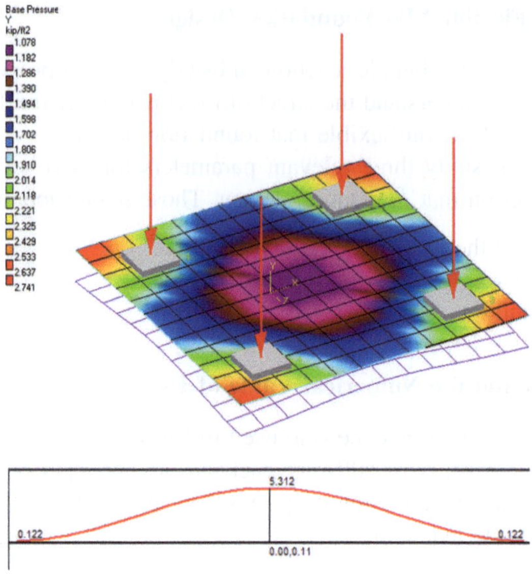

Fig. 9. A simple Mat foundation for a case study

Table 1. Key value comparisons for two uniform Ks values

Ks (psi/in)	40	80	Difference (%)
σ_{min} kip/ft^2	1.078	0.761	41.66%
σ_{max} kip/ft^2	2.741	3.156	13.15%
Moment (kip-ft/ft)	5.312	3.927	35.27%

1.10 Parametric Study by Horvilleur and Patel [4]

Mr. Horvilleur and Mr. Patel studied and published a paper (SP-152: Design and Performance of Mat Foundations) on different soil and structural parameters and their effects on mat foundation design including dishing effect. The study was accompanied with parametric studies on two actual mat foundations. The purpose was to study the effect of different design parameters on the magnitude and distribution of soil pressure and bending moment. The most important parameters studied were as followings:

1. Mat Flexural Properties EI
2. Effect of Mat Shear Deformations
3. Modulus of Subgrade Reaction "Ks"

We will limit our discussion on Ks because of its relevancy with the current topic.

1.11 NCNB Corporate Center

The first building was the NCNB Corporate Center, which was a 60-story tall concrete building located in downtown Charlotte, North Carolina. Two foundation systems were used in the building. The perimeter columns were supported by deep concrete caissons bearing on 150 ksf rock. The core columns were supported on a core mat having dimensions 84ft × 93ft × 8ft. The mat was founded on partially weathered rock with an average modulus of subgrade reaction of 290 pci (290 psi/in).

Two analyses were conducted. First using K = 290 psi/in and then K = 580 psi/in. Figures below show variations for pressures and moments for two K values. The maximum change in pressures was 7%. There was a 30% change in positive moment and a 43% change in negative moment. It is interesting to note, that the variations of results are in line with our earlier simple case study (Figs. 10 and 11).

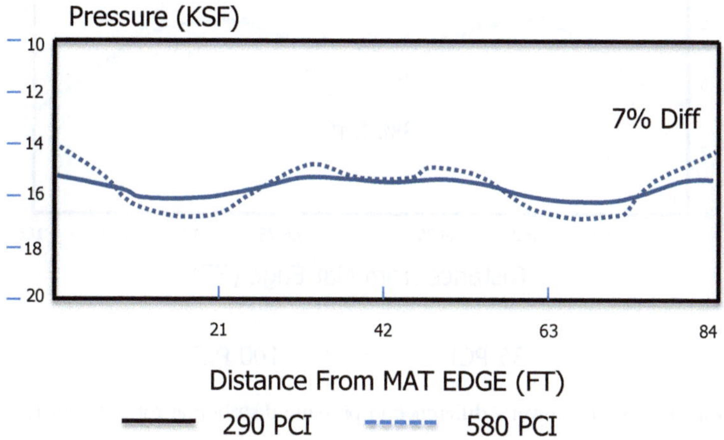

Fig. 10. NCNB Corporate Center - difference in pressure distribution for different Ks

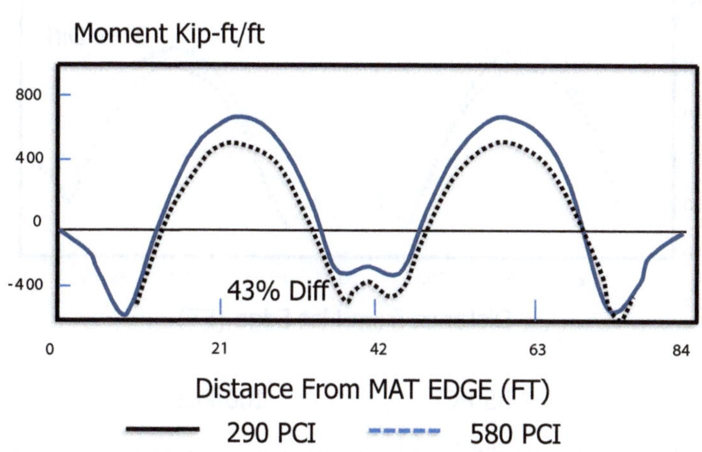

Fig. 11. NCNB Corporate Center - difference is bending moment diagrams for different Ks

1.12 Smith Tower

The Next study site was Smith Tower. It is a 30-story tall concrete building located in the Texas Medical Center, Houston, Texas. In this case, the entire building was supported by a rectangular mat having dimensions 106 ft. × 228 ft. × 5′–6″. The mat was founded on Houston clays with an average modulus of subgrade reaction of 35 psi/in.

As in NCNB Corporate Center case study, two analyses were conducted. First using K = 35 psi/in and then K = 100 psi/in. Figures below show variations for pressures and moments for two K values. The maximum change in pressures was 9% and maximum change in the moment was 21%.

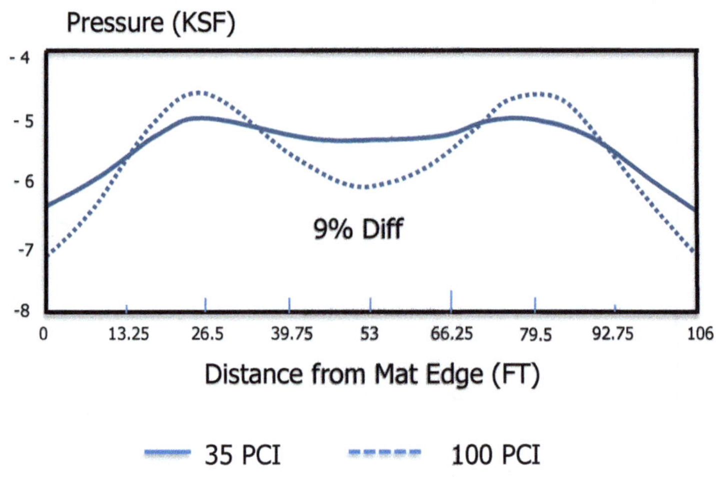

Fig. 12. Smith Tower - difference in pressure distribution for different Ks

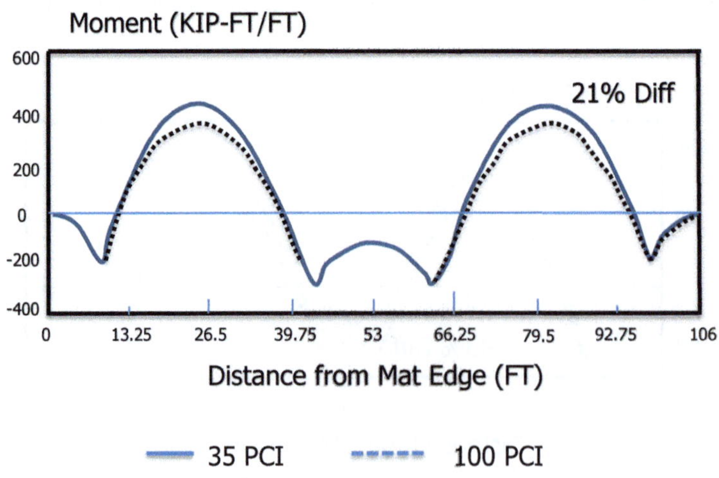

Fig. 13. Smith Tower - difference is bending moment diagrams for different Ks

From both the studies, it can be noted that moment diagrams are more sensitive to the magnitude of Ks than the pressure values. It is evident from these case studies that all efforts must be made for the selection of a proper Ks value and is probably the single most important factor responsible for the analysis of a mat foundation (Figs. 12 and 13).

1.13 Distribution of Ks

As much as the value of Ks is important, distribution of Ks and its effect on the foundation is even more important. It is a complex subject and often not used for routine foundation design because of the non-availability of any simple mechanism. For large scale projects, engineers often collaborate and use sophisticated software for soil-structure-interaction.

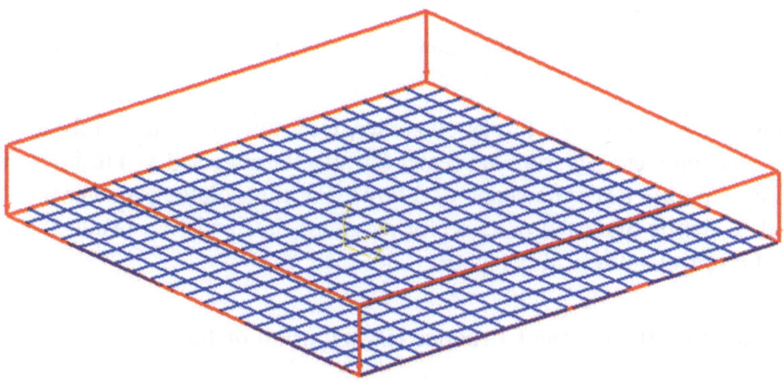

Fig. 14. A case study of a uniformly loaded mat foundation

From the theory of subgrade reaction, we know that the settlement profile of a uniformly loaded flexible foundation will take the shape of a bowl or trough. We would like to study if it holds true for a supporting soil medium having uniform Ks. As in earlier case study, we will try a simple case study using a standard commercial software package (Fig. 14).

Modeling -

Mat geometry: Square Mat 12 ft × 12 ft × 0.5 ft

Soil Bearing Capacity: 4 kip/ft^2

Loading: 1 Kip/ft^2

Solution –

Ks can be estimated as [3, 6],

$$Ks = \frac{Iq_a}{\delta} = \frac{3 \times {}^4\!/_{144}}{1} = 0.0833\, kip/in^2/in$$

Where,
I = Safety factor = 3.0 (assumed)
q_a is the allowable bearing capacity (given 4.0 kip/ft^2)
δ is the allowable soil settlement = 1 inch (assumed)

As discussed earlier, the displacement profile is expected to take the shape of a bowl or trough. However, foundation settled uniformly (Fig. 15) and displacement profile didn't match our expected behavior.

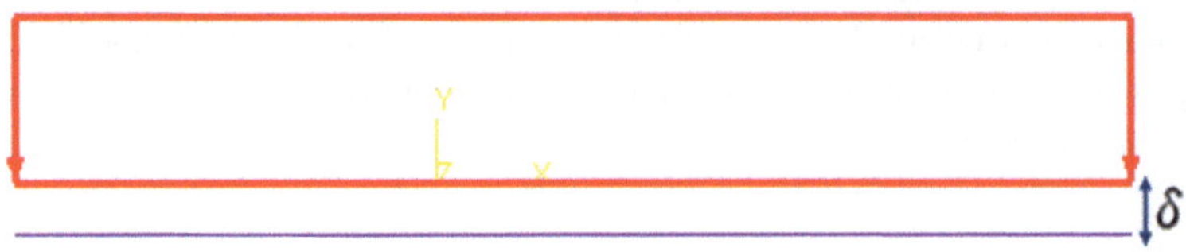

Fig. 15. Vertical deflection diagram of the uniformly loaded Mat foundation

It is hardly a surprise. As discussed earlier, we know, for a uniformly loaded mat, subgrade reaction increases from the center towards the edge and as a result, foundation displacement profile takes the shape of a bowl. For this solution, a constant Ks value was used (which is often the practice for its simplicity) and as a result, the program computed erroneous physical behavior.

1.14 Extending the Current Practice (Distribution of Ks)

Fortunately, it is not too difficult to manipulate a computer program to get the expected physical behavior. We can extend the model to create varying soil properties from the center of the foundation towards the edge. In other words, we can use variable Ks to depict the desired settlement profile of the supporting soil medium.

Solution

As shown in Fig. 16, divide the supporting soil medium in several bands (five for this example) ranging from K_{s1} to K_{s5}. Also, let's assume subgrade reaction increases as much as 100% from the center towards the edge.

K_{s1} = 0.0833 kip/in^2/in
K_{s2} = 0.09996 kip/in^2/in
K_{s3} = 0.119952 kip/in^2/in
K_{s4} = 0.143942 kip/in^2/in
K_{s5} = 0.172731 kip/in^2/in

After running the analysis, the deflection profile takes the shape of a bowl or trough and matches our expected soil settlement profile. Also, it is interesting to notice the soil pressure contour. Unlike the case for a uniform Ks, base pressure contour now shows varying pressure from the center towards the edge.

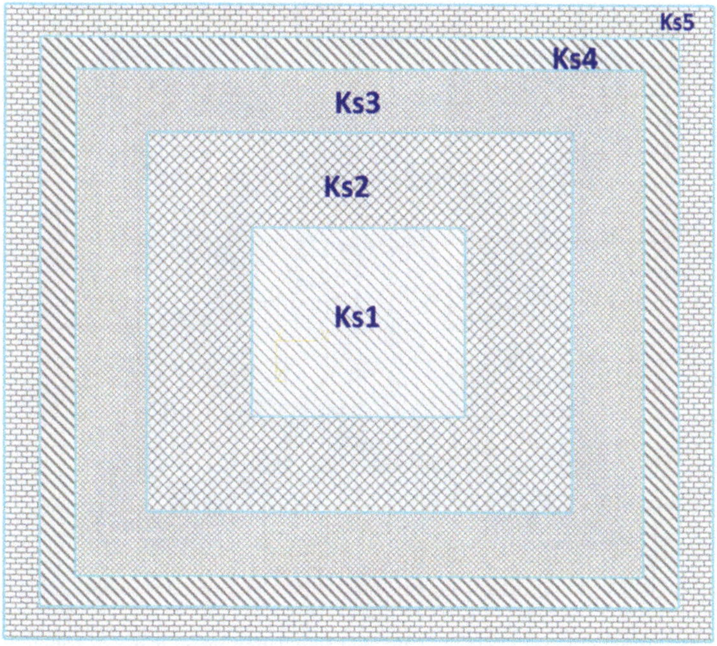

Fig. 16. Banded Ks distribution from lowest at the center to the highest towards the edge

So, it is natural to conclude that a flexible mat foundation should always be analyzed using a variable modulus of subgrade reaction. It predicts more accurate physical behavior and hence the results are more accurate. This approach was implemented for several case studies and briefly discussed below (Figs. 17 and 18).

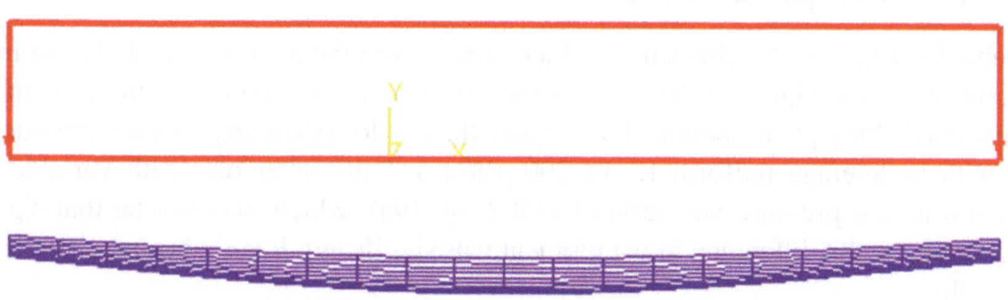

Fig. 17. Vertical displacement profile of a mat foundation when using distributed Ks

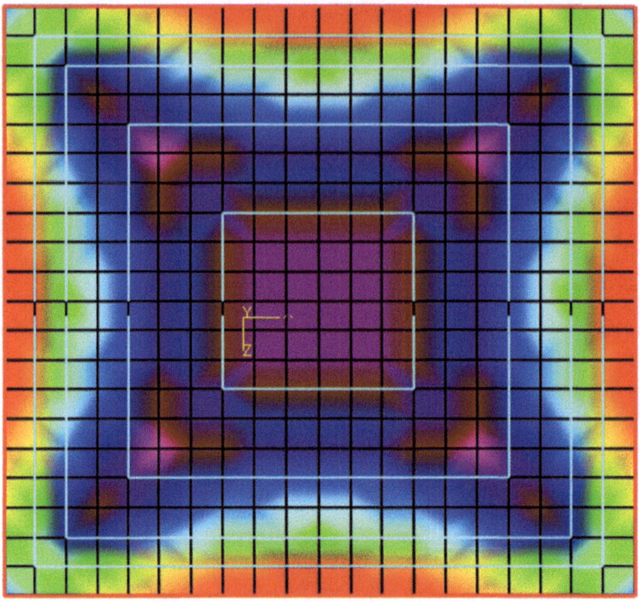

Fig. 18. Base pressure contour of the mat foundation when using distributed Ks

1.15 Parametric Study by Horvilleur and Patel [4]

As discussed earlier, Horvilleur and Patel studied the effect of variable Ks on two sites and documented the effects of those parameters on soil pressure distribution and bending moments.

1.16 NCNB Corporate Center

Variable Ks was used for this study, which varied from the lowest value (181 psi/in) at the centroid to the highest value (548 psi/in) at the edge. As expected, the researchers observed a dishing phenomenon. To compare the results, two analyses were conducted. One with an average uniform Ks of 290 psi/in and the other one with variable Ks. Variation in soil pressure was around 11% (Fig. 19a), which may not be that significant. However, the difference in the moment was significant. It varied as much as 120% (Fig. 19b).

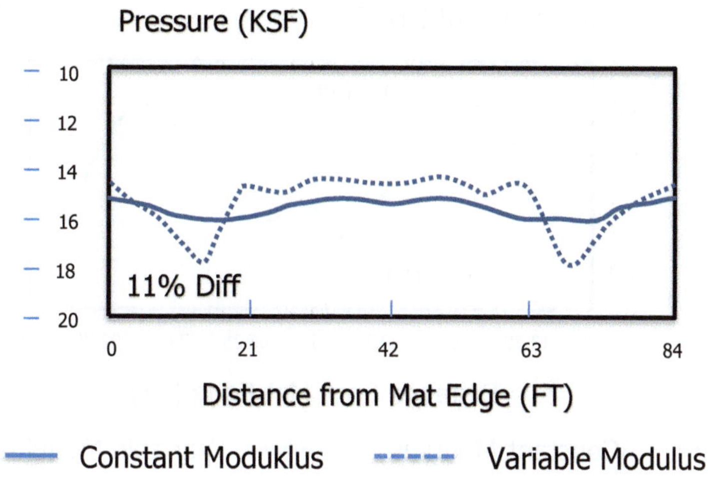

Fig. 19a. NCNB Corporate Center – base pressure comparison for constant vs variable Ks

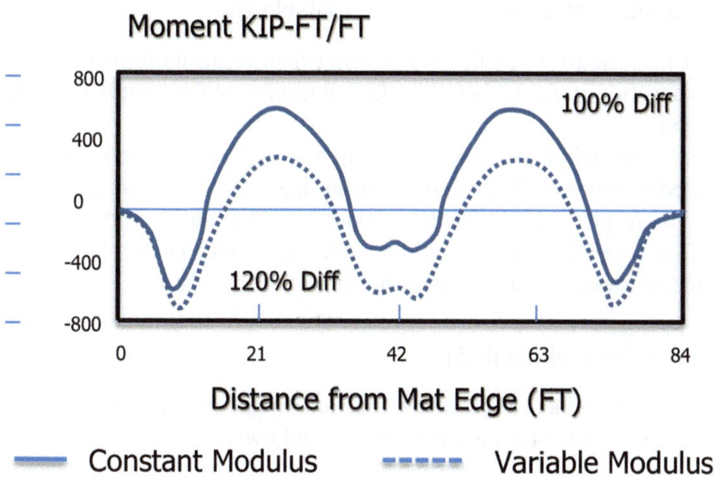

Fig. 19b. NCNB Corporate Center – moment comparison for constant vs variable Ks

1.17 Smith Tower

Like the earlier case study, two analyses were performed. One with an average uniform Ks and the second analysis used variable Ks, ranging from a low 24 psi/in to 38 psi/in. Interestingly the observations were very similar, there was a difference of 13% in soil pressure but a significant difference in the moment (Fig. 20).

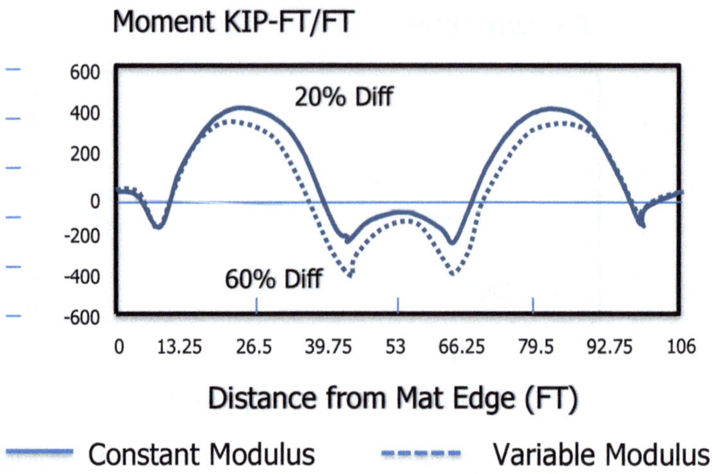

Fig. 20. Smith Tower – moment comparison for constant vs variable Ks

1.18 An Iterative Method to Estimate Variable Ks

It is evident that use of a variable Ks is better than using uniform Ks. However, the value of Ks depends on many factors including rigidity of the foundation, soil stiffness, and settlement.

As an example, initial estimated values of Ks for a given zone were derived based on a predicted settlement. After finite element analysis, calculated displacement for a given zone (more precisely for a given node) may not match with the predicted displacement. So, a new Ks value should be calculated and run the analysis again. This iterative process should continue until the solution converges.

1.19 Discrete Area Method [5]

This is an iterative method to achieve deflection compatibility between mat deflection and soil settlement. The steps are described as follows:

1. Create a finite element model for mat foundation and analyze it using a geotechnical engineer's best-estimated uniform Ks value.
2. Using base pressure from step 1, geotechnical engineer calculates soil settlement at each node of the FEA model and a new set of Ks value at corresponding nodes.
3. Input a new set of subgrade modulus in the structural finite element model and obtain a new pressure distribution and settlement.
4. Using the pressure profile from step 3, the geotechnical engineer calculates settlement at each node and corresponding Ks at each node.

Repeat

Repeat step 3 and step 4 until a convergence is achieved. This happens when the displacements predicted by the structural engineer's finite element analysis match the settlements predicted by the geotechnical engineer. The resulting coefficient of subgrade modulus may differ significantly from the initial estimated values and, as a result, will significantly influence mat design. It is obvious that the convergence may require multiple iterations and as the process cannot be automated, use of this method demands close collaboration between structural and geotechnical engineers.

1.20 A New Approach: Settlement Profile Convergence Method

Base pressure, settlement, stiffness, and modulus of subgrade reaction are all intertwined. Dependency among these parameters adds to the complexities and makes it difficult to achieve a quick definitive solution.

A new computer-based approach has been developed and proposed by the author. It is an iterative solution to converge foundation displacement and soil settlement. The most important aspects of the solution are an initial estimation of soil settlement and normalization of the settlement profile which minimizes the effect of various factors used in different proposed equations for soil stress calculation.

The steps are described as followings:

(1) Generate a mesh to create a finite element analytical model of the foundation slab.
(2) Use standard Boussinesq's equation or any other method to calculate soil stress under each meshed node for a given load condition.
(3) Multiply soil stress values with calculated Borowicka stiffness factor and calculate soil settlement for each node.
(4) Normalize settlement values between 1 and the ratio between maximum over the minimum settlement.
(5) Use a surface fitting technique to create a soil settlement profile.
(6) Calculate Ks for each node from allowable bearing capacity and the calculated soil settlement from step 5.
(7) Calculate nodal spring constants by multiplying Ks with the nodal tributary area. Assign spring constant as compression-only springs.
(8) Run the finite element analysis.
(9) Extract nodal displacement values from the analysis.
(10) Repeat step 4 through step 9.

(11) Compare the new set of displacement values with displacement values from the last analysis.
(12) Repeat steps 10 and 11 until two consecutive analyses converge or fall within a reasonable tolerance limit.

1.21 The Effectiveness of the New Method

The new method is significant for several reasons.

- It focuses on the shape and relative displacements rather than the absolute displacements, which minimizes the effect of initially estimated soil settlement.
- The process is automated and complementary to the current FEA based mat foundation analysis.
- It considers structural rigidity.

1.22 Comparison with the Zonal Ks Distribution Method

We can compare the example from the section, "distribution of Ks" to study the effectiveness of the above-mentioned method. As expected, displacement profile from both the methods takes the shape of a bowl. However, the shape produced by the "settlement profile convergence method", is smooth. It is even more apparent from the resulting base pressure diagrams from each analysis. As discussed earlier, Ks is expected to vary radially from the center towards the corner. The resulting contour diagram as shown in Fig. 22 correctly depicts that expected radial distribution. It is a major improvement over any existing method (Fig. 21).

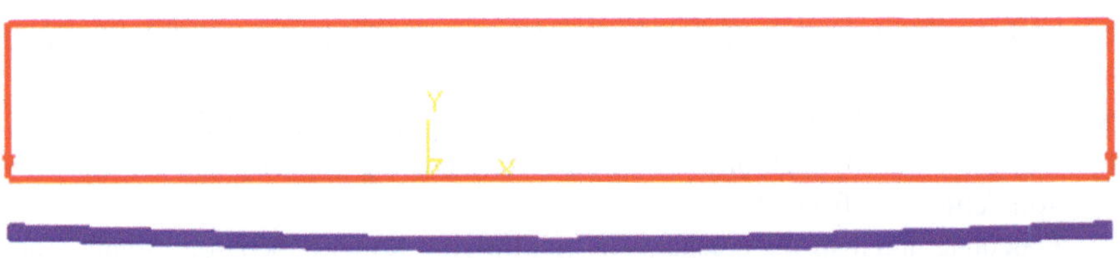

Fig. 21. Revised displacement diagram for settlement profile convergence method

1.23 Effect on Foundation Rigidity

In theory, if we run the same example with a different thickness, base pressure distribution should vary. So, let us run the same example for a slightly thicker footing. After changing the footing thickness to 0.75 ft., we get a base pressure contour as shown in Fig. 23. As expected, it varies from the diagram for 0.5 ft. thick footing. Also, it demonstrates narrowing down of pressure variation because of the increase of foundation thickness (or rigidity) (Fig. 24).

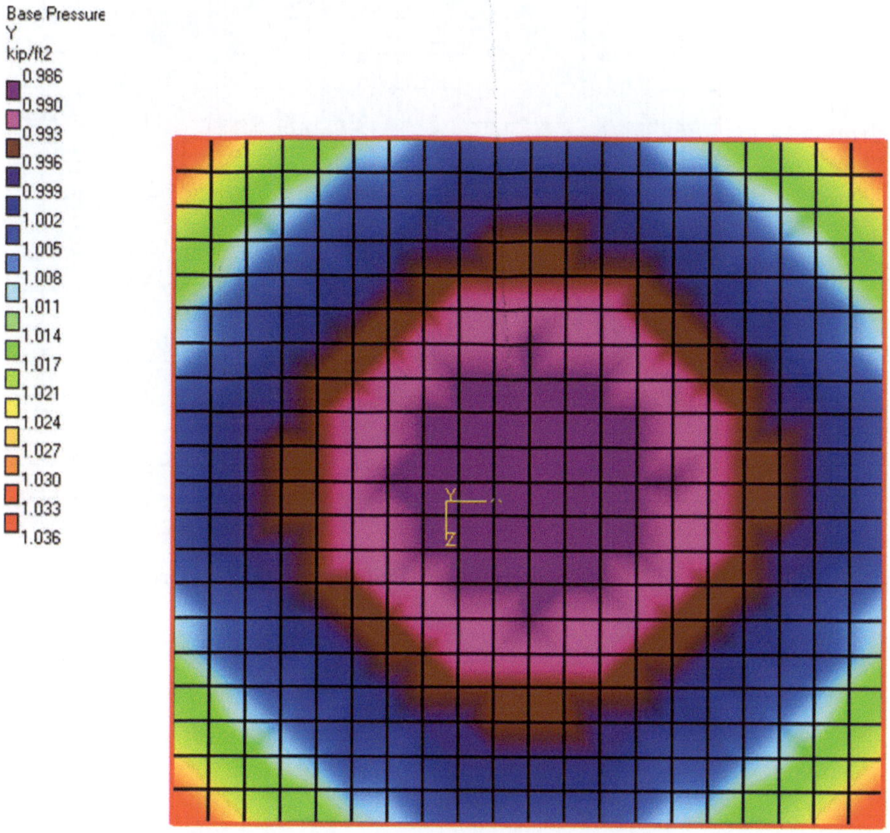

Fig. 22. Revised base pressure contour for 0.5 ft. thick mat foundation

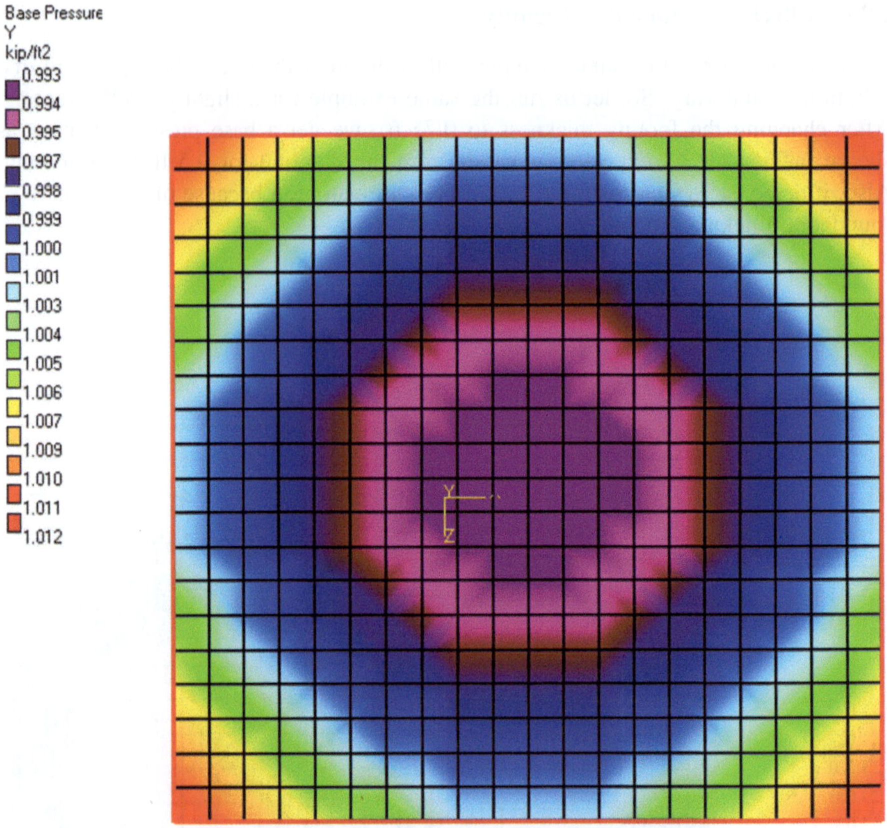

Fig. 23. Revised base pressure contour for 0.75 ft. thick mat foundation

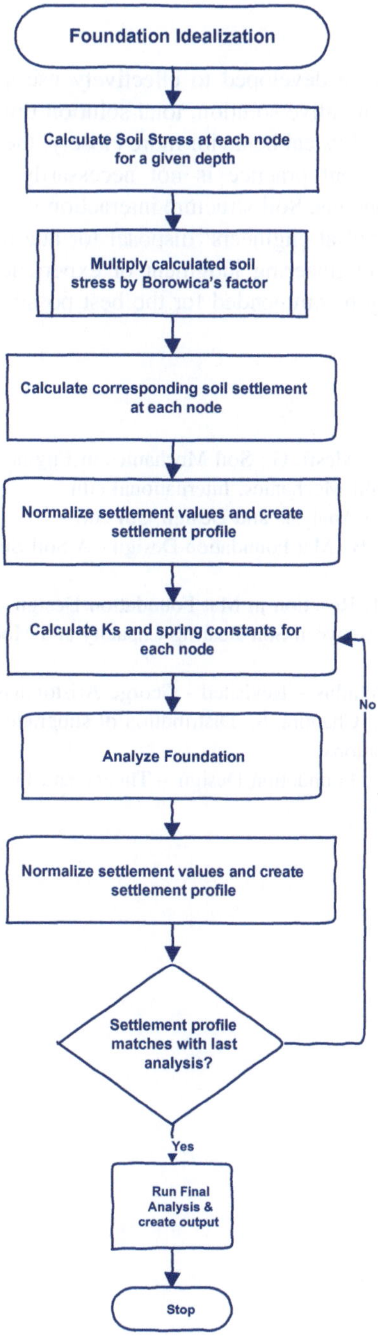

Fig. 24. A computer program for displacement profile convergence method

2 Conclusion

A computer program can be developed to effectively use settlement profile convergence method. As it is an iterative solution, total solution time will extend marginally. However, as it depicts the physical behavior more closely, the final results are expected to be more accurate. Current practice is not necessarily conservative. It is oversimplified and can be erroneous. Soil-structure interaction is a complex subject and this method is an additional tool at engineers' disposal for automating a foundation analysis. It does not replace engineering judgment or experience. A close collaboration among engineers is highly recommended for the best possible outcome.

References

1. Terzaghi, K., Peck, R.B., Mesri, G.: Soil Mechanics in Engineering Practice, 3rd edn
2. Das, B.M.: Advanced Soil Mechanics, International edn
3. Bowles, J.E.: Foundation Analysis and Design, 5th edn
4. Horvilleur, J.F., Patel, V.B.: Mat Foundation Design - A Soil-Structure Interaction Problem, 1 March 1995
5. Ulrich Jr., E.J.: Subgrade Reaction in Mat Foundation Design, 1 March 1995
6. Tribedi, A.: Correlation Between Soil Bearing Capacity and Modulus of Subgrade Reaction (2013)
7. Gómez, J.: Subgrade Modulus – Revisited - George Aristorenas (2014)
8. Dey, A., Basudhar, P.K., Chandra, S.: Distribution of subgrade modulus beneath beams on reinforced elastic foundations
9. Kameswara Rao, N.S.V.: Foundation Design – Theory and Practice

Environmentally Friendly Building Materials in Ground and Special Foundation Engineering on the Example of a Fully Biological Soft Gel for Subsoil Injection

Arne Kindler[1(✉)], Karolina Nycz[1], and Michael Rembe[2]

[1] Department for Research and Development, Stump Spezialtiefbau GmbH, Valeska-Gert-Straße 1, 10243 Berlin, Germany
arne.kindler@stump.de
[2] Rembe Consulting PartG mbB, Alte Leipziger Straße 50, 99734 Nordhausen, Germany

Abstract. The Climate Change Conference in Paris in 2015 supported by all present countries decided to reduce the CO_2 emissions in order to make a substantial contribution to reducing global warming. In this context in 2016 a new fully biological ground injection medium on the basis of the bionics was developed by Stump Spezialtiefbau GmbH Germany. In the following, the most important material parameters of the developed organic soft gel as well as the results of related numerical simulations are presented.

1 Introduction

Soil injections are being carried out for more than 200 years, with the first injections using water-cement suspensions (Kutzner 1991) as the injections medium. Over the years the injection technique developed steadily. In Germany the injection technique experienced a rapid development only after the invention of the chemical injection by Joosten in 1926, although the pure chemical injection has been already invented at the beginning of the 20th century (Kutzner 1991). The injection technique in geotechnical engineering reached its highest point with increased planning of deep structures in high groundwater levels. Nowadays such buildings still require a temporary draining of the excavation pits until a buoyancy-safe stage of construction is reached.

Due to the existing fine grain of sands in conjunction with the size of the solid particles of a cement suspension in order to achieve a successful injection, the use of ultrafine cements and/ or soft gels is required.

In the years leading up to 1995 the injection of soft gels was of a large popularity, particularly in Berlin. In the same year the responsible Senate Department of the city of Berlin decided to withhold the allowance to use soft gels in order to, among other things; clarify the question of the chemical influence of the injection medium on the groundwater.

H. Shehata and B. Das (Eds.): GeoMEast 2018, SUCI, pp. 185–196, 2019.
https://doi.org/10.1007/978-3-030-01923-5_14

Since the year 2000 the German Institute for Building Technology (DIBt) of the Federal Republic of Germany has released a directive on "Evaluation of the effects of construction products on soil and groundwater" (currently: DIBt 2009a). The directive specifies in detail which trials and boundary conditions for an assessment of the environmental impact of construction products in the groundwater are of importance.

2 History

The chemical injection mediums used mainly for large-scale seals on high ground water levels in Germany were at the time mostly so-called soft gels. The main components of those soft-gels were water, water glass and a reactive. The used water glass is an aqueous solution of sodium silicates as a three-phase system in combination of alkali oxides, silica and water (Eiswirth et al. 1998). Often as reactive sodium aluminate was used.

At the time the order to stop the usage of soft-gels imposed by the Senate of Berlin was justified by the alkaline reaction of the water glass solutions (pH-values $\geq$ 10) as well as by a possible release of the aluminate of the reactive during the setting process.

In 1997 the numerical model investigations commissioned by the DIBt showed various worst-case-studies (Brauns et al. 1997) of time-independent constant discharge of reaction products from the soft gel sole in to the ground. These studies were proved wrong by Eiswirth et al. (1998) nevertheless they found sufficient consideration in the approval procedures of the competent senate administration. In the following years, other German cities followed the precedent decision of the Senate of Berlin.

3 Directive on the "Assessment of the Effects of Construction Products on Soil and Groundwater"

As a result of this situation a directive on the assessment of the effects of construction products on soil and ground water was created by the DIBt, which forms the basis for any injection medium, and on the basis of which an environmental impact of the injection medium is evaluated and issued by the DIBt, as general technical approval (abZ).

Published for the first time in 2001 the DIBt guideline describes control parameters for the ingredients of the previous so-called soft gels which are still to be investigated even for modern injection medium. This includes the possible inorganic water components such as aluminum, sodium and silicon, as well as the physio-chemical parameters such as electrical conductivity, pH-value and temperature.

As in all in-put products of today's soft gels also traces of As, Cu, Fe, Mg, Hg, Ti, and Zn (Brauns et al. 1996/1997) can be present, the soft gels have to be investigated also regard to their heavy metal contamination.

For this purpose an investigation of the injection substances through the evaluation of inverse column elution tests according to DIN 19631:2016-07 is being made.

The aim of the research is to establish findings with regard to mobilizable ingredients as well as the ecotoxicological effects of these substances. The most important criteria of the DIBt is compliance with a pH value <10.5 in the reactive mixture (DIBt 2009b) of the injection medium since it is possible that by a change of the pH-value in the soil a mobilization of a tied medium (for example, a humic substances) and related secondary effects may occur.

The inverse column experiment is essentially marked by its procedure. The experiment set-up and the implementation in the laboratory can be described by the following points:

- standard acrylic glass cylinder
- filling of the cylinder through a standardized sand
- conditioning of the fill sand with drinking water
- injection of soft gels
- passing through the filled and injected cylinder with drinking water
- collecting of the eluate in defined time intervals

In a first step, general parameters of the eluate (smell, turbidity, a tendency to foam) are to be identified and assessed organoleptically.

In the next step, the eluates are analyzed in terms of their material parameters. This includes the determination of pH-value, electrical conductivity and TOC content. In addition, determination of further material parameters can be required by DIBt.

In Germany the results of this experiment are to be compared in terms of their electrical conductivity and the heavy-metal pollution with the limits issued by the Country Working Community for Waste (LAGA 2002). For further limits not covered by the LAGA material parameter the limits of the Drinking Water Ordinance (TrinkwV 2011) of the Federal Republic of Germany are relevant.

If the sum of the determined TOC parameter excess the value of 20 mg/l further ecotoxicological studies (biological test) are required.

For the verification of the test results, as well as to gain a better understanding of the injection mobilization in the ground all test are being calculated numerically modeled (see Fig. 1).

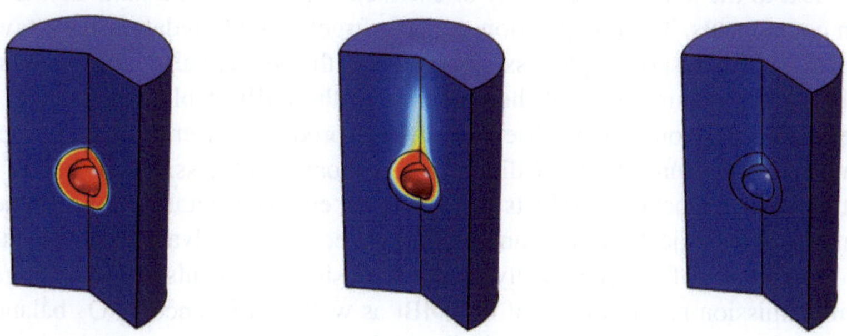

Fig. 1. Numerical simulation of an inverse column experiment for the verification of the mobilization on the example of the chloride effluent (a) at time t = 000 h after the injection (b) at time t = 200 h after the injection (c) at time t = 168 h after the injection

Figure 1 shows fragmentary a time study for the mobilization of chloride from an injection body. The time points are:

- t = 0 h for the time after the injection,
- t = 2 h after the injection,
- t = 168 h after the injection.

As expected it is shown that the mobilization due to the fall in the total concentration with the experiment time progress decreases.

Through the simulation of the material transport as well as their validation with the laboratory results, there is a growing in the understanding of the material transport in soils and their impact on the injection techniques. The numerical simulation was carried out using the commercial software COMSOL Multiphysics.

4 Environmentally Friendly Soft Gels

As a result of this historical development in the chemical injection technique foundation engineering companies in Germany began with the development of new environmentally friendly and therefore the certified soft gels. In this context, the first general building inspectorate approval for environmentally friendly soft gel was granted to the company Bauer (silicate-gel), followed by the company Züblin/ Insond (acrylate-gel).

4.1 House-Own Environmentally Friendly Soft Gels

In the light of the above-mentioned historical development the Stump Spezialtiefbau GmbH has also developed its own environmentally friendly soft gels. These include three different products, which differ from each other in their reaction mechanisms, as well as in their composition and thus in their possible application areas.

The currently officially approved silicate-gel “Stump-Silikatgel SIG 60” is characterized by the fact that all reaction components in confined construction site conditions as well as for silo storage are available in unsolved condition. The recipe consists next to the main part of water of chemically produced and naturally occurring reaction components. The composition of the soft-gel is the foundation for a favorable CO_2 balance. The response process associated with the pH-value change is significantly below the requirements of the guideline of the DIBt of pH-value <10.5.

The second environmentally-friendly soft gel product, currently in the process of the admission procedure “Stump Silitight 69” is more of a “classic” soft gel, in which all of the silicate reaction products are made of environmentally-friendly chemical components. This injection medium has some economic advantage over the first silicate-gel. In spite of the chemically produced basic components this silicate-gel also meets all admission requirements of the DIBt as well as a balanced CO_2 balance.

The third soft gel product is a fully biological soft gel currently also in the admission procedure. It consists of a biopolymer with corresponding natural reaction components. This soft gel is distinguished by its pH-value neutrality as well as a fully neutral CO_2 balance. Because of its material components this soft gel will be distributes under its own registered trademark:

The advantage of this product is mainly the applicability in natural protected areas as well as in groundwater protection areas.

5 The Bio-Soft Gel

The development of the Bio-Soft Gel has a special history behind it. Unlike the other two soft gels, the development of the Bio-Soft Gel did not take place in a laboratory. A lot more the development of the Bio-Soft-Gel has its origins in bionics, in particular, in the "excruciating" TV program.

The origin of the Bio-Soft Gel (floats) in about 30 m deep tropical shallow water regions, such as the Indian Ocean. There is a native species that due to their way of life live in self-dug caves in sandy seabed. In order to stabilize the cave walls, as well as to avoid the trickling of sand, they self-produce a secretion. This secretion is according to its chemical composition an animal biopolymer, which is also available as an herbal biopolymer on the world market.

Building on this, the Bio-Soft Gel was developed. It stands out due to its composition as well as the origin of its individual components through a genuine environmental compatibility.

Investigations have shown that at an average total mixing ratio of water/ biopolymer of 1:0.03 various drips and setting times are possible. The bandwidth is depending on the mixing ratio between a few seconds up to well over 180 min. In this way, all technically relevant processing times are covered.

The following section presents the main results of the inverse column elution tests of all three soft gels to be compared with the demands of the DIBt. It should be noted that the start value for the trial period t = 0 h is a blind value (BV). The blind value represents the start value, which is used in the drinking water.

At this point it is worth mentioning that the Bio-Soft-Gel was developed fully by the civil engineers without the help of other scientific disciplines.

6 Results of the Inverse Column Elution Tests

6.1 Experiment Results of the "Stump-Silikatgel SIG 60"

The investigated silicate gel distinguished itself in terms of its general parameters as inconspicuously. The DIBt demanded because of the individual components, among other things, the investigation of the chloride injection discharge from the injection body. Following the pH-value, in addition to determining the TOC content as well as the chloride content are graphically shown in the Figs. 2, 3 and 4.

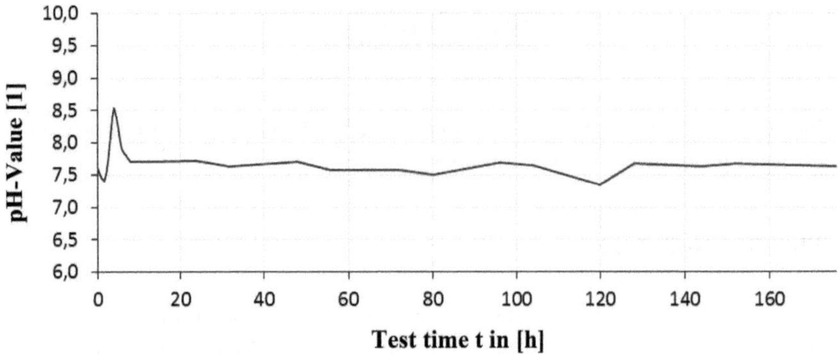

Fig. 2. pH-value during the course of the *"Stump-Silikatgel SIG 60"* inverse column elution experiment over a total test time of 168 h

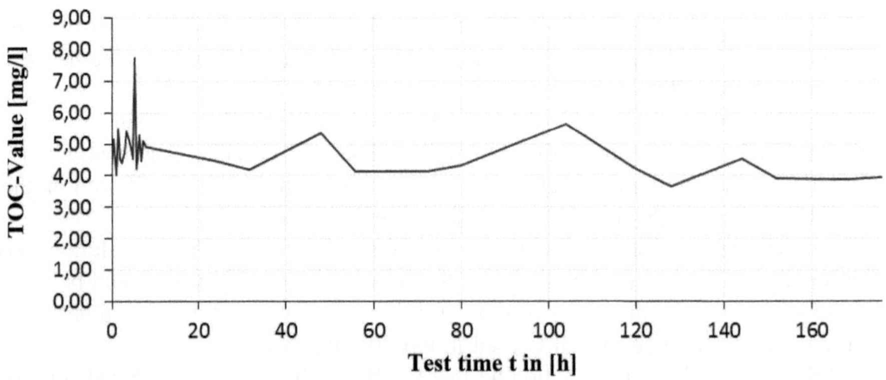

Fig. 3. TOC-value during the course of the *"Stump-Silikatgel SIG 60"* inverse column elution experiment over a total test time of 168 h

The results of the series of experiments in the laboratory showed only a transient exceeding of the permissible values in the chloride values and were therefore considered as critical by the DIBt. Considering all the experimental results, it can be stated that all graphs are marked by a first significant peak.

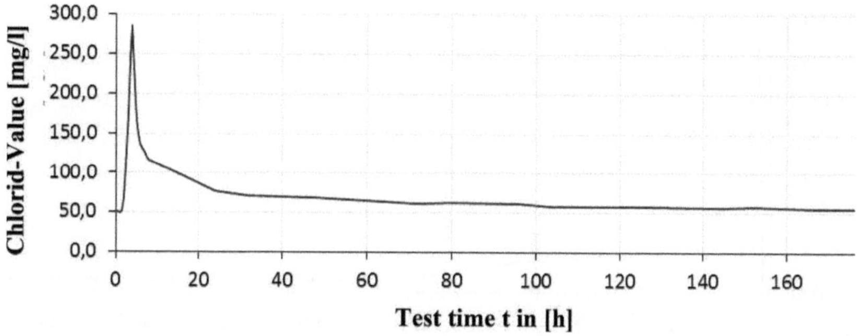

Fig. 4. Chloride-value during the course of the *"Stump-Silikatgel SIG 60"* inverse column elution experiment over a total test time of 168 h

The numerical simulations showed that by the injection of the test soil with two-phase-flow two different injection areas were formed. Area 1 (inner area) is marked by a complete change of the free water, whereas the area 2 is marked by a lack of change of the free water. This leads to a dilution of the injection material, with the result that the drip time in the second area through the dilution is greatly increased. With the beginning of the flow through the column, the material is carried out in the eluate and leads to a short-term increase in the measured values.

6.2 Experiment Results of the "Stump-Silitight 69"

The examined silicate-gel distinguished itself in terms of its general parameters as inconspicuously, although due to the chemical reaction a significant pH-value and TOC values increase in the first trial period range up to 60 h after the injection (see Figs. 5 and 6) can be noted. The ecotoxicological studies, however, showed no abnormalities.

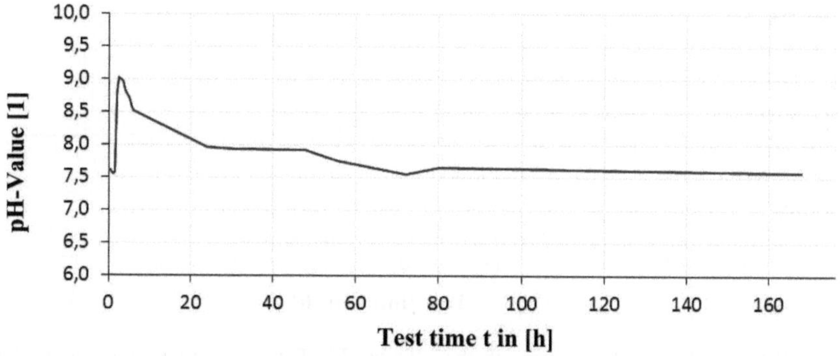

Fig. 5. pH-value during the course of the *"Stump-Silitight 69"* inverse column elution experiment over a total test time of 168 h

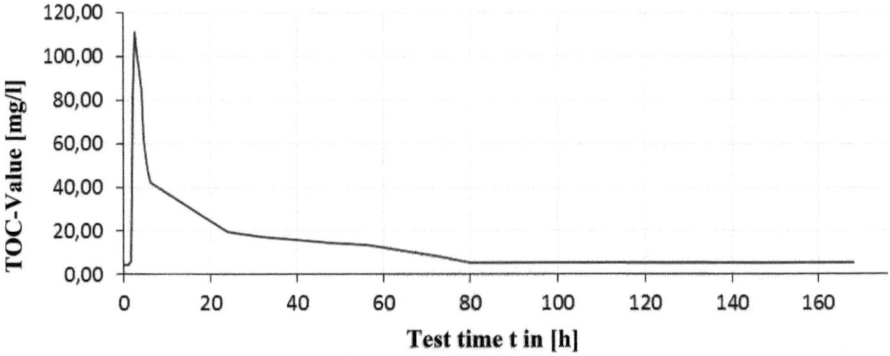

Fig. 6. TOC-value during the course of the *"Stump-Silitight 69"* inverse column elution experiment over a total test time of 168 h

Following the pH-value and the TOC content are displayed graphically.

The characteristic course of the graph is as previously described to attribute to the procedures described previously. However, a larger buffer capacity of the injection good with regard to the pH and TOC value by the chemical reaction of the individual components can be noticed.

6.3 Experiment Results of the "Stump-Bio-Soft Gel"

The new "Stump-Bio-Soft Gel" is characterized in particular by its pH value neutrality. As shown in Fig. 7 to the pH-value is only marginally affected. This ensures that any mobilization of tied substances due to pH-value change cannot occur. In addition, the TOC values are shown in Fig. 8. These results are also as expected.

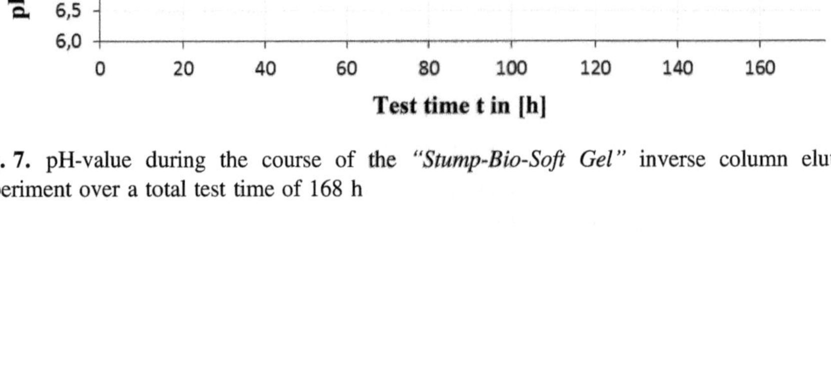

Fig. 7. pH-value during the course of the *"Stump-Bio-Soft Gel"* inverse column elution experiment over a total test time of 168 h

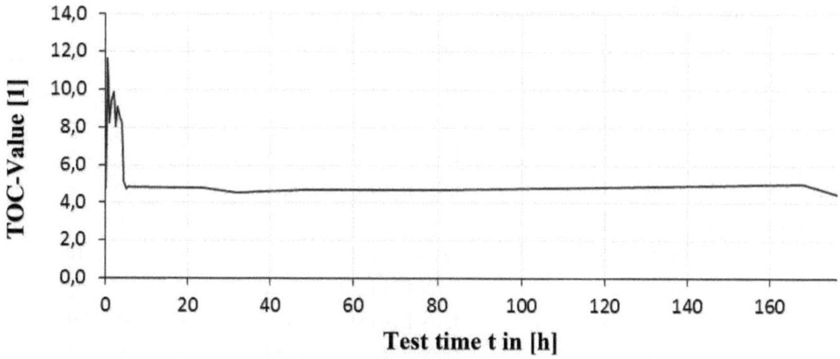

Fig. 8. TOC-value during the course of the *"Stump-Bio-Soft Gel"* inverse column elution experiment over a total test time of 168 h

After the injection, due to the two-phase flow a short-term increase in the TOC content in the eluate occurs. The increase in the TOC value is in relation to the measurement results in Fig. 6 rather marginal. Notwithstanding, the TOC value increase must occur on the basis of purely biological components. It is evident however, that the injected medium builds a stable injection compound because after a short period of time the TOC value falls back to its initial state.

Tables 1 and 2 show the results of the heavy-metal pollution of the basic component from the analysis of the dry mass as well as the analysis of the eluate.

Table 1. Results of heavy metal analysis of the dry mass of the biopolymer

Parameters	Unit	Measuring value	Method	Z0	Z1	Z2
		Dry substance		Limits		
Arsenic	mg/kg TS	<5	DIN EN ISO 11885 (E 22)	10	45	150
Cadmium	mg/kg TS	<0.1	DIN EN ISO 11885 (E 22)	0.4	3	10
Chromium (total)	mg/kg TS	<0.5	DIN EN ISO 11885 (E 22)	30	180	600
Copper	mg/kg TS	27	DIN EN ISO 11885 (E 22)	20	120	400
Nickel	mg/kg TS	0.7	DIN EN ISO 11885 (E 22)	15	150	500
Lead	mg/kg TS	<2	DIN EN ISO 11885 (E 22)	40	210	700
Zinc	mg/kg TS	4.7	DIN EN ISO 11885 (E 22)	60	450	1500
Mercury	mg/kg TS	<0.1	DIN EN ISO 12846 (E 12)	0.1	1.5	5

The only abnormality is an elevated copper value in the dry earth, and the eluate. This abnormality is limited to the natural boundary conditions of the growth location of the biopolymer as no substances were added to the biopolymer.

Table 3 shows the results of the heavy metal contamination on the solid material and the eluate of all three soft gels. The maximum values of the analysis are highlighted for better comparability.

Table 2. Results of the analysis of heavy metals in the eluate biopolymer

Parameters	Unit	Measuring value	Method	ZO	Z1.1	Z1.2	Z2
		Eluat		Limits			
Arsenic	µg/l	<4	DIN EN ISO 11885 (E 22)	14	14	20	60
Cadmium	µg/l	<0.4	DIN EN ISO 11885 (E 22)	1.5	1.5	3	6
Chromium (total)	µg/l	<2	DIN EN ISO 11885 (E 22)	12.5	12.5	25	60
Copper	µg/l	20	DIN EN ISO 11885 (E 22)	20	20	60	100
Nickel	µg/l	<4	DIN EN ISO 11885 (E 22)	15	15	20	70
Lead	µg/l	<4	DIN EN ISO 11885 (E 22)	40	40	80	200
Zinc	µg/l	<40	DIN EN ISO 11885 (E 22)	150	150	200	600
Mercury	µg/l	<0.4	DIN EN 1483 (E 12)	0.5	0.5	1	2

Table 3. Comparison of the results of heavy metal analysis in the solid material and the eluate of the three soft gels

Parameters	*Bio-Weichgel*	*Silitight 69*	*Silikatgel SIG 60*	*Bio-Weichgel*	*Silitight 69*	*Silikatgel SIG 60*
	Measuring value dry substance mg/kg			Measured values Eluat in µg/l		
Arsenic	<5	–	<5	<4	<4	**19**
Cadmium	<0.1	–	<0.1	<0.4	**0.68**	<0.4
Chromium (total)	<0.5	–	**0.77**	<2	<2	<2
Copper	**27**	–	**16**	**20**	0.51	**82**
Nickel	**0.7**	–	0.7	<4	<4	<4
Lead	<2	–	<2	<4	**8.8**	**4.6**
Zinc	**4.7**	–	**5.6**	<40	<40	<40
Mercury	<0.1	–	<0.1	<0.4	<0.4	<0.4

As a result of the inverse column elution experiments a numerical analysis with regard to the material effluent of injection from a sole is simulated.

In the Fig. 9 an example of the TOC effluent of an injection sole in flowing ground waters for the Stump Silitight 69 over three periods of time is shown. Figure 9(a) shows the initial state at time t = 0 days (injection start). Figure 9(b) shows the TOC effluent, taking into account a groundwater recharge after a period of t = 365 days. Finally, it is shown in Fig. 9(c) effluent state for the time t = 3650 days.

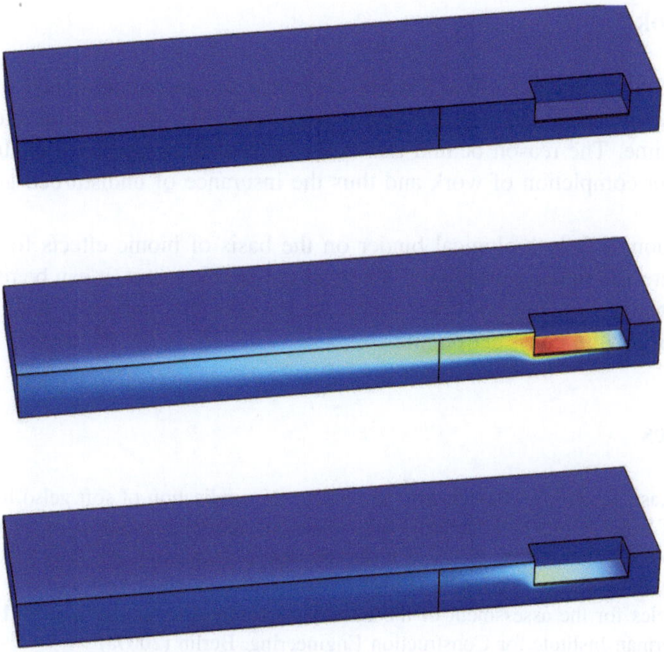

Fig. 9. Numerical simulation model of TOC effluent of the *Stump-Silitight 69* from an injection sole over time (a) at time t = 0 d output condition (b) at time t = 365 d after injection completion (c) at time t = 3650 d after injection completion

7 Conclusions

At the 2015 Climate Conference in Paris it was decided to significantly minimize CO_2 emission in order to make a contribution to reducing global warming. On the other hand, cements or cement slurries are used worldwide in special civil engineering. The development of a fully biological soft gel based on bionics has made a significant contribution to protecting the environment and reducing CO_2 emissions. It was shown that the essential parameters of the organic soft gel, which could lead to the mobilization of bound substances in the soil, occur only partially or does not occur at all. Furthermore, it is shown that the use of the organic soft gel does not lead to the introduction of harmful substances into the soil and thus has no lasting negative effect on the soil. In the course of development, various material compositions were examined and evaluated. Their effects on the subsoil were additionally investigated using numerical simulation models. With the development of the first fully biological soft gel, it was shown that the use of naturally occurring biological resources opens the way for the development of genuine environmentally compatible building materials.

8 Outlook

Currently, Stump Spezialtiefbau GmbH examines the extent to which a biological injection additive can influence the dissolution of the injection medium after a certain amount of time. The reason behind this is the defined decomposition of the injection medium after completion of work and thus the insurance of undisturbed groundwater conditions.

In addition, a fully biological binder on the basis of bionic effects for the use of building materials in the geotechnics is now in the development. It can be reported that a natural model has already been found and works are currently taking place to implement it for the construction site.

References

Brauns, J., Kast, K., Hotzl, H., Eiswirth, M.: Select the application of soft gelso for horizontal sealing of building pits by injection soles - questions about the influence of the groundwater quality. Verification on the basis of a demonstration excavation pit. Unpublished Report; University of Karlsruhe Karlsruhe (1997)

DIBt: Principles for the assessment of the effects of construction products on soil and ground water. German Institute for Construction Engineering, Berlin (2009a)

DIBt: Recommendations for the evaluation of special construction products regarding possible changes in pH in the soil and ground water. German Institute for Construction Engineering, Berlin (2009b)

DIN 19631: Leaching of construction products - percolation method for the examination of the leaching behavior of material injected into the soil. Beuth Verlag, Berlin (2016)

DIN EN ISO 11885: Water quality - determination of selected elements by inductively coupled plasma optical emission spectrometry (ICP-OES) (ISO 11885:2007); German version EN ISO 11885:2009. Beuth Verlag, Berlin (2009)

DIN EN ISO 12846: Water quality - determination of mercury - method using atomic absorption spectrometry (AAS) with and without enrichment (ISO 12846:2012); German version EN ISO 12846:2012. Beuth Verlag, Berlin (2012)

Eiswirth, M., Ohlenbusch, R., Schnell, K.: Groundwater affected by soft gel injection soles. Series of Applied Geology Karlsruhe, vol. 50, pp. 117–134 (1998)

LAGA: LAGA PN 98 - guidelines for the procedure for physical, chemical and biological investigations in connection with the recycling/disposal of waste. Release 32. Country Working Community waste, Munich (2002)

TrinkwV: Drinking water regulations. Federal Ministry of Health, the Federal Ministry for Consumer Protection, Food and Agriculture, in agreement with the Federal Ministry of Economics and Technology and the Federal Ministry for Environment, Nature Conservation and Nuclear Safety, with the consent of the Senate, Berlin (2011)

Effect of Shape and Depth of Shallow Foundations on Failure Mechanism and Wedge Angle of Sandy Soil

W. Nashaat Abd El Samee(✉)

Civil Engineering Department, Beni-Suef University, Beni-Suef, Egypt
Waelnashat@eng.bsu.edu.eg

Abstract. In the present study, finite element analysis was selected to develop numerical models to study the failure mechanism of sandy soil and the wedge angle. The sandy soil is simulated by a semi-infinite element isotropic homogeneous elastic material. The analysis program consists of five footings with different L/B ratio (Ratio between length and width of footing) ranged between 1 to 8. Each footing was analyzed with different foundation depth. In addition, also different angle of internal friction of sandy soil was taken. It was concluded that, the obtained failure mechanism in the present study is identical with the conventional failure mechanism. In addition, it was concluded also that the wedge angle values obtained in the present study decrease with increasing depth of footing and increasing angle of internal friction of the soil as well as increasing L/B ratio while the wedge angle value of sandy soil obtained from previous theoretical approaches is constant [$\Psi = 45 + (\text{Ø}/2)$].

Keywords: Failure mechanism · Wedge angle · Sandy soil · Depth of footing Ratio

1 Introduction

The redistribution of stresses within the soil mass, often referred to as arching, has been analyzed using physical and analytical modeling of active and passive zone systems Terzaghi [1]. The movement of an active zone causes a reduction of soil stresses immediately and increase of stresses in the adjacent soil mass.

McManus and Burdon investigated three shallow foundations each 4.25 m wide 4.6 m long consisting of a 100 mm thick slab "on-grade" with two foundation beams 600 mm wide embedded 450 mm constructed in coarse granular material. The tests were supplemented with several, simpler interface sliding tests performed on 2 m wide × 3 m long concrete slabs constructed "on-grade" using one or two layers of polymer damp-proof membranes. Lateral loading of the slab and beam foundations caused a wedge type of failure mechanism with significant passive soil pressures acting against the vertical faces of the foundation beams. The passive soil wedge developing against the trailing beam lifted one side of the structure vertically leaving hollow space beneath the floor slab. For the somewhat narrow structures tested, significant rotations of the structure occurred. A simple method of analysis was developed to give

H. Shehata and B. Das (Eds.): GeoMEast 2018, SUCI, pp. 197–216, 2019.
https://doi.org/10.1007/978-3-030-01923-5_15

predictions for the experimental results while accounting for all of the main parameters. The analysis predicts that lateral load capacity was highly sensitive to the eccentricity of the applied lateral load [2].

Kholdebarin et al. presented a rigid footing model with specified properties and dimensions on a sandy-clay soil with Mohr-coulomb material. This elasto-plastic soil foundation model was analyzed dynamically with finite difference 2D FLAC software under vertical component of ground excitations. Then the soil was improved with cement grouting and analyzed again. Consequently, the vertical stress under a strip footing, due to vertical component of ground accelerations through the underlying soil was plotted. Also the stress distribution and dynamic bearing capacity of natural and soil cemented foundation was presented [3].

Kentaro et al. presented a various types of foundations suitable for a variety of soils and applications as an alternative to conventional flat foundations for the increase of bearing capacity and the savings of materials. The concept of shells was not new in foundation design, considering the construction with inverted brick arch foundation. Various types of foundations with different geometrical shapes have been extensively investigated in the structural design. However, the corresponding studies on the geotechnical design in terms of bearing capacity and deformation are scare. As such the advantages of various of foundations have not yet been clarified in terms of the geotechnical design. The objective of the study was to examine the overall geotechnical performance of various types of foundation on sand using model loading tests and numerical limit analysis. The general superiority of various types of foundations has been revealed by comparing the loading tests with the analysis results [4].

Nguyen and Merifield presented a prediction of the undrained bearing capacity of strip, square, rectangular and circular footings embedded in clay by using finite element analysis. From these analyses, rigorous shape and depth factors have been derived and were compared with previous numerical and empirical solutions in the literature. The bearing capacity behaviour was discussed and the bearing capacity factors were given for various cases involving a range of embedment depths and footing shapes [5].

Conte et al. presented the analyzed of the response of shallow foundations on soils with strain-softening behavior by using finite element approach. In these soils, a progressive failure can occur owing to a reduction of strength with increasing the plastic strains induced by loading. The theoretical approach allows this failure process to be properly simulated by using a non-local elasto – viscoplastic constitutive model in conjunction with a Mohr – Coulomb yield function in which the shear strength parameters were reduced with the accumulated deviatoric plastic strain. Another advantage of the method was that it requires few material parameters as input data, with most of these parameters that can be readily obtained from conventional geotechnical tests. To assess the reliability of the proposed approach, some comparisons with experimental results from physical model tests were presented. A fairly agreement was found between simulated and observed results. The progressive failure process that occurs in a dense sand layer owing to loading was analyzed and the main aspects concerning the associated failure mechanism were highlighted [6].

Zhang et al. presented a method for progressive failure analysis of strain-softening slopes based on the strength reduction method and strain-softening model. The mutation was more pronounced in strain-softening analysis, and the mutation of displacement at slope crest was taken as critical failure criterion. This method was applied to a cut slope in an industry site. The results were as follows: (i) The thickness of the shear zone considering strain softening behaviour is narrower than that with non-softening analysis. (ii) The failure of slope is the process of the initiation, propagation and connection of potential failure surface. The strength parameters were mobilized to a non-uniform degree while progressive failure occurs in the slope. (ii) The factor of safety increases with the increase of residual shear strain threshold and elastic modulus. The failure mode of slope changes from shallow slip to deep slip. Poisson's ratio and dilation angle have little effect on the results [7].

Qiming and Murad presented some analytical solutions for estimating the ultimate bearing capacity of strip footings on RSFs. A general failure mode for RSFs was first proposed based on previous studies conducted by the authors and test results from literature study. A limit equilibrium stability analysis of RSFs was performed based on the proposed failure mechanism. New bearing capacity formulas, which consider both the confinement and the membrane effects of reinforcements on the increase in ultimate bearing capacity, were then developed for strip footings on RSFs. Several special cases of RSFs were presented and discussed. The proposed model was verified by the experimental data reported in the published literature. The predicted ultimate bearing capacity was in good agreement with the results of model tests reported in the literature. The study showed that the depth of the punching shear failure zone (D_P) depends on the relative strength of the reinforced soil layer and the underlying unreinforced soil layer [8].

El-Samny et al. presented the effect of footing shape and size on the wedge angle. The wedge angle of cohesionless soil at surface with and without surcharge has been determined in field for graded sand samples. The tests have been conducted in field by using the plate load test. The relationships between the wedge angle and the angle of internal friction have been obtained. An empirical formula to determine the wedge angle for circular, square and rectangular plates at different depths has been presented. The relationships between the wedge angle and the ratio (D/B) of footing depth and width have been also obtained [9].

Acharyya and Dey presented the effect of foundation placed on a slope on the ultimate bearing capacity by its vicinity to the slope face, which offers substantially lesser passive resistance as compared to a footing resting on a semi-infinite medium. Conventional bearing capacity theories fail to address the behavior of such foundations. However, deformation along the slope plays a major role in governing the failure of such foundations, thus requiring a coupled stress-deformation based failure analysis. With the aid of 3-D finite element modelling, employing coupled stress deformation analysis. Addresses the failure mechanism and the bearing capacity (q_u) of a square footing located on a dry cohesionless slope. The effect of various parameters, namely the angle of internal friction of soil, setback distance, slope inclination, footing width and the depth of embedment of the footing, have been investigated. Variations parameters were found to noticeably alter the bearing capacity estimate and the observed failure mechanism. The unit weight and modulus of elasticity of the soil material was found to have negligible effect on the bearing capacity [10].

The main aim of the present study is to determine, the effect of shape and depth of shallow foundations on wedge angle and failure mechanism of soil. In addition, determine, the effect of angle of internal friction of the soil as well as increasing L/B ratio on wedge angle and failure mechanism of soil.

2 Failure Mechanism

The true theoretical failure mechanism is determined using theory of plasticity concepts and is illustrated conceptually in Fig. 1.

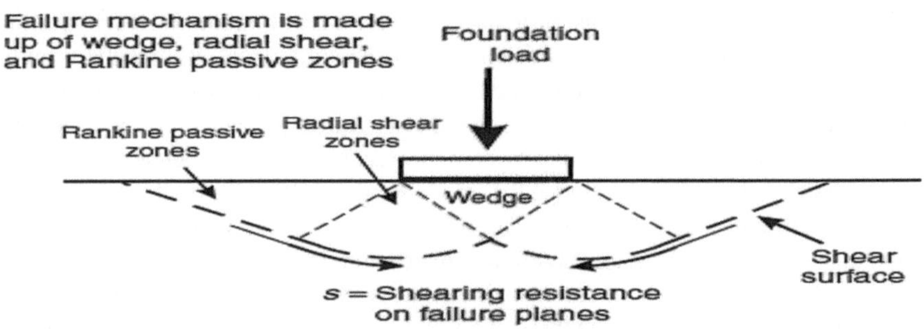

Fig. 1. Failure mechanism for bearing capacity [11].

Where:

$$\Psi = \text{wedge angle} = 45 + \Phi/2 \quad \alpha = 45 - \Phi/2 \quad \text{B : Width of footing.}$$

Prandtl [11] developed an equation based on his study of penetration of long hard metal puncher into softer materials for computing the ultimate bearing capacity. The mechanisms of failure of soil under plastic equilibrium are:

i. Failure occurs along definite slip surfaces and symmetrical with respect to the centerline of the puncher or footing in the case of soil for symmetrical vertical loading.
ii. Under ultimate, load Q_{ult} acting on the footing of width B as shown in Fig. 2, a triangular wedge of soil abc is formed below the footing and pushed down as a rigid body. The stresses acting on abc will be in active stage.

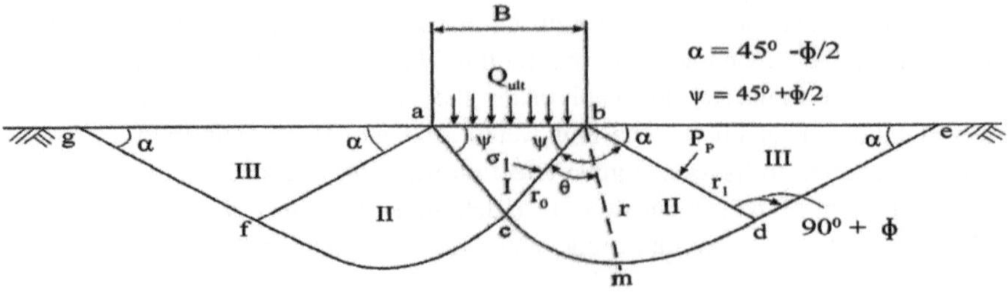

Fig. 2. Failure surface [11].

iii. Due to gradual downward push of the wedge abc, the soil mass on its left and right will be pushed outward and upward. The failure takes place along the slip surfaces bcde and acfg. These surfaces comprise triangular wedges bde and afg, and sectors bcd and acf.
iv. The mass of soil will be in a plastic state above the failure surface gfcde and in elastic state below the surface.
v. The failure surfaces can be divided into three Zones I to III. Zone I will be in active state, Zone II in radial shear and Zone III in plastic state.
vi. The lower boundaries of zone II are parts of logarithmic spirals with their poles coinciding with the corresponding edge of the footing. The straight lines de and fg are tangential to spirals at points d and f respectively and meet the horizontal surface.

Terzaghi studied change in the properties of soil behavior beneath and around the foundation [12, 13]. It is based ultimate bearing capacity theory for strip footing. Failure mechanism and width footing explain this theory. Later on, Meyerhof and Hansen (1963) expanded Terzaghi theory. Terzaghi in (1943) gave a general bearing capacity theory for a strip foundation. For the first time, he developed his theory by incorporating weight of failure wedge in the analysis. Terzaghi considered a continuous footing of width B placed at a depth of D below the ground surface as shown in Fig. 2. The soil is assumed to fail along the surface (aedcgfb). The failure surface consists of 5 zones. The Zone I, abc is an elastic zone. Zone II, which comprises wedges beg and acd is the zone of radial shear. The wedges bfg and ade comprise Zone III and are known as passive Rankine's zone and will be in a plastic state. When the footing is loaded the wedge abc sinks into the ground as an integral part of the footing and remains in the elastic state due to the cohesion and adhesion between the base of the footing and the soil. The straight boundaries ac and bc of this zone are inclined at an angle with the base of the footing. In Zones II and III, shear patterns develop. Zone II, the zone of radial shear constitutes a set of radial lines emerging from the outer edges b and a of the footing.

3 Finite Element Analysis

Finite element method was selected in the present study to develop numerical models to study the failure angle in soil (wedge angle). The finite element analysis was performed using two finite element software (Optum & Limitstate) [14, 15], limit analysis was performed in this study including upper and lower bound. A semi-infinite element isotropic homogeneous elastic material simulates the soil and the material model used is Mohr-Coulomb, while the concrete simulated as rigid material.

3.1 Numerical Program

The analysis program consists of five footings including square footing and rectangular footings with L/B ratio (Ratio between length and width of footing) ranged between 1 to 4 and 8. Each footing was analyzed with foundation depth ranged between 0.0 B,

0.10 B. 0.15 B, 0.25 B, 0.50 B and 1.0 B. The angle of internal friction of sandy soil used was also ranged between 30°, 35°, 38° and 40°. The analysis program is shown in Table 1. The soil is simulated by a semi-infinite element isotropic homogeneous elastic material. The type of soil is sand. The used soil properties are listed in Table 2.

Table 1. Investigated cases of study by numerical analysis program.

Footing shape	L/B ratio	Angle of internal friction (Ø)	Depth of foundation (D_f)
Square footing (B * B)	1	30°, 35°, 38° and 40°	0.00 B
			0.10 B
			0.15 B
			0.25 B
			0.50 B
			1.00 B
Rectangular footing (L * B)	2	30°, 35°, 38° and 40°	0.00 B
			0.10 B
			0.15 B
			0.25 B
			0.50 B
			1.00 B
Rectangular footing (L * B)	3	30°, 35°, 38° and 40°	0.00 B
			0.10 B
			0.15 B
			0.25 B
			0.50 B
			1.00 B
Rectangular footing (L * B)	4	30°, 35°, 38° and 40°	0.00 B
			0.10 B
			0.15 B
			0.25 B
			0.50 B
			1.00 B
Strip footing (L * B)	8	30°, 35°, 38° and 40°	0.00 B
			0.10 B
			0.15 B
			0.25 B
			0.50 B
			1.00 B

Where:
D_f: Depth of footing. L: Length of footing. B: Width of footing.
L/B: Ratio between length and width of footing.

Table 2. Soil properties.

Parameters	Soil
Material model	Mohr - Coulomb
Type of material	Fine to med sand
Thickness (m)	10
Unit weight, γ (kN/m^3)	17.5
Young's modulus, E_s (kN/m^2)	15000
Poisson's ratio, υ	0.25
Friction angle, Ø (deg)	30, 35, 38 and 40
Dilatancy angle, ¥ (deg)	0, 5, 8 and 10

3.2 Failure Mechanisms by Finite Element

The failure mechanism in the soil beneath large footings is observed internally thus movement of soil around the foundation is seen (the passive zone is not distinguished clearly). This mechanism is identical from what Terzaghi has proposed. Increasing lateral pressure (confining) the ground located under the footing will increase the pressure bearing in this region. According to the conventional failure mechanism, stress is distributed in the soil beneath and around the footing, so, the stress penetration depth decreases (formation of stress bubbles). According to the proposed models in the present study, in order to predict and interpret the results obtained from numerical analysis, the effect of shape and depth of shallow foundations on wedge angle and failure mechanism of soil have been obtained. Figures 3, 4, 5, 6, 7, 8, 9, 10 and 11 show examples of failure mechanism for the modeled footings.

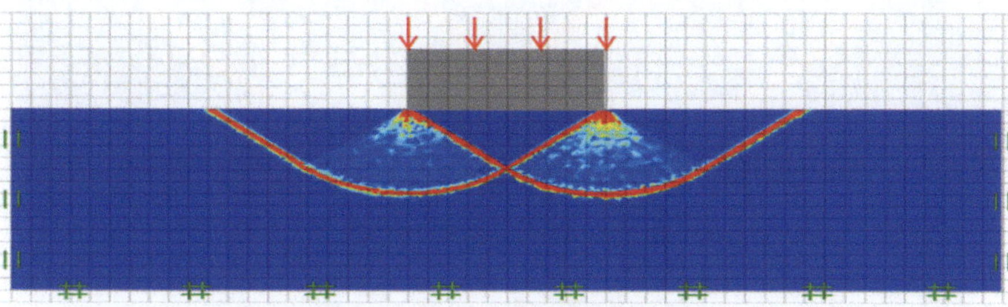

(3-a) Failure mechanism of square footing L/B = 1.00.

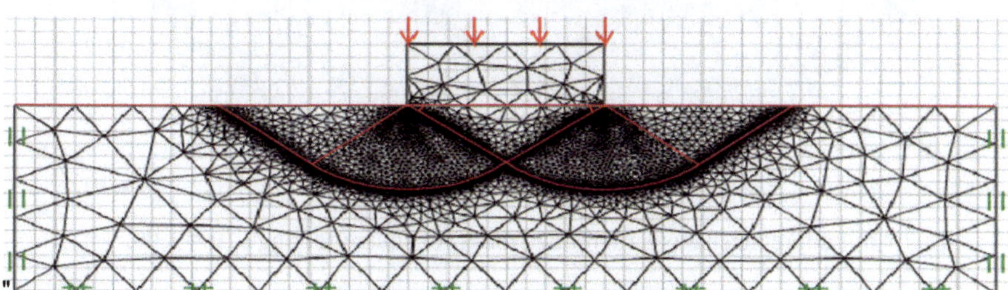

(3-b) Failure mechanism (deformed mesh) of square footing L/B = 1.00.

Fig. 3. Failure mechanism of square footing L/B = 1.00, at Ø = 30° and at D_f = 0.00 B.

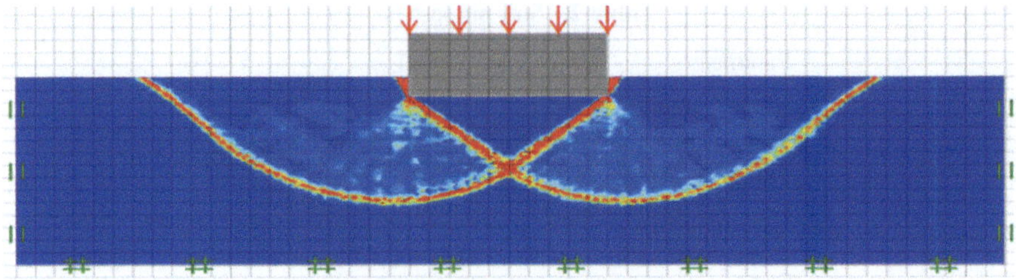

(4-a) Failure mechanism of square footing L/B = 1.00.

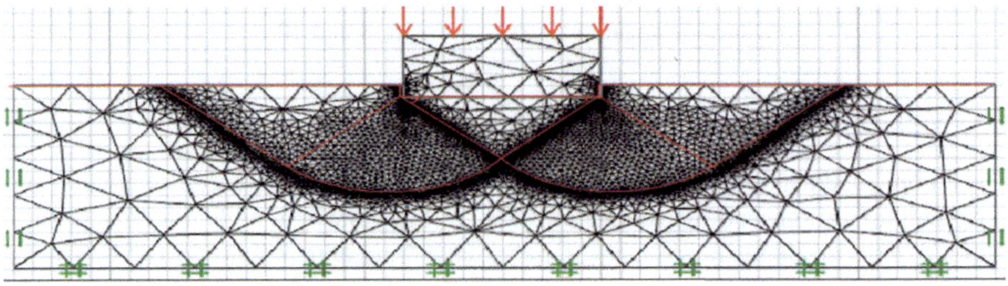

(4-b) Failure mechanism (deformed mesh) of square footing L/B = 1.00.

Fig. 4. Failure mechanism of square footing L/B = 1.00, at Ø = 30° and at Df = 0.25 B.

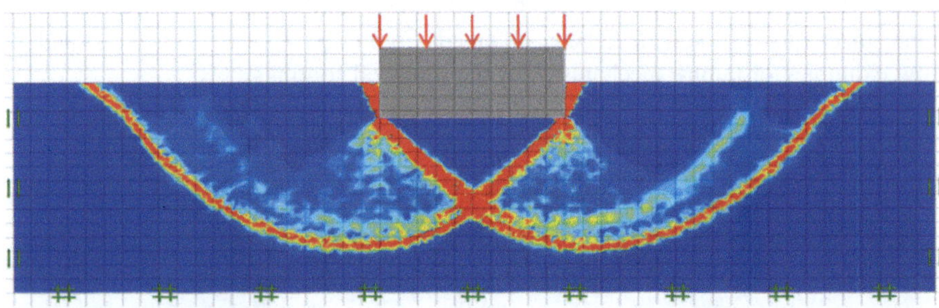

(5-a) Failure mechanism of square footing L/B = 1.00.

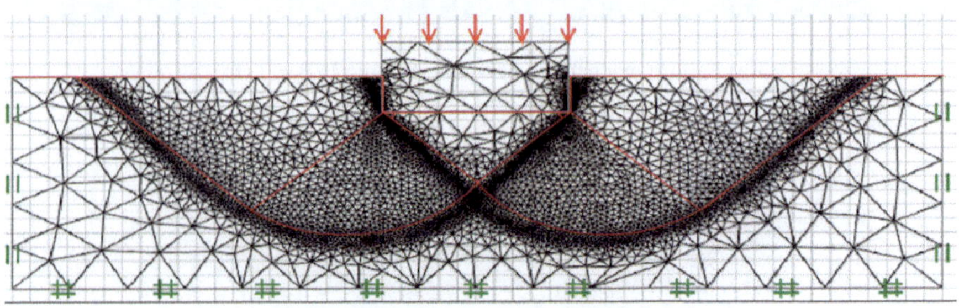

(5-b) Failure mechanism (deformed mesh) of square footing L/B = 1.00.

Fig. 5. Failure mechanism of square footing L/B = 1.00, at Ø = 30° and at D_f = 0.50 B.

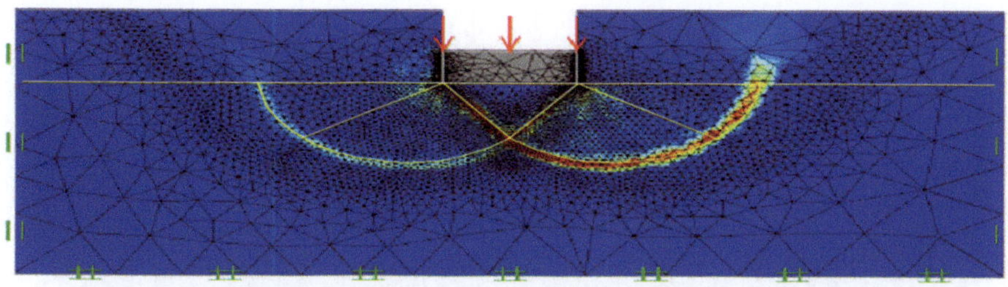

Fig. 6. Failure mechanism of square footing L/B = 1.00, at Ø = 30° and D_f = 1.00 B.

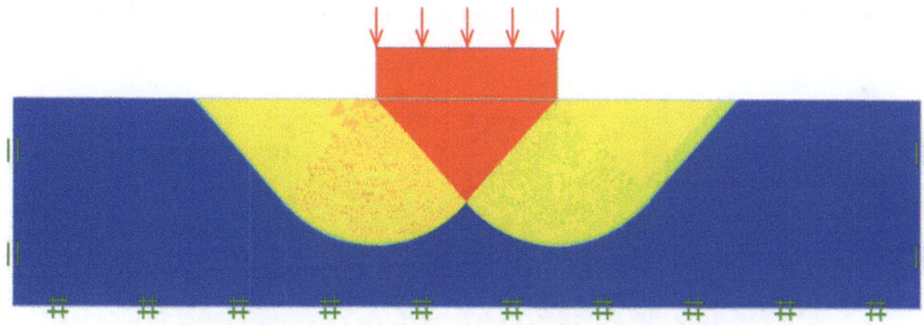

(7-a) Failure mechanism of rectangular footing L/B = 2.00.

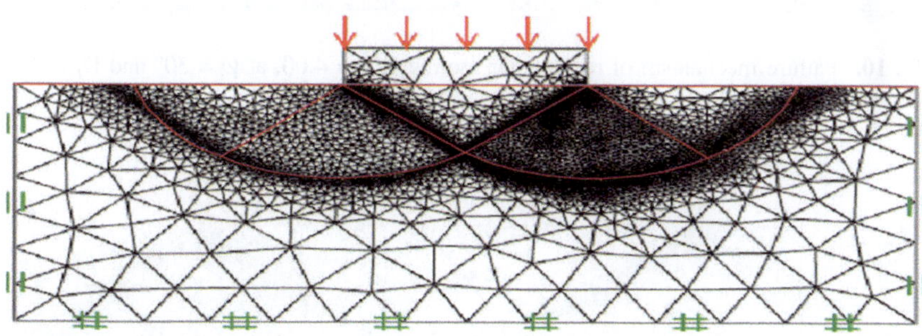

(7-b) Failure mechanism (deformed mesh) of rectangular footing L/B = 2.00.

Fig. 7. Failure mechanism of rectangular footing L/B = 2.00, at Ø = 30° and D_f = 0.00 B.

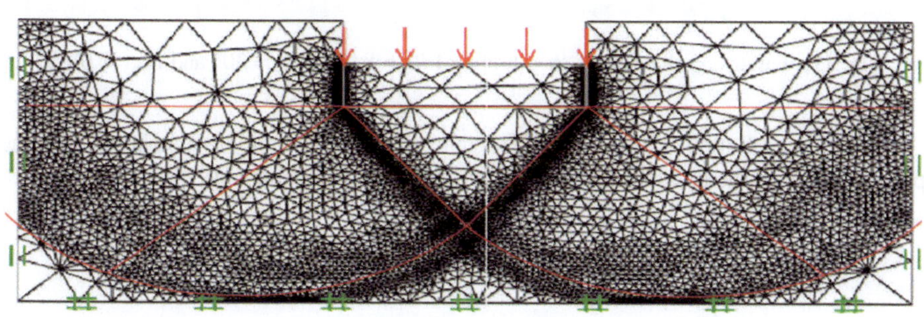

Fig. 8. Failure mechanism of rectangular footing L/B = 2.00, at Ø = 30° and D_f = 0.50 B.

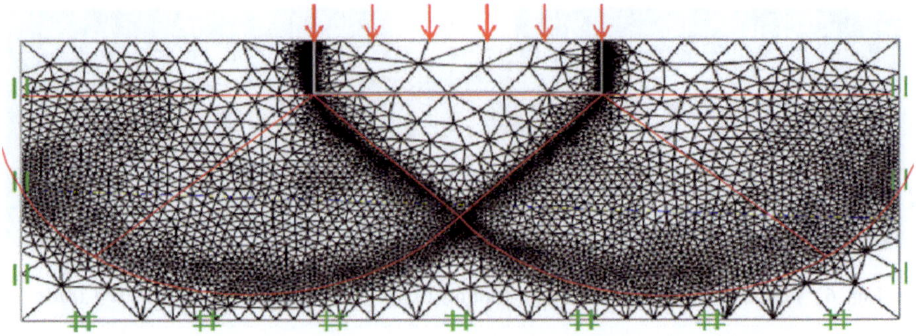

Fig. 9. Failure mechanism of rectangular footing L/B = 3.00, at Ø = 30° and D_f = 0.25 B.

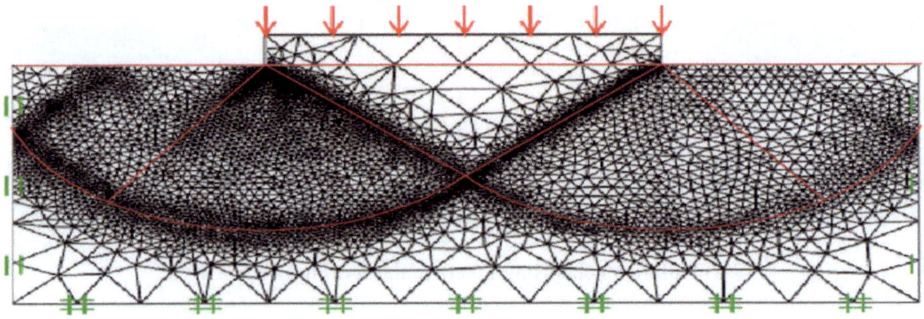

Fig. 10. Failure mechanism of rectangular footing L/B = 4.00, at Ø = 30° and D_f = 0.00 B.

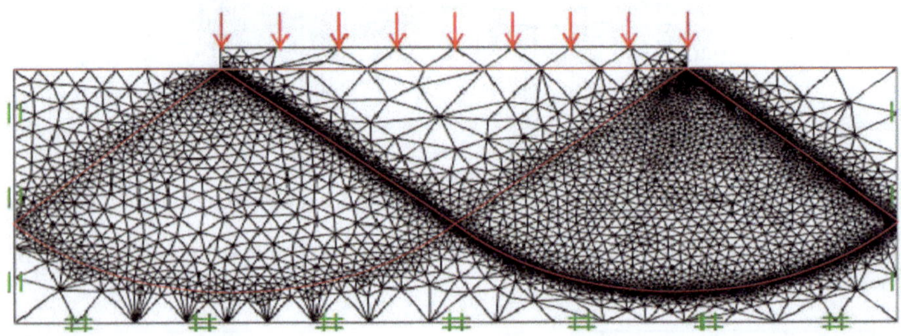

Fig. 11. Failure mechanism of strip footing L/B = 8.00, at Ø = 30° and D_f = 0.00 B

From these figures it can be noticed that, the obtained failure mechanism in the present study is identical with the conventional failure mechanism.

4 Finite Element Results

The obtained results from the numerical analysis are described as follows:

4.1 Determination of Failure Angle (Wedge Angle)

The failure angle (wedge angle) in soil has been obtained using finite element models through locating the points of deformation along the axis of the triangle of failure which is the active zone, then by drawing a line intersecting these points it results in a triangle shape. By knowing the triangle axis lengths, the angle between the triangle sides are then determined as shown in Fig. 12.

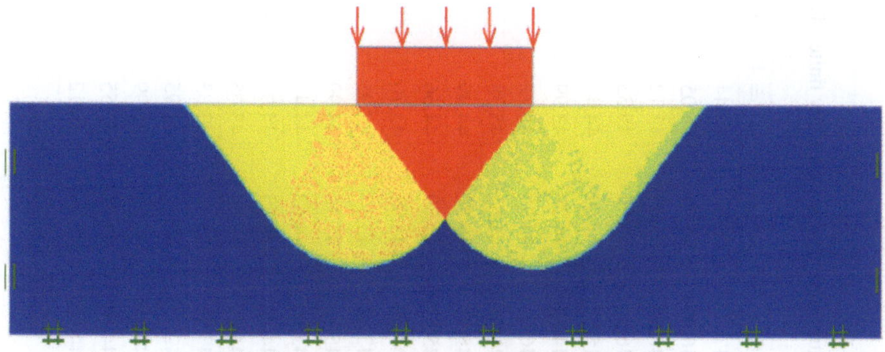

(12-a) Failure mechanism of rectangular footing L/B.

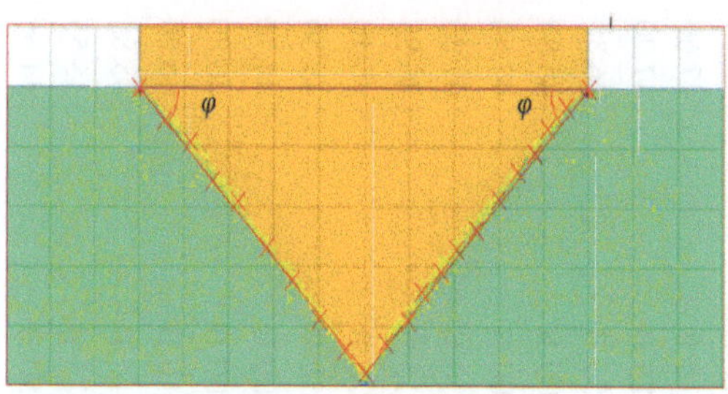

(12-b) (wedge angle of failure mechanism for rectangular footing L/B

Fig. 12. Determination of failure angle (wedge angle).

4.2 Effect of Ratio L/B on Failure (Wedge) Angle of Soil

The values of failure (wedge) angle of sandy soil have been obtained by the numerical analysis and compared with values of wedge angle obtained by theoretical approaches using Terzaghi and Meyerhof are presented in Table 3. A comparison between the values of wedge angle of sandy soil with the ratio L/B (shape of shallow foundation) is presented in Figs. 13, 14, 15 and 16.

Table 3. The obtained values of wedge angle by numerical analysis program and compared with values of wedge angle obtained by theoretical approaches using Terzaghi and Meyerhof at Ø = 30°, 35°, 38° and 40°.

Footing shape	L/B ratio	Depth of foundation (D_f)	Obtained values of wedge angle (Ψ) by finite element				Terzaghi and Meyerhof Ψ = 45 + (Ø/2)			
			(Ø) = 30°	(Ø) = 35°	(Ø) = 38°	(Ø) = 40°	(Ø) = 30°	(Ø) = 35°	Ø) = 40°	(Ø) = 45°
Square footing (B * B)	1	0.00 B	48.35	47.38	46.67	46.44	60.00	62.50	64.00	65.00
		0.10 B	47.95	46.99	46.29	46.05				
		0.15 B	47.60	46.65	45.95	45.72				
		0.25 B	47.20	46.26	45.56	45.33				
		0.50 B	46.30	45.37	44.69	44.47				
		1.00 B	45.80	44.88	44.21	43.99				
Rectangular footing (L * B)	2	0.00 B	47.60	46.65	45.95	45.72				
		0.10 B	47.25	46.31	45.61	45.38				
		0.15 B	46.85	45.91	45.22	44.99				
		0.25 B	46.50	45.57	44.89	44.66				
		0.50 B	45.50	44.59	43.92	43.70				
		1.00 B	44.70	43.81	43.15	42.93				
Rectangular footing (L * B)	3	0.00 B	46.80	45.85	45.16	44.99				
		0.10 B	46.20	45.30	44.62	44.47				
		0.15 B	45.85	44.90	44.23	44.03				
		0.25 B	45.55	44.50	43.83	43.65				
		0.50 B	44.65	43.81	43.15	42.83				
		1.00 B	43.65	42.75	42.11	41.95				
Rectangular footing (L * B)	4	0.00 B	45.80	44.88	44.21	43.99				
		0.10 B	45.35	44.44	43.78	43.55				
		0.15 B	44.90	44.00	43.34	43.12				

(*continued*)

Table 3. *(continued)*

Footing shape	L/B ratio	Depth of foundation (D_f)	Obtained values of wedge angle (Ψ) by finite element				Terzaghi and Meyerhof Ψ = 45 + (Ø/2)			
			(Ø) = 30°	(Ø) = 35°	(Ø) = 38°	(Ø) = 40°	(Ø) = 30°	(Ø) = 35°	Ø) = 40°	(Ø) = 45°
		0.25 B	44.68	43.66	43.00	42.79				
		0.50 B	43.78	42.92	42.28	42.07				
		1.00 B	42.55	41.70	41.07	40.87				
Strip footing (L * B)	∞	0.00 B	44.90	44.00	43.34	43.12				
		0.10 B	44.45	43.56	42.91	42.69				
		0.15 B	44.10	43.22	42.57	42.35				
		0.25 B	43.80	42.92	42.28	42.07				
		0.50 B	42.90	42.04	41.41	41.20				
		1.00 B	41.30	40.47	39.87	39.66				

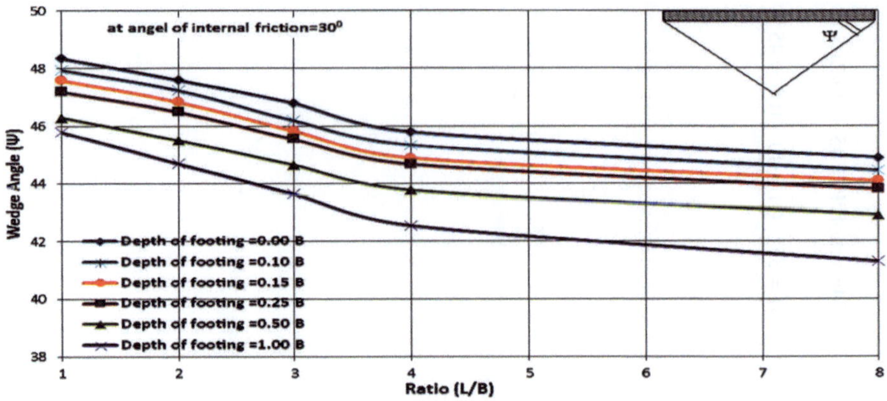

Fig. 13. Comparison of the values of wedge angle and the ratio L/B at different of depth of footing (D_f) and at angle of internal friction (Ø) = 30°.

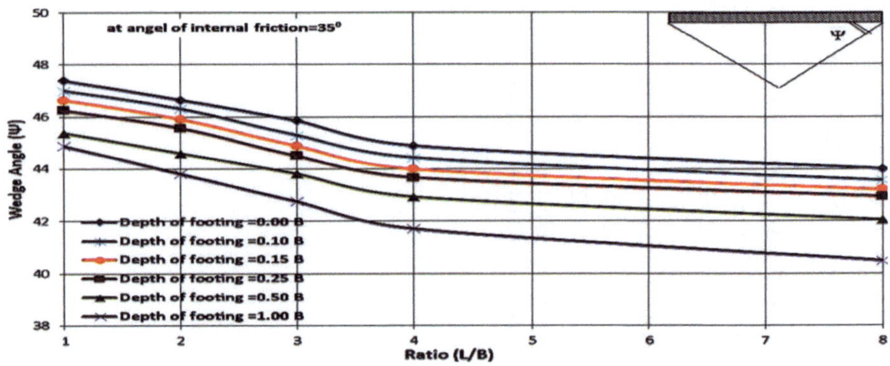

Fig. 14. Comparison of the values of wedge angle and the ratio L/B at different of depth of footing (D_f) and at angle of internal friction (Ø) = 35°.

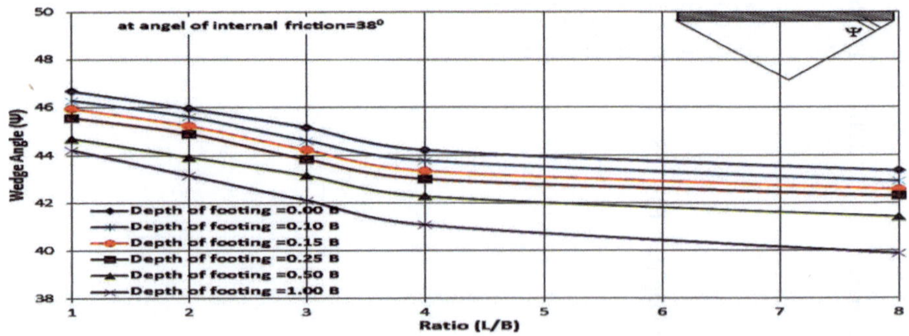

Fig. 15. Comparison of the values of wedge angle and the ratio L/B at different depths of footing (D_f) and at angle of internal friction (Ø) = 38°.

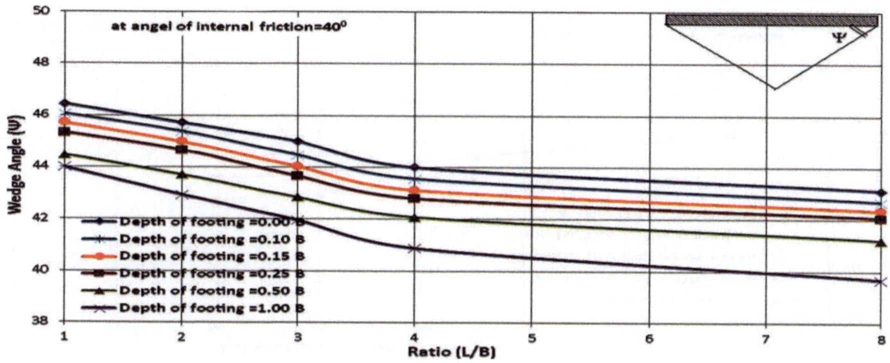

Fig. 16. Comparison of the values of wedge angle and the ratio L/B at different of depth of footing (D_f) and at angle of internal friction (Ø) = 40°.

4.3 Effect of Angle of Internal Friction on Wedge Angle of Soil

A comparison between the values of wedge angle of sandy soil with the angle of internal friction is presented Figs. 17, 18, 19, 20, 21 and 22. show the relationships between angle of internal friction of sandy soil and the wedge angle under different shape of shallow foundation (ratio L/B) for different depths of footings.

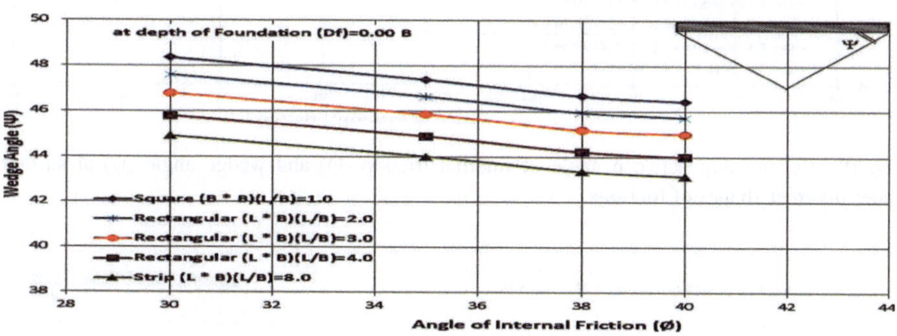

Fig. 17. Relationship between angle of internal friction (Ø) and wedge angle (ψ) at surface under different shapes of footings.

From Table 3 and Figs. 13, 14, 15, 16, 17, 18, 19, 20, 21 and 22, it can be concluded that the wedge angle values of sandy soil decrease with increasing depth of footing and increasing of angle of internal friction of the soil as well as increasing L/B ratio (Ratio between length and width of footing).

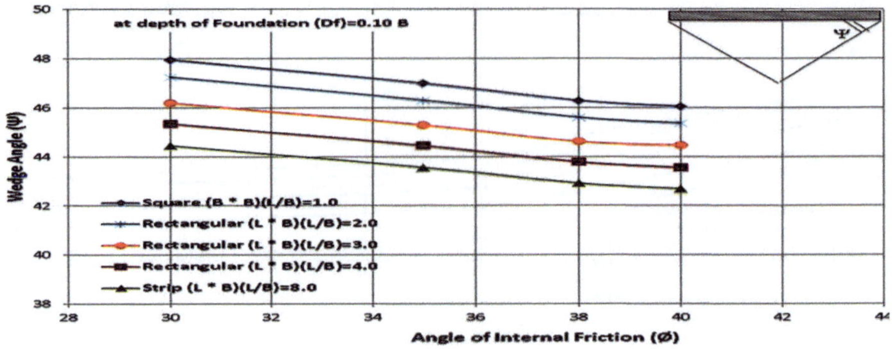

Fig. 18. Relationship between angle of internal friction (Ø) and wedge angle (ψ) at surface under different shapes of footings.

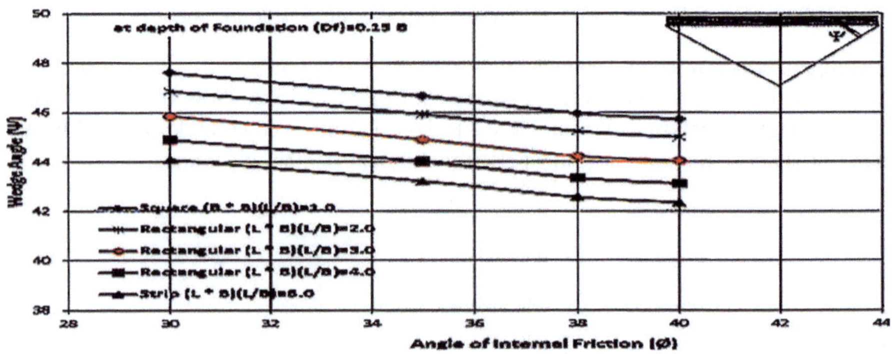

Fig. 19. Relationship between angle of internal friction (Ø) and wedge angle (ψ) at surface under different shapes of footings.

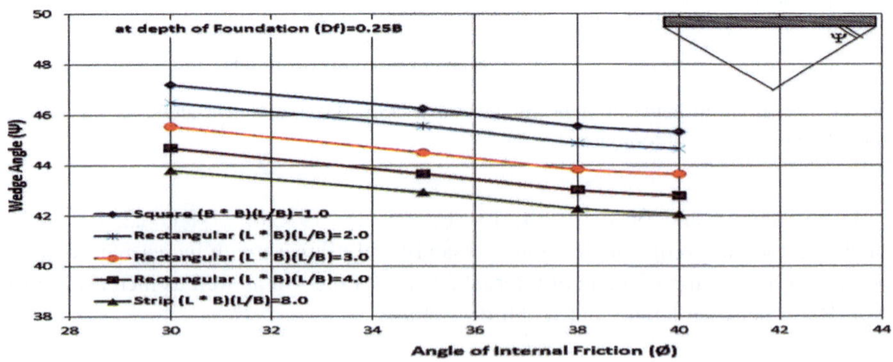

Fig. 20. Relationship between angle of internal friction (Ø) and wedge angle (ψ) at surface under different shapes of footings.

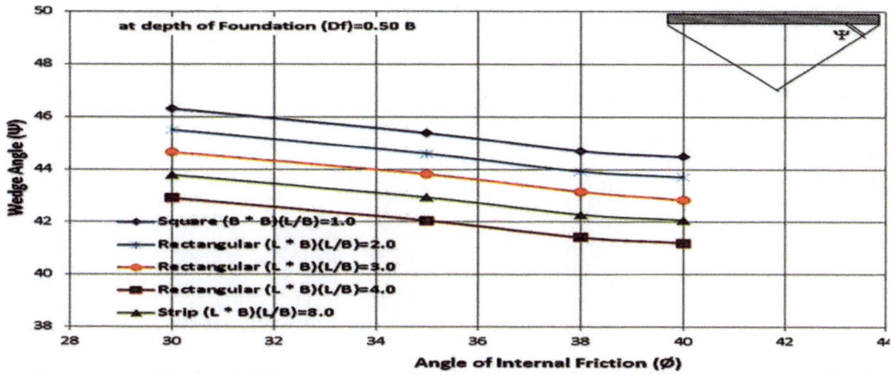

Fig. 21. Relationship between angle of internal friction (Ø) and wedge angle (ψ) at surface under different shapes of footings.

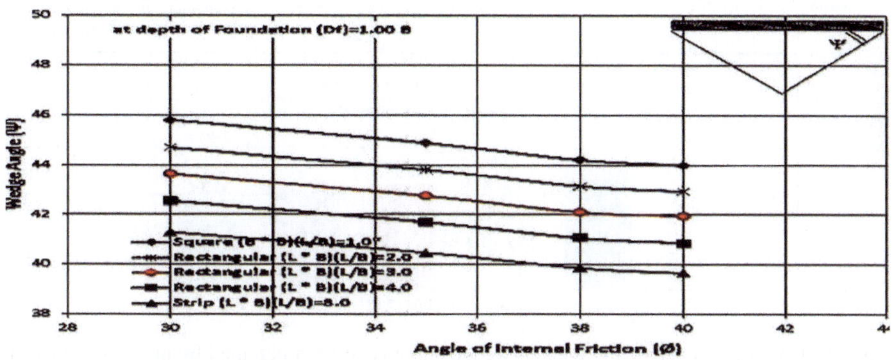

Fig. 22. Relationship between angle of internal friction (Ø) and wedge angle (ψ) at surface under different shapes of footings.

4.4 Comparison Between Values of Wedge Angle Obtained by Numerical Analysis and Theoretical Approaches

A comparison between values of wedge angle obtained by the present numerical analysis and theoretical approaches is presented in Figs. 23, 24, 25, 26, 27 and 28.

From these figures, it can be concluded that the wedge angle values obtained in the present study decrease with increasing depth of footing and increasing angle of internal friction of the soil as well as increasing L/B ratio while the wedge angle value of sandy soil obtained from previous theoretical approaches is constant [$\Psi = 45 + (Ø/2)$].

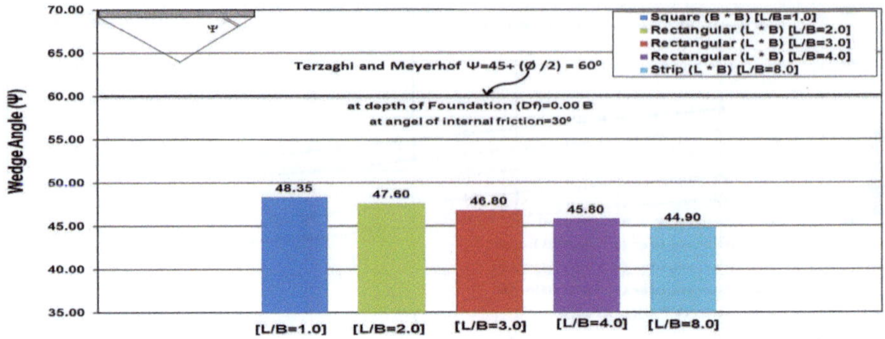

Fig. 23. Comparison between values of wedge angle have been obtained by numerical analysis and by using Terzaghi and Meyerhof at surface (0.00 B).

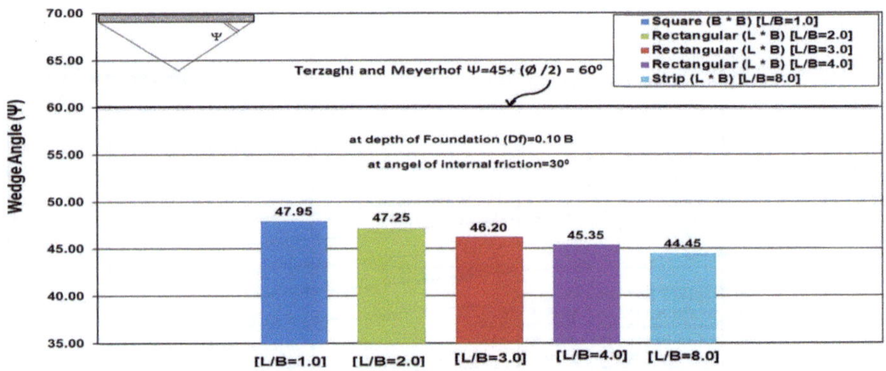

Fig. 24. Comparison between values of wedge angle have been obtained by numerical analysis and by using Terzaghi and Meyerhof at depth of footing = (0.10 B).

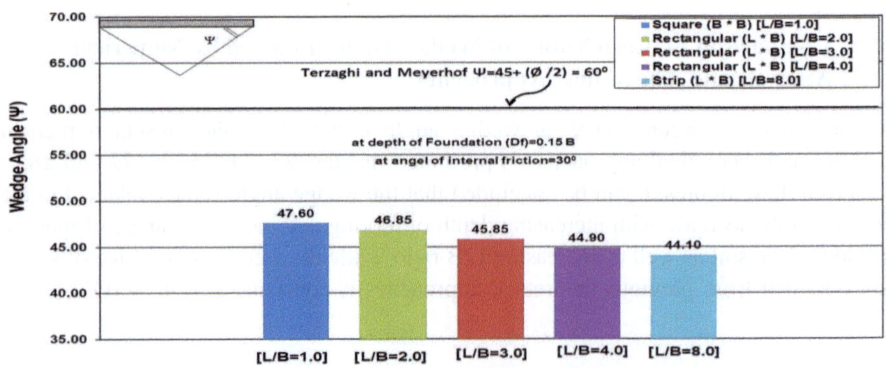

Fig. 25. Comparison between values of wedge angle have been obtained by numerical analysis and by using Terzaghi and Meyerhof at depth of footing = (0.15 B).

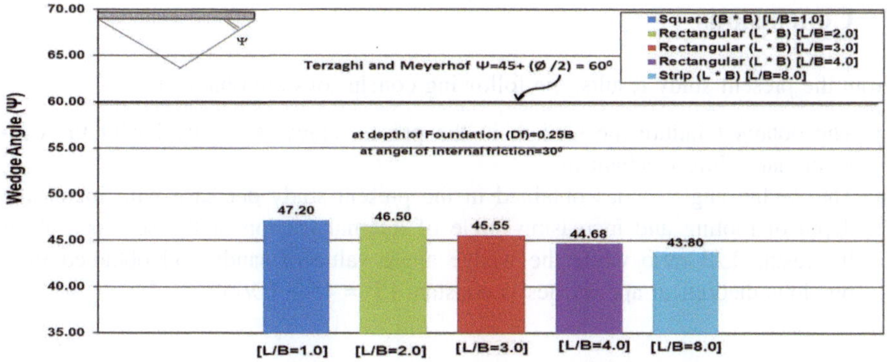

Fig. 26. Comparison between values of wedge angle have been obtained by numerical analysis and by using Terzaghi and Meyerhof at depth of footing = (0.25 B).

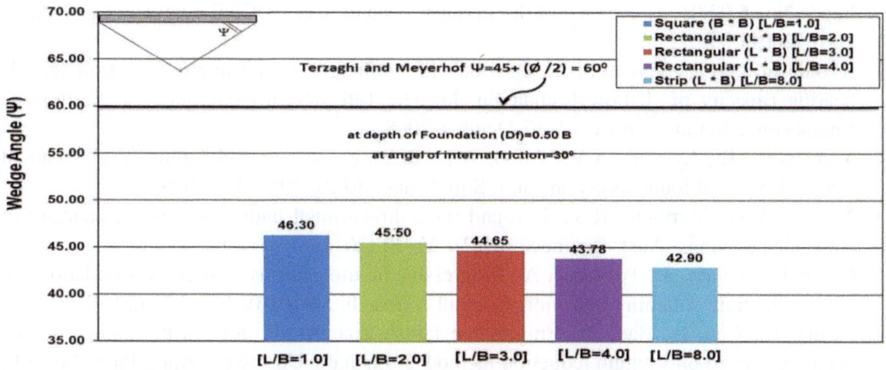

Fig. 27. Comparison between values of wedge angle have been obtained by numerical analysis and by using Terzaghi and Meyerhof at depth of footing = (0.50 B).

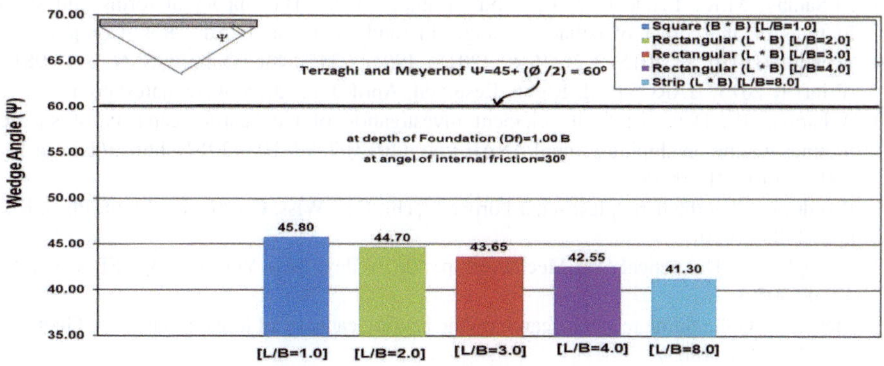

Fig. 28. Comparison between values of wedge angle have been obtained by numerical analysis and by using Terzaghi and Meyerhof at depth of footing = (1.00 B).

5 Conclusions

From the present study results, the following conclusions are obtained:

i. The obtained failure mechanism in the present study is identical with the conventional failure mechanism.
ii. The wedge angle values obtained in the present study decrease with increasing depth of footing and increasing angle of internal friction of the soil as well as increasing L/B ratio while the wedge angle value of sandy soil obtained from previous theoretical approaches is constant [$\Psi = 45 + (Ø/2)$].

References

1. Terzaghi, K.: The actual factor of safety of foundations. Struct. Engr. **13**, 126–160 (1935)
2. McManus, K.J., Burdon, N.R.R.: Lateral resistance of shallow foundations. In: Conference Paper No. 6.03.01. University of Canterbury, Department of Civil Engineering, NZSEE 2001 (2001)
3. Kholdebarin, A.R., Massumi, A., Davoodi, M.: Influence of soil improvement on seismic bearing capacity of shallow foundation. In: The 14th World Conference on Earthquake Engineering, Beijing, China, 12–17 October 2008
4. Yamamoto, K., Lyamin, A.V., Abbo, A.J.: Bearing capacity and failure mechanism of different types of foundations on sand. Soil Found. **49**(2), 305–314 (2009)
5. Nguyen, V.Q., Merifield, R.S.: Two-and three-dimensional undrained bearing capacity of embedded footings. Aust. Geomech. **47**(2), 25–40 (2012)
6. Conte, E., Donato, A., Troncone, A.: Progressive failure analysis of shallow foundations on soils with strain-softening behavior. Comput. Geotech. **54**(2013), 117–124 (2013)
7. Zhang, K., Cao, P., Bao, R.: Progressive failure analysis of slope with strain-softening behaviour based on strength reduction method. J. Zhejiang Univ.-Sci. (Appl. Phys. Eng.) **14** (2), 101–109 (2013). ISSN: 1673-565X (Print), ISSN: 1862-1775 (Online), www.zju.edu.cn/jzus, www.springerlink.com, E-mail: jzus@zju.edu.cn
8. Chenn, Q., Abu-Farsakh, M.: Ultimate bearing capacity analysis of strip footings on reinforced soil foundation. Soils Found. J. **55**, 74–85 (2015). www.sciencedirect.com
9. El-Samny, M.K., Elsideek, M.B., Abd Elsamee, W.N.: The impact of forms and sizes footings on the corner of collapse triangle in sandy soil. Int. J. Adv. Res. Comput. Sci. Technol. (IJARCST 2015) **3**(2), 76–85 (2013). ISSN: 2347-8446 (Online) ISSN: 2347-9817 (Print) © 2013, IJARCST All Rights Reserved, April–June 2015. www.ijarcst.com
10. Acharyya, R., Dey, A.: Finite element investigation of the bearing capacity of square footings resting on sloping ground. INAE Lett. (2017) **2**, 97–105 (2017). https://doi.org/10.1007/s41403-017-0028-6
11. Prandtl, L. Über die härte plastischer körper. Nachr. Ges. Wiss. Goettingen, Math.-Phys. Kl., pp. 74–85 (1920)
12. Terzaghi, K.: Theoretical Soil Mechanics, p. 528. Wiley, New York (1943). ISBN: 978-0-471-85305-3
13. Meyerhof, G.G.: Some recent research on the bearing capacity of foundations. Can. Geotech. J. **1**, 224–229 (1963)
14. Optum G2, v2, 2018.02.27 (www.optum.com)
15. Limitstate GEO v3.4 (www.limistate.com)

3D FE Analysis Model for Geotechnical Problems of Historic Masonry Structures

Sayed Hemeda[1,2(✉)]

[1] Conservation Department, Faculty of Archaeology, Cairo University, Cairo, Egypt
sayed.hemeda@cu.edu.eg

[2] Doctor of Civil Engineering, Aristotle University of Thessaloniki, Thessaloniki 12613, Greece

Abstract. Geotechnical modeling of complex soil problems requires an advanced 3D discretization and a sophisticated soil models. The aim of the present study is to accurately analysis the geotechnical and induced structural failure mechanisms observed in the field and experimental studies. In order to provide a contribution to quantify the stress and deformation analysis and settlement of soft clay subsoil, a geotechnical analytical model is presented in details. The study presents an attempt of a PLAXIS 3D FE analysis model to simulate the subsoil problems, deformation and stress analysis of the multi-layer masonry structures of Abu Serga church in Cairo, Egypt which is loaded in plane. Advanced soil plastic modeling has been used throughout the different phases of the numerical analysis. The numerical and experimental analysis allowed us to define the pathology of this unique architectural heritage and to estimate the subsoil deformation as well as the stress-strain analysis of its superstructure, also to define the ultimate load that they can survive under their present conditions. The results are discussed with respect to the stress-strain and volumetric behavior of the soil. Finally, the paper presents the design and implementation of the complete intervention retrofitting project of Abu Serga church.

Keywords: Abu Serga church · Geotechnical problems · Masonry structures
Soil settlement · 3D FE models · PLAXIS 3D · Predicting model
Vertical displacements

1 Introduction

Construction degradation is a slow process of structural and operational performance changes. Diagnosis of a historical buildings involves several aspects: general organization of the building, the nature of the establishment soil, the general strength of the site, the transmission of dead loads and flat loads to the establishment soil, the nature of the materials, the development methods, the support level, the impact of mediations after some time and so on (Ilies and Popa 2013).

The structural damage of architectural heritage is very often due to displacement of the bearing subsoil and differential settlement or rotation of their foundations, or other effect of subsoil-structure interaction. Although it is necessary to investigate both base

H. Shehata and B. Das (Eds.): GeoMEast 2018, SUCI, pp. 217–241, 2019.
https://doi.org/10.1007/978-3-030-01923-5_16

shear resistance and settlements for any structure, very limited researches which presents contributions to development of a numerical methods to assess and classify the consequences of differential settlements on historic masonry structures (Nasser et al. 2014; Müthing et al. 2018). In many cases settlement criteria will control the allowable bearing capacity (Brinkgreve et al. 2011; Klai et al. 2015).

Because of late advances in computational innovation, numerical examination has turned into a standard apparatus to explore the geotechnical issues. In any case, numerical recreations and their contributing physical-mechanical parameters are ending up progressively hard acquiring. On one hand the satisfactory outcomes for numerical investigation of the entangled geotechnical issues can be procured just by producing a powerful three dimensional numerical model; then again the conduct of soils subjected to confused pressure ways must be anticipated by utilizing fitting progressed constitutive models (Duncan and Chang 1970; Vakili et al. 2013).

Commercial package PLAXIS 3D (PLAXIS v.b, 2018) was utilized for computing stresses and settlements. It is a limited component program produced for geotechnical building plan and inquires about and has been as of late utilized as a part of the examination of geo-structures. The Mohr – Coulomb display is utilized both for static stacking and solidification examination. The code has a helpful methodology for programmed stack venturing, called Load Advancement, which we made use here (Hemeda and Pitilakis 2010a, b; Yamamoto et al. 2001).

Keeping in mind the end goal to enhance the nature of the outcomes from the numerical recreation, the adequate site examination had been performed. In the present contextual analysis we played out a few sorts of geotechnical and geophysical field and in-situ tests, for example, SPT tests, encompassing clamor estimations, and geotechnical boreholes with necessary inspecting tests, borehole disfigurement tests and lab testing on soils.

2 Geotechnical Conditions

The subsoil of Abu Sega church comprises of delicate soaked plastic silty clay layers blended with variable rates of sand. Fine to medium sand layers show up at profundity 6 m beneath the inside chapel's floor level. The groundwater table shows up at a profundity of 1.8 m (Hemeda and Pitilakis 2017). Figure 1 presents the electrical resistivity tomography inversion model for profile ERT-3 in Mr. Girgis Main Street, 5 m above the Abu Serga church ground floor level. The geotechnical properties of the soil layers underneath the Abu Serga church are summarized in Table 1. For Clay Layers: liquid limit (WL) = 100, Shrinkage limit (WS) = 17, Plasticity limit (Wp) = 45, Dry Unit weight = 17 kN/m^3, Saturated unit weight = 20 kN/m^3, Specific gravity GS = 2.7, Initial void ratio e_i = 1.84, Compression Index Cc = 0.3. Young's modulus E = 1500 kN/m^2, Poisson's ratio υ = 0.4, Cohesion (c) = 50 kN/m^2, friction angle ϕ = 27.

For Sand Layers: liquid limit (WL) = 20, Plasticity limit (Wp) = 0, Dry Unit weight = 18 kN/m^3, Saturated unit weight 21 kN/m^3. Young's modulus E = 6000 KN/m^2, Poisson's ratio υ = 0.35, Cohesion (c) = 0 kN/m^2, friction angle ϕ = 34.

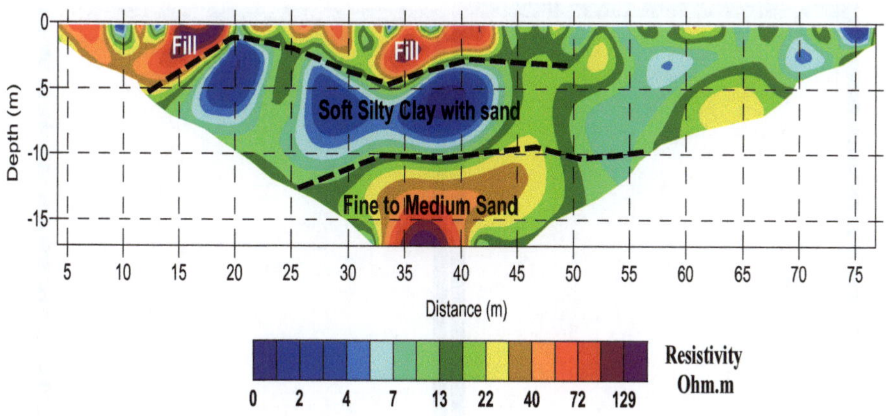

Fig. 1. Presents the electrical resistivity tomography inversion model for profile ERT-3 in Mr. Girgis main street, 5 m above the Abu Serga church ground floor level.

3 Monitoring

The old Cairo area has historically water flooded periodically before the construction of Aswan High dam on the Nile River in 1971. These water floods caused loading and unloading of the subsoil for at least decades up to hundreds of years, as the vertical effective stresses in the clay soil and its bearing capacity also changed with the rising and falling water table. After the termination of the ground water lowering, differential settlements were observed and recorded between parts of the super structures of the church. Settlements have been calculated up to 9 mm at the areas of filter wall and perforated shafts.

4 Geotechnical and Structural Problems of Abu Serga Church

Abu-Serga church stands among the most prestigious historic churches and monasteries in old Cairo area in Egypt. This church gained its high prestige from the fact that it was built on top of the holly Crypt.

The Church is about 27 m long, 17 m wide and 15 m high. Its floor level is about 1.5 m below the surrounding alleys and 4.5 below the Mar Girgis main street. It has a typical Basilicas design with a narthex leading to the nave with two side aisles, as shown in Figs. 2, 3, 4, 5a and 5b (Hemeda 2014).

Cracks and local failure were observed mainly during the historical fluctuations of Nile River (floods cause loading and unloading of the soil) and during the dewatering process.

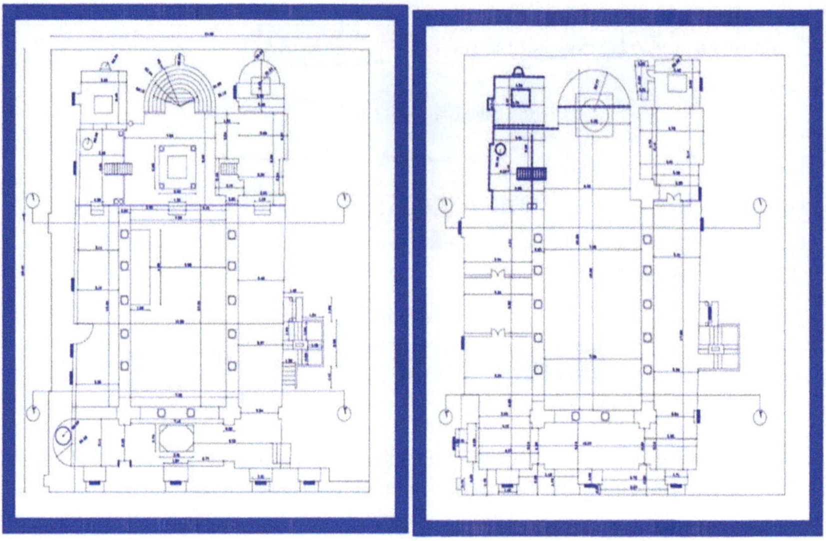

Fig. 2. Layout of the ground floor and first floor of Abu Serga church

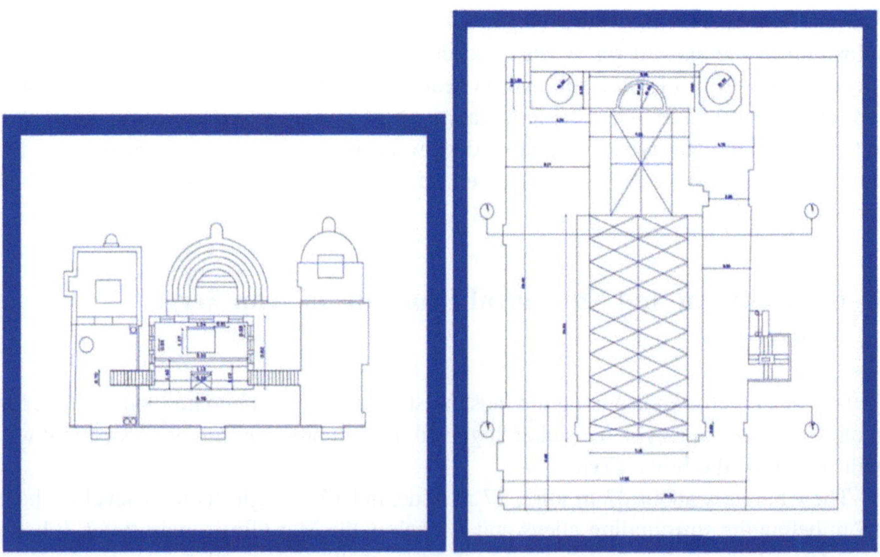

Fig. 3. Layout the holly crypt underneath the church, and the layout of the roof of the Abu Serga church.

Fig. 4. A cross section 1-1 inside Abu Serga church. B Cross section 2-2 inside Abu Serga church.

Fig. 5a. Sketch up 3D model of Abu Serga church, shows the western facade.

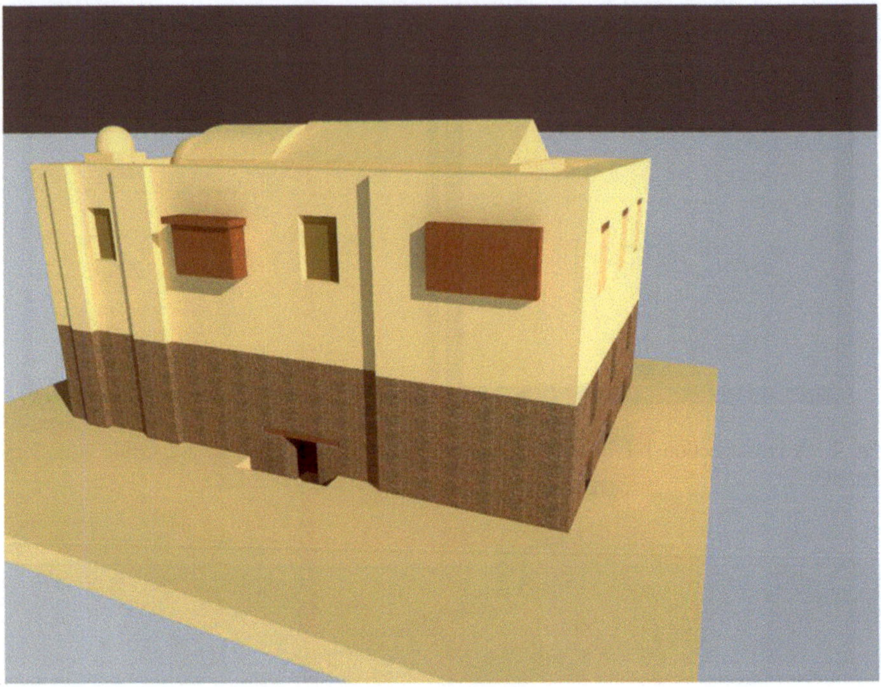

Fig. 5b. Sketch up 3D model of Abu Serga church, shows the northern facade.

The main geotechnical problems and structural deficiencies of the vertical component of the structural system are:

The superstructure of the church has suffered considerable damage due to extensive cracking associated with soil settlement and foundations movements undergoing a shear failure accompanying with large settlement both towards to the western and northern directions.

Cracks widths were 5–10 cm and exhibiting periodical widening. The cracks had a preferential orientation in two principal directions namely E-W and S-N as shown in Fig. 4, which includes typical views of major cracking in the masonry walls and foundations. Such effects were attributed to earthquake damages, permanent deformations and large settlement of the bearing soil, natural wear of construction materials and the construction history of the complex;

The success of structural restoration project depends mainly upon the accuracy of the diagnosis.

Almost causes of damage and deterioration appear to be three of the problems that affect the church, namely:

(a) Soil differential settlement due to the poor geotechnical properties of the soft clay subsoil and the variation of ground water table in addition the dewatering project in the surrounding area (contract 102).
(b) Seismic actions, and

(c) Deterioration of materials where humidity often plays a prominent role due to water table rising a long period of time. The in-plane deformations (vertical and horizontal cracks) of the main facades of Abu Serga church are obvious, as shown in Fig. 6. The differential leaning and tilting of the supporting marble columns, and the vertical cracks in arches inside Abu Serga church is presented in Fig. 7.

Fig. 6. In-plane deformations (vertical and horizontal cracks) of the main facades of Abu Serga church.

Fig. 7. Differential leaning and tilting of the supporting marble columns, and the vertical cracks in arches inside Abu Serga church.

All the starting moment of the technical assessment, almost all masonry walls (especially those in the holly crypt, ground and first floors) presented a general state of shear and out of plan failure; this aspect, together with the presence in the building of a large amount of priceless archaeological exhibits, determined the owner to start the rehabilitation works.

5 Constitutive Modeling

The geometry and FE discretization of the 3D Model are displayed in Fig. 8. In this examination plastic soil show with various multifaceted natures is chosen for the depiction of the constitutive soil conduct. Before utilizing the soil model in the investigation of versatile displaying, an alignment of the soil models and their soil parameters, particularly in view of the diverse solidness' and Elastic moduli are required. Therefore, the trial soil tests (triaxial and Oedometer test) are directed.

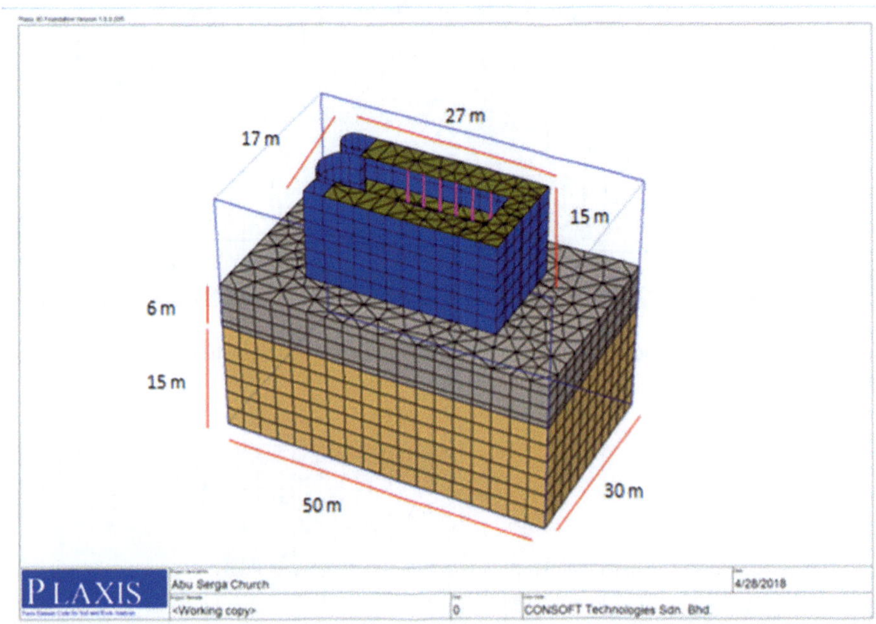

Fig. 8. The geometry and FE discretization of the 3D Model.

The numerical simulation was run by PLAXIS 3D code (version 2018). A plane strain numerical modeling was adopted. It is composed by 15 nodes triangular finite elements with medium precision mesh to minimize the computation time. All material characteristics of soil layers and construction material of Abu Serga church are given in Tables 1 and 2. The Mohr Coulomb constitutive law was selected to describe the

Table 1. Determined physical and mechanical properties of the soil layers underneath the Abu Serga church.

Friction Angle (ϕ) o	Soil Cohesi on (c) kN/m²	Poisson's ratio	Young's Modulus kN/m²	Compression Index Cc	Initial void ration (e)	Specific gravity (Gs)	Saturated Unit Weight (γ_{sat})	Dry Unit Weight (γ_{unsat})	Plasticity limit (WP)	Shrinkage limit (WS)	Liquid limit (WL)	Soil type
27	50	0.4	1500	0.3	1.84	2.7	20	17	45	17	100	Clay
34	0	0.35	6000	–	–	–	21	18	0	–	20	Sand

behavior of soft saturated clay behavior. Related to the boundary conditions of the numerical model, vertical and horizontal displacements are zero at the rigid stratum level; the horizontal displacement along vertical borders is also assumed zero. The water table is located at 1.8 m depth. The settlement was monitored in accordance with the time given to the actual settlement (Müthing et al. 2018).

Table 2. Determined physical and mechanical properties of the building materials; fired brick, marble, and timber.

G (MPa)	E_{dy} (MPa)	V_P (km/s)	E (GPa)	σ_t (MPa)	σ_c (MPa)	n (%)	Wa (%)	Gs (g/cm^3)	Sample
917	2443	1.71	8.4	1.8	5	0.27	20.1	1.8	Brick
11095	10921	2.87	30	6	20	0.32	0.12	2.7	Marble
–	–	–	7	3	10	–	30	0.64	Wood

Gs = specific gravity, **Wa** = Water absorption, **n** = Porosity, $\boldsymbol{\sigma_c}$= Uniaxial compressive strength, $\boldsymbol{\sigma_t}$= Brazilian tensile strength, **Vp** = primary wave velocity, **E** = static Young's modulus, **Edy** = Dynamic Young's modulus, **G** = Shear modulus

The soil behavior is reported and discussed. It follows by a detailed explanation of the general concept for the adaption of constitutive soil models and its corresponding influences on the soil parameters and on the plasticity properties of soil.

6 Numerical Analysis Results and Discussion

Results of the numerical analysis of the Abu Serga church as originally designed shows that some surface subsidence was induced during construction on soft silty clay thick layers (6 m) and due to the dewatering project in 2000 (contract 102). The displacement developed at the surface, right above reached a maximum value of about 17.4 cm which is the total consolidation settlement and local shear failure of the bearing soil due to the in-plane loading. Figure 9 presents the deformed mesh and the calculated vertical displacements of the bearing soft clay saturated soil and its distribution underneath the superstructure of the church is shown in Figs. 10, 11 and 12. The distribution of the volumetric strain of the soil is shown in Fig. 13, where the extreme value is 7.3%. The value of the initial vertical effective compressive stress at the mid depth of the clay layer has been determined, where σ_v = 60 kN/m^2. The effective mean stresses P'on the subsoil underneath the structure where the extreme value is 263.2 kN/m^2 are shown in Figs. 14 and 15. The effective mean stresses P'also calculated and the extreme value is 263 kN/m^2, as shown in Fig. 16. The distribution of the active pore pressure P_{active} in the subsoil, where the extreme value reached 240 kN/m^2 are shown in Fig. 17.

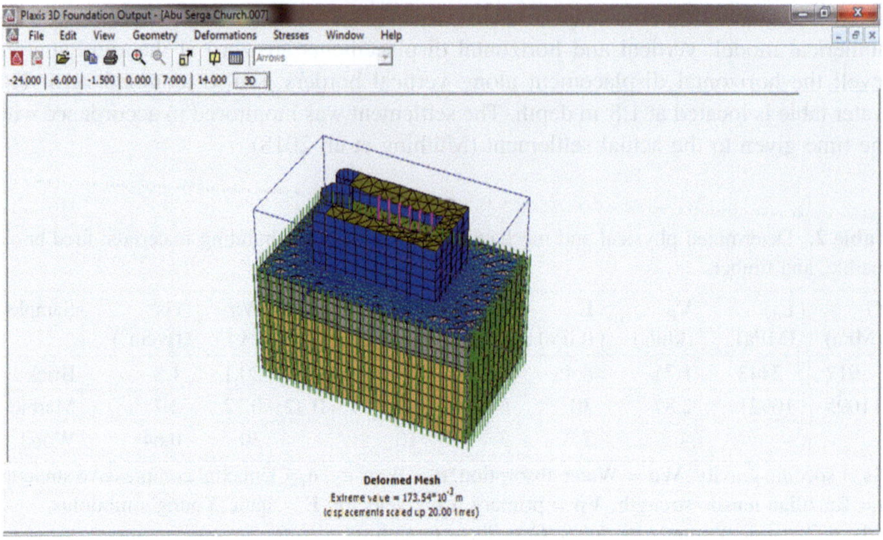

Fig. 9. PLAXIS 3D FE deformed mesh.

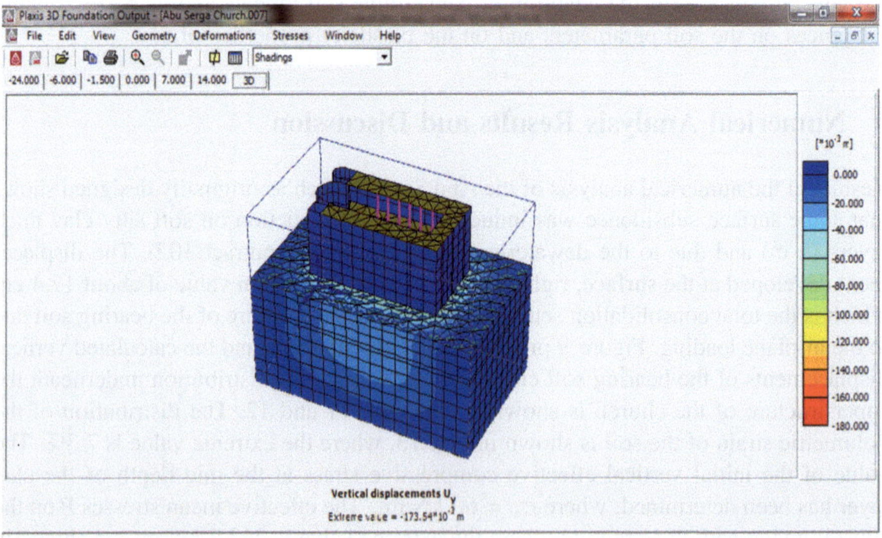

Fig. 10. Vertical displacement distribution in subsoil underneath the structure of Abu Serga church. The extreme value is 173.54 mm

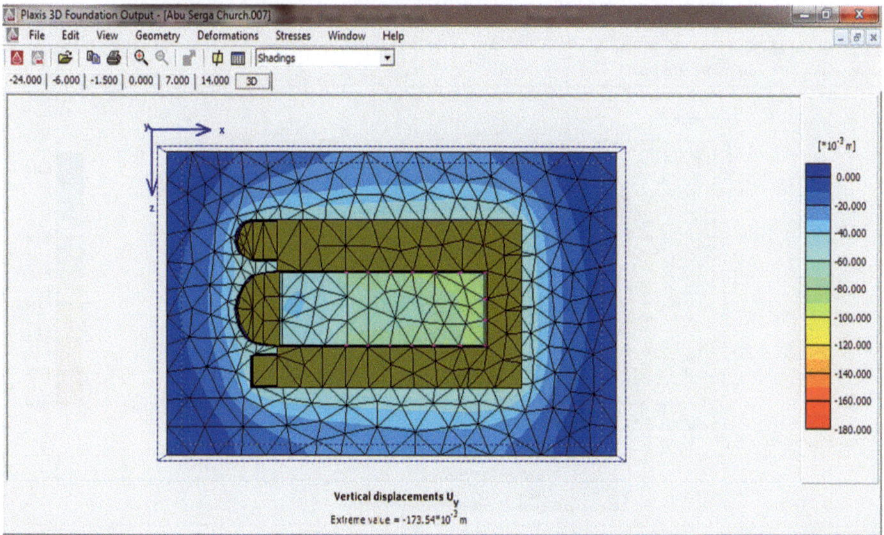

Fig. 11. Vertical displacement distribution in subsoil underneath the structure of Abu Serga church.

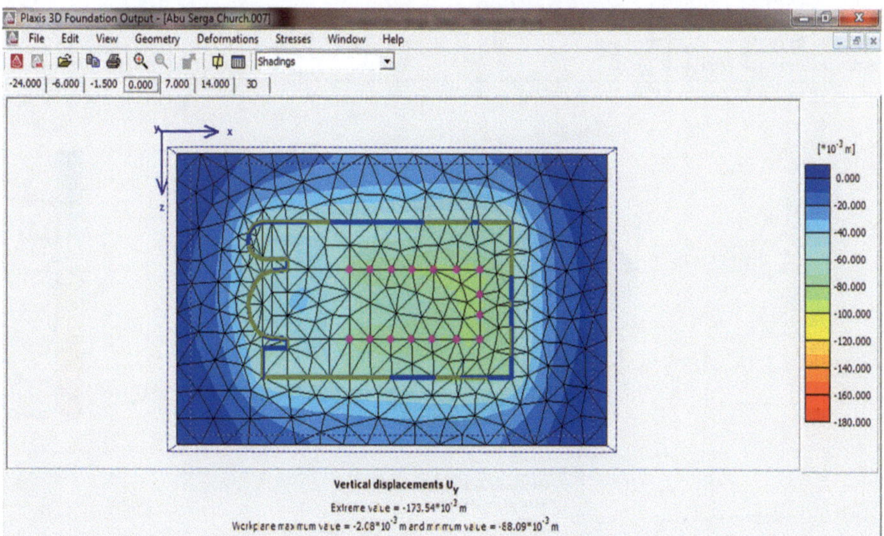

Fig. 12. Vertical displacement distribution in subsoil underneath the structure of Abu Serga church.

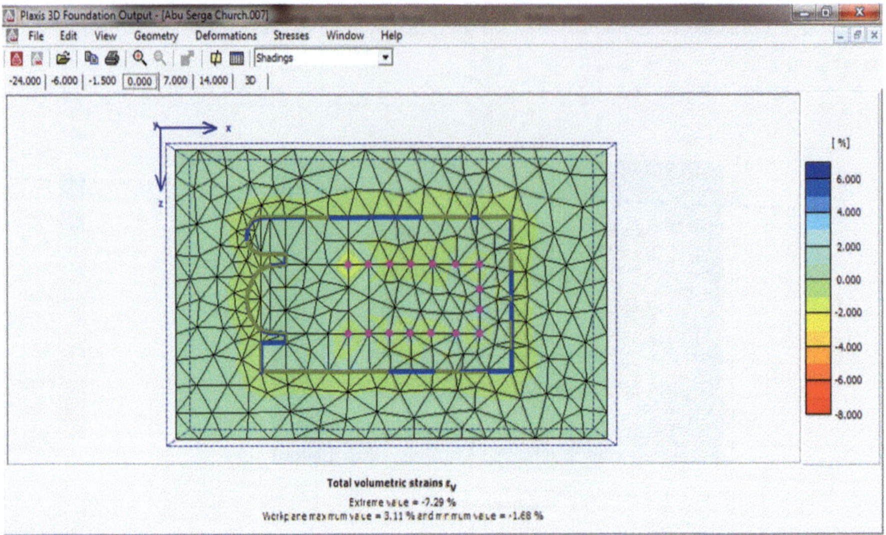

Fig. 13. Total volumetric strain distribution in subsoil underneath the structure of Abu Serga church. The extreme value is 7.29%

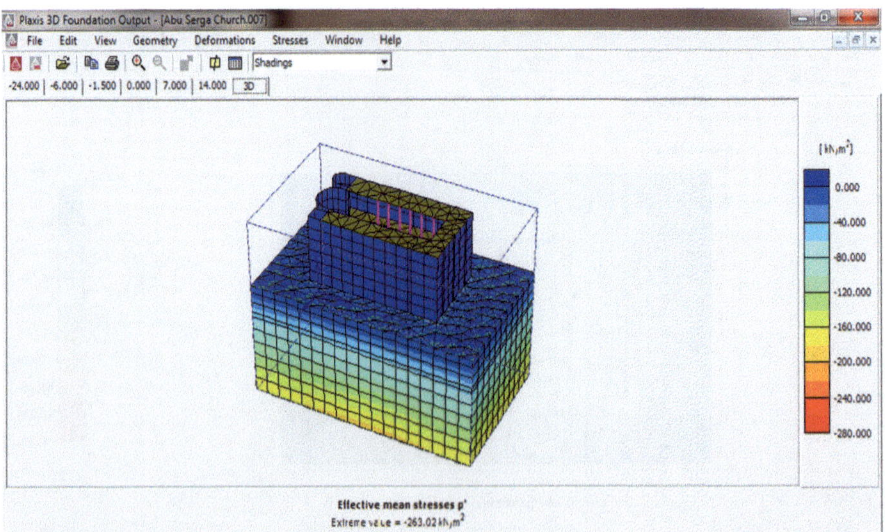

Fig. 14. Effective mean stresses distribution in subsoil underneath the structure of Abu Serga church. The extreme value is 263 kN/m^2.

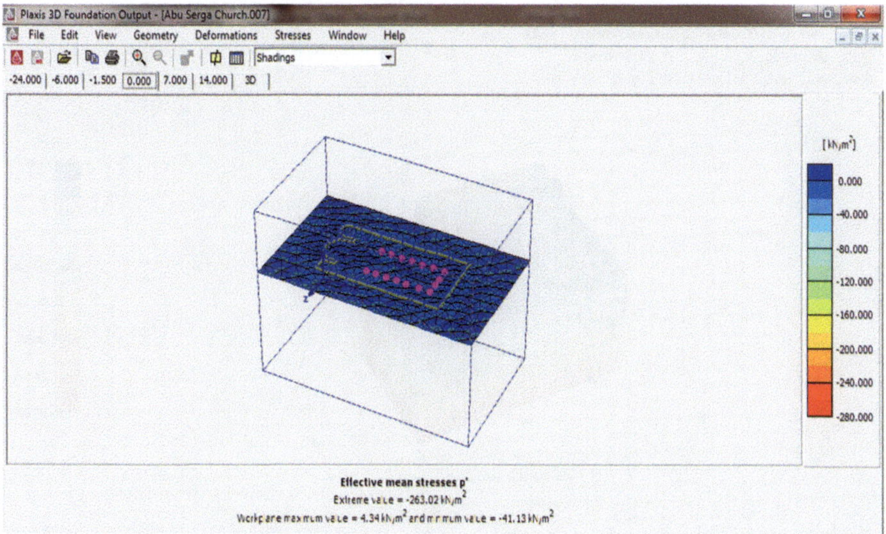

Fig. 15. Vertical effective mean stresses distribution in subsoil underneath the structure of Abu Serga church. The extreme value is 263 kN/m^2.

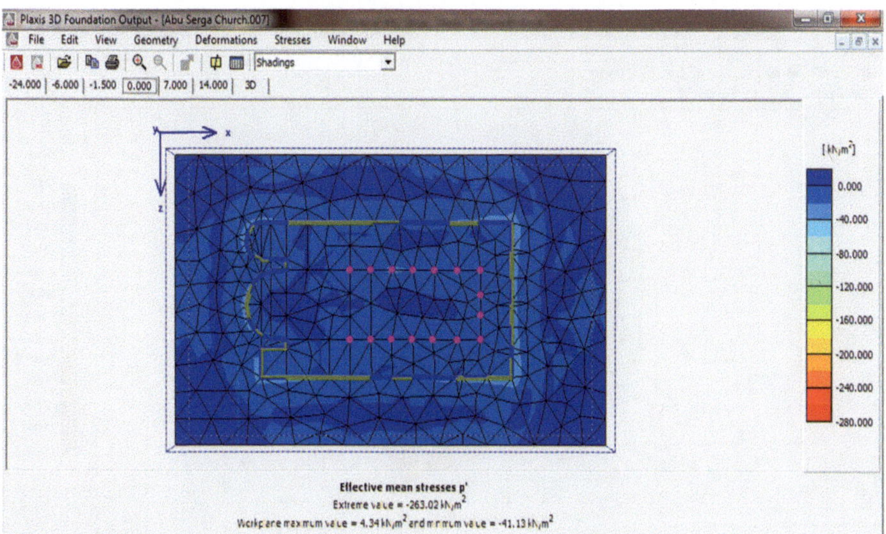

Fig. 16. Vertical effective mean stresses distribution in subsoil underneath the structure of Abu Serga church. The extreme value is 263 kN/m^2.

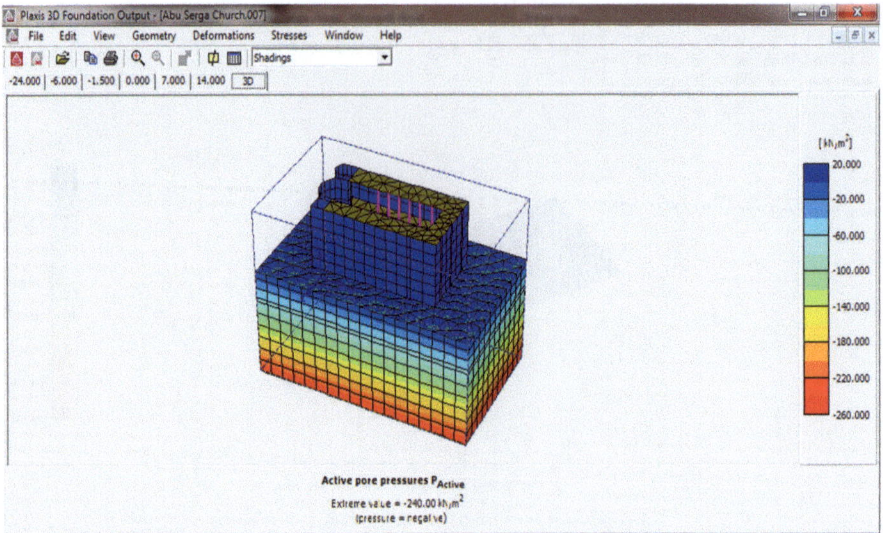

Fig. 17. Active pore pressures distribution in subsoil underneath the structure of Abu Serga church. The extreme value is 240 kN/m^2.

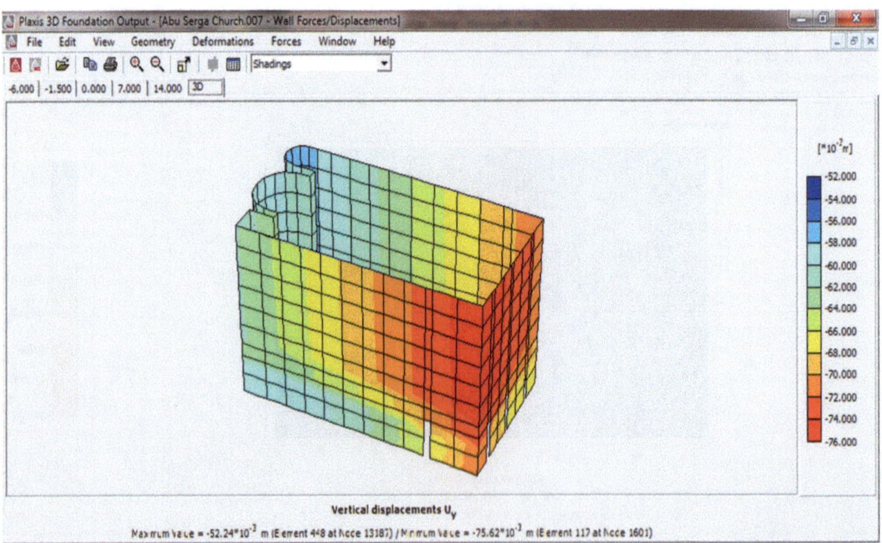

Fig. 18. Vertical displacements of structure of Abu Serga church due to the differential soil settlement. The extreme value is 76 mm.

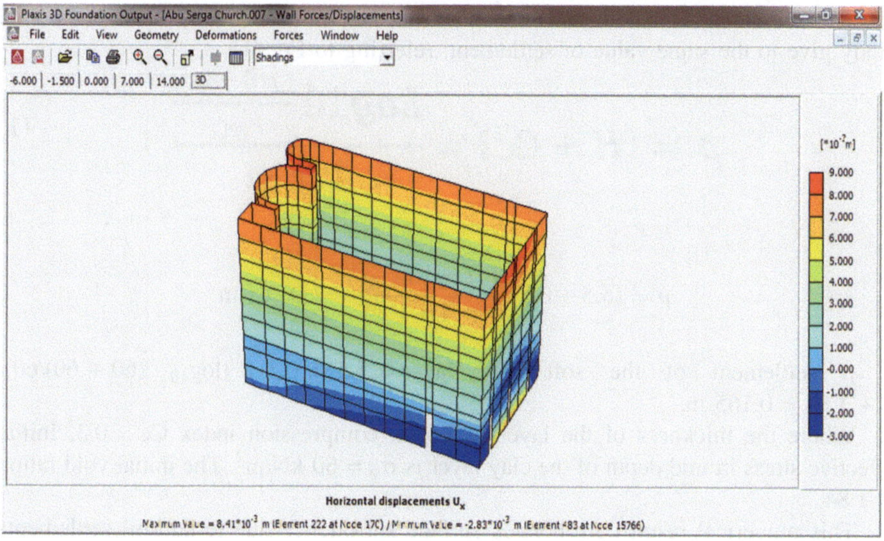

Fig. 19. Horizontal displacements of structure of Abu Serga church due to the differential soil settlement. The extreme value is 9 mm.

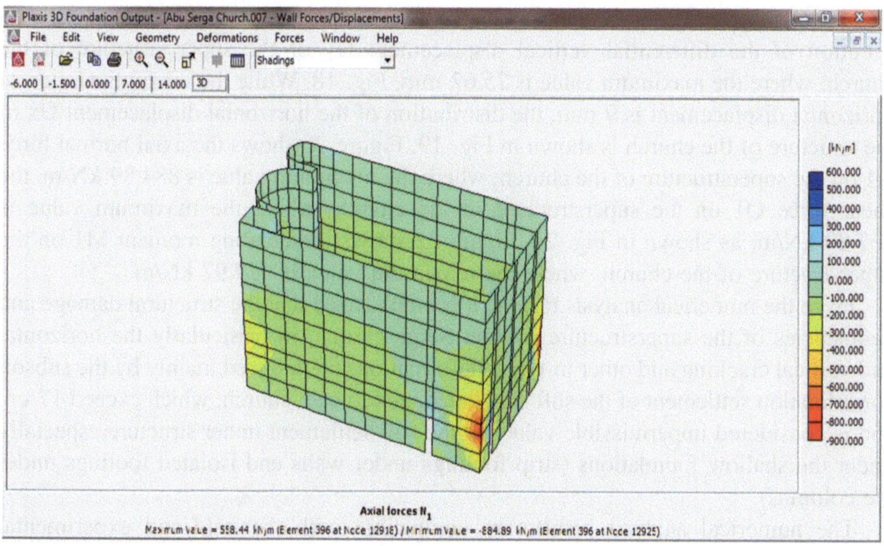

Fig. 20. Axial normal force on the structure of Abu Serga church due to the differential soil settlement. The extreme value is 885 kN/m.

The settlement calculations which have been determined from the experimental study give to the same value of settlement, referring to Eq. (1).

$$\rho = (H + Cc) = \frac{Log10\,\frac{(\sigma 0 - \Sigma\sigma)}{\sigma 0}}{1 + ei} \tag{1}$$

$$\rho = (5.5 + 0.3) = \frac{Log10\,\frac{(60-60)}{60}}{1 + 1.84} = 0.165\ \text{m}$$

ρ settlement of the soft clay layer = 5.5 × 0.3 × ($\log_{10}$ (60 + 60)/60)/ 1 + 1.84 = 0.165 m.

Where the thickness of the layer is 5.5 m, compression index Cc = 0.3, Initial effective stress in mid depth of the clay layer is σ_v = 60 kN/m^2. The initial void ration is 1.84.

This numerical result corroborates surface subsidence due to subsoil settlements and the deformations and straining reactions observed at the superstructure of the church.

Subsoil subsidence and settlement may lead to cracking of the superstructures of the church. The primary causes are consolidation. Also the shear failure characteristics of subsoil may lead to subsidence. The numerical analysis results indicated the distribution of the differential vertical displacement Uy of the super structure of the church, where the maximum value is 75.62 mm, Fig. 18. While the maximum value of horizontal displacement is 9 mm, the distribution of the horizontal displacement Ux of the structure of the church is shown in Fig. 19. Figure 20 shows the axial normal force N1 on the superstructure of the church; where the maximum value is 884.89 kN/m. the shear force Q1 on the superstructure of the church, where the maximum value is 467.89 kN/m, as shown in Fig. 21. Figure 22 shows the bending moment M1 on the superstructure of the church, where the maximum value is 127.97 kN/m.

From the numerical analysis results, it is emphasized that the structural damage and deficiencies of the superstructure of Abu Serga church, in particularly the horizontal and vertical cracking and other in-plane deformations are induced mainly by the subsoil consolidation settlement of the soft clay layers underneath church, which exceed 17 cm and is considered impermissible value for the soil settlement under structure especially under the shallow foundations (strip footings under walls and isolated footings under the columns).

The numerical analysis results are compatible with the field and experimental observations, where the critical and deformed zones are imposed by different high values of normal force (884.89 kN/m), shear force (467.89 kN/m), and bending moments (127.97 kN/m) and the uniaxial compressive strength of the brick is low in range of 5 MPa.

Consequently it can be concluded that modeling the behavior of subsoil soft clay underneath Abu Serga church by the Numerical analysis is suitable for understanding of the studied behavior of this kind of problematic soil.

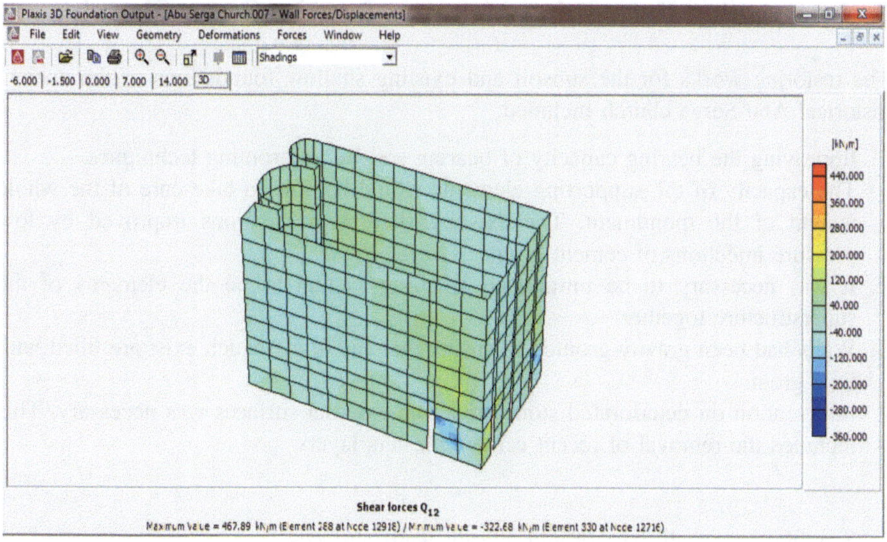

Fig. 21. Shows the shear forces on the structure of Abu Serga church due to the differential soil settlement. The extreme value is 468 kN/m.

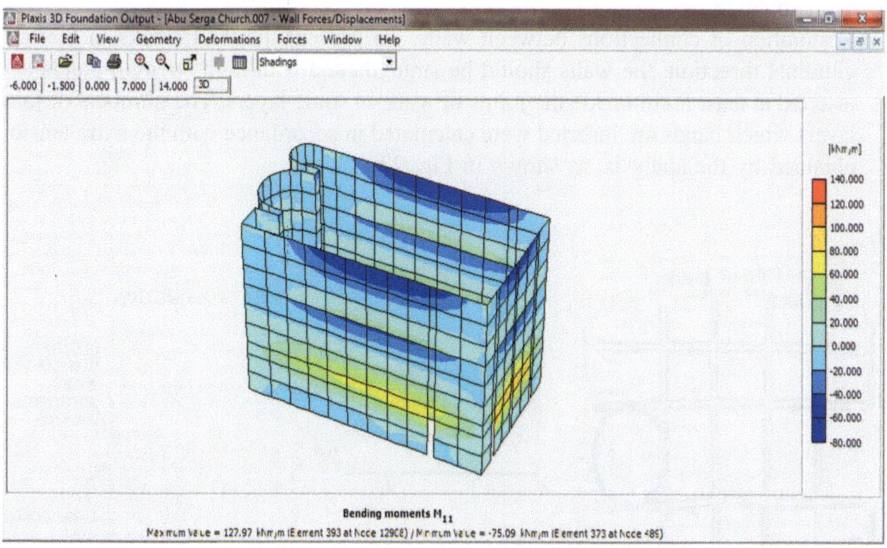

Fig. 22. Bending moment on the structure of Abu Serga church due to the differential soil settlement. The extreme value is 128 kN/m.

7 Intervention Retrofitting Measures for Abu Serga Church

The restoring works for the subsoil and existing shallow foundations of the ancient historical Abu Serga church included:

1. Improving the bearing capacity of bearing soil by jet grouting techniques.
   The capacity of the supporting elements would be able to take care of the whole weight of the monument. The existing shallow foundations improved by low pressure injections of cement grout.
2. It was necessary to be introduced stitching system to tie the elements of the superstructure together.
3. Walls had been gravity grouted to ensure that any voids which exist are filled with lime grout.
4. Intervention on deteriorated stone/brick and external surfaces was necessary. This included the removal of recent cement plasters layers.

- Sandblasting of all exterior and interior walls surfaces.
- Repointing of all surfaces with lime mortars and removal of previous interventions with cement paste.

5. Wall stitching
6. Restoration of connections between walls. To prevent the wall cracks in the longitudinal direction, the walls should be strengthened with high strength bands and inserted at least 5 cm inside the joints of walls in some layers. The numbers of joint layers which bands are inserted were calculated in accordance with the extra tension obtained by the analysis, as shown in Fig. 23.

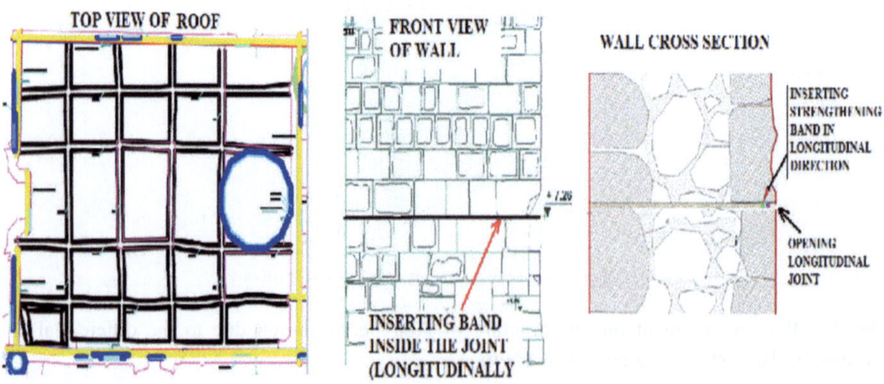

Fig. 23. Left: Applying high tension bands in the roof saddle to prevent widening of the structure. Middle: Inserting band in the longitudinal direction of the wall to prevent wall cracking. Right: The bands that were shown in cross section of the wall. One band is used for each 1 m.

To prevent the double skin walls from separation in transverse direction, due to the existence of filling garbage which is not stuck to the skin of walls, special type of stirrup elements such as carbon fiber band or high strength stainless steel were introduced to be used. This will prevent the walls from extra ridging and enables them to be more resistant in the future.

Fig. 24. Extensive deterioration of the internal walls of Aub Serg church, induced by rising water table and salt weathering mechanisms, before restoration measures.

Fig. 25. Leaning and decay of the supporting marble columns of Aub Serg church, before restoration measures.

Fig. 26. Aub Serg church, during the restoration measures, removal of cement renders, stitching walls, and treatment of the fired bricks.

Fig. 27. Aub Serg church, during the intervention retrofitting of the supporting marble columns.

Fig. 28. Aub Serg church after the restoration and strengthening measures.

Fig. 29. Aub Serg church after the restoration and strengthening of the marble supporting columns.

It was recommended to insert lead shim plates of about a few millimeters thick in the joints in each 1–1.5 m above the ground up to the roof in double skin walls. The widths of these plates should be the same as the skin of walls. For the rubble stone part which is within them is not necessary to be isolated. This causes the shakings due to minor earthquakes that are low in magnitude but numerous in several centuries be stopped and prevent to go upwards and damage the structure.

7. Stone stitching
8. Reinforcing of marble columns with the steel ring beams
9. Restoration of mural paintings and Icons
10. Restoration of wooden elements, doors and windows

Figures 24 and 25 show the state of preservation of the church before the restoration project.

Fig. 30. Aub Serg church after the restoration and strengthening measures. Reinforcing of marble columns with steel ring beams.

The state of preservation of the Abu Serga church after the fully installation of remedial intervention retrofitting measures are presented in Figs. 26, 27, 28, 29 and 30, in particularly the walls cleaning, salt removal and stitching processes, and supporting marble columns strengthening.

8 Conclusions

The calculation and assessment of soft clay subsoil settlement and its induced damage of Abu Serga Church were investigated. A 3-dimensional finite element model was used to perform several types of static plastic and consolidation analyses. The structure's long term creep deformation during the last several centuries was analyzed to find a suitable method for its strengthening. A numerical model was implemented based on assumptions and simplifications regarding the geometrical conditions and staged construction of the structure, the initial and boundary conditions as well as the constitutive modeling. The calibration of the numerical model was conducted based on in situ measurement data to identify the corresponding model parameters. Different methods for parameter optimization were investigated. Consequently it can be concluded that modeling the behavior of subsoil soft clay underneath Abu Serga church by the Numerical analysis is suitable for understanding of the studied behavior of this kind of problematic soil.

Various types of strengthening techniques were investigated to best suit to this structure. The results of the analyses concluded that the decayed cement plasters which had been covered the interior brick walls during 1960s should be removed due to the rising of water table and humidity also due to asymmetric deformation during earthquakes shaking. The result of analysis of deformation shows that the effect of impermissible subsoil settlements is obvious which exceed 17 cm, also the settlement induced structural damage has highlighted that the in-plane collapse of the masonry walls is the main way of failure in old masonry buildings, especially due to ineffective connection with floors.

References

Brinkgreve, R.B.J., Engin, E., Swolfs, M.W.: Material models manual. In: Plaxis 3D Plaxis bv, Delft, The Netherlands (2011)

Duncan, J.M., Chang, C.: Nonlinear analysis of stress and strain in soils. J. Soil Mech. Found. Div. ASCE. **96**(SM5), 1629–1654 (1970)

Hemeda, S., Pitilakis, K.: Serapeum temple and the ancient annex daughter library in Alexandria, Egypt: Geotechnical–geophysical investigations and stability analysis under static and seismic conditions. Eng. Geol. **113**, 33–43 (2010a)

Hemeda, S., Pitilakis, K., Bakasis, E.: Three – dimensional stability analysis of the central rotunda of the catacombs of Kom El-Shoqafa, Alexandria, Egypt. In: 5th International Conference in Geotechnical Earthquake Engineering and Soil Dynamics, 24–29 May 2010, San Diego, California, USA (2010b)

Hemeda, S.: Seismic hazard analysis for preservation of architectural heritage: the case of the Cairo's oldest Abu Serga church. J. Civ. Eng. Sci. **3**, 2 (2014)

Hemeda, S., Pitilakis, K.: Geophysical investigations at Cairo's oldest, the church of Abu Serga (st. sergius), Cairo, Egypt. Res. Nondestruct. Eval. **28**(3), 123–149 (2017). https://doi.org/10.1080/09349847.2016.1143991

Ilies, N.M., Popa, A.: Geotechnical problems of historical buildings from Transylvania. In: Bilota, et al. (eds.) Geotechnical Engineering for the Preservation of the Monuments and Historic Sites. Taylor & Francis Group, London (2013)

Klai, M., Bouassida, M., Tabchouche, S.: Numerical modeling of tunis soft clay. Geotech. Eng. J. SEAGS AGSSEA **46**(4 December), 87–95 (2015)

Müthing, N., Zhao, C., Hölter, R., Schanz, T.: Settlement prediction for an embankment on soft clay. Comput. Geotech. **93** (2018). https://doi.org/10.1016/j.compgeo.2017.06.002

Nasser, H., Marawan, H., Deck, O.: Influence of differential settlements on masonry structures. In: Conference: Computational Modelling of Concrete Structures vol 2. At: St Anton am Arlberg, Austria (2014). https://doi.org/10.1201/b16645-93

Vakili, K.N., Barciago, T., Lavason, A.A., Schanz, J.: A practical approach to constitutive models for the analysis of geotechnical problems. In: The 3rd International Conference on Computational Geomechanics (ComGeo III), vol. 1, Krakow, Poland (2013)

Yamamoto, K., Tabata, K, Kitamura, R.: Finite Element Analysis of Seepage and Deformation Properties in Shirasu Ground for the Situations of Sheet Pile Excavation. Elsevier BV (2001)

Hypotheses of the Tilting of Nevyansk Tower

Vladimir Lushnikov[1], Askar Zhussupbekov[2], Abdulla Omarov[2], and Gulzhanat Tanyrbergenova[2(✉)]

[1] GeoStroiExpert Ltd., Saint-Petersburg, Russia
[2] Department of Civil Engineering, L.N. Gumilyov Eurasian National University, Astana, Kazakhstan
astana-geostroi@mail.ru, tanyrbergen_guldy@mail.ru

Abstract. The Nevyansk Tower is a unique monument of history and culture in the Ural region of Russia, which was constructed at the beginning of the 18th Century. At the article along with the description of the history of construction of this monument are reported the information on the known and many unknown facts related to it, data on the tilting of the Tower. Also it shows the reconstruction of the history of inclination of the Tower and the prediction of its development for 100 and 300 years. In the case of possibility of a dangerous progress of its tilt there must provide a detailed engineering investigation of the Nevyansk Tower and develop the projects for stabilization of tower position similarly to what has been made on the worldwide famous monument of Pisa tower.

1 Introduction

Nevyansk tower are located in the Nevyansk town, Yekaterinburg Region, Russia (see in Fig. 1). It is one of the most unique historical monuments of the 18th Century. Tower is famous not only its inclination (not leaning), but also its history and its myths and legends. The Tower of 57.5 meters height was built about 1725 and not later than 1742 according to different historical sources on the territory of the metallurgy plant on the orders of mine owner Akinfii Demidov.

Since 1930 at various times the investigations of the Nevyansk tower have been carried out. The first measurements of the Tower and the study of building structures were carried out by P. Podolsky in 1936. From the data of geodetic observation which have been made since 1962 till 1965 the inclination of top of Tower from the base center of square section is 1.8 m (Skulsky 1965). In this observation five settlement marks were installed on the basement of the tower: No. 2, 3 and 4 - in the corners, No. 1 - in the corner of addition of Tower, No. 5 - nearby the connection of Tower and addition. The accuracy of the measurements was ensured within 1 mm (leveling of the 2nd class). For three years of observation have been fixed the settlement of marks ΔS = 2.1 to 3.8 mm.

However, this investigation enables to get the first information on the Tower tilt, its curved axis, the settlement to be gained as well as the first prediction of the deformation development. The data obtained were as follows: deflection of Δ = 1.8 m; average tilt 2°; the tilt of independent floors decreases from 3.6 at the bottom to 0.48 at the top,

H. Shehata and B. Das (Eds.): GeoMEast 2018, SUCI, pp. 242–248, 2019.
https://doi.org/10.1007/978-3-030-01923-5_17

Fig. 1. General view of the Nevyansk Tower

i.e., the axis is curved so that it obtained a specific shape (Fig. 2b); the rate of inclination of the opposite edges of the "chetveric" (an average relative settlement per year over the period of observations) was 0.9 mm per year. This value was confirmed by observations of a research group of the Ural Polytechnic Institute (Lushnikov et al. 1979). Using the settlement marks of V. Skulsky, there was recorded an increasing in the total settlement of the foundations of the Tower for 15 years (from 1962 to 1977) ΔS = 13.9–14.5 mm, which roughly corresponds to the given by V. Skulsky, but velocity of settlement less than 0.5 mm per year.

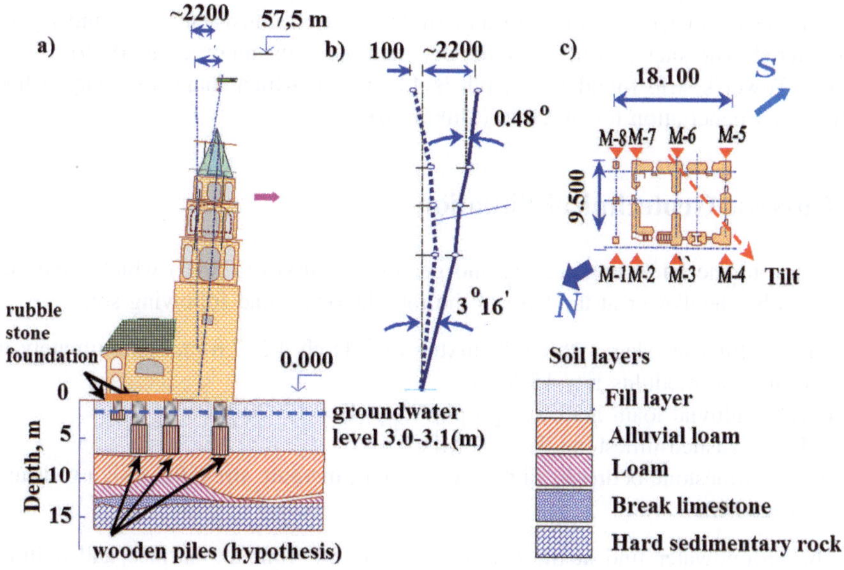

Fig. 2. The Nevyansk Tower with soil layers (a); a curved axis of the Nevyansk Tower (b); Tower's 1st level plan with settlement marks locations (c)

Certainly, these data were constantly reflected through the subsequent measurements.

In the calculations of prediction (Zaretskii et al. 1981) the following initial data were used: by the end of construction works, the axis of tent has been vertical and its tilt of 0°48′ if the tower were a rigid. It is the possibility for determination of the eccentricity of the mass center at the beginning of the of 18th Century and compare it with 1979 (mc = 1.1 m, height of 23 m from the foundation level); the most dangerous recognizes the eccentricity Δmc = 1.7 m, equal to 1/6 the diameter of the foundation (10.5 m), when under its edges there As a result of the Tower is turned off by 3.50 m. According to this prediction, the Nevyansk Tower would work out the stability yield by 2200–2250, i.e., by this time, the coefficient of stability would reach = 1 (according to the standardized allowance the safe values for such buildings are k_{st} = 1.2–1.3).

During 1977–1979 the researchers of the Ural Polytechnic Institute were carry out a detailed investigation of tower and foundation. It was found that the walls of the Tower are made of red clayey bricks and have a thickness from 1.8 m at the bottom to 0.77 m at the top. The Tower was made of brick and has a square cross section of 9.5 × 9.5 m (in Russian “chetveric”) up to the height of 24 m, and further it has three octahedral levels (in Russian “vosmeric”). Six spans are built as brick arches, while the remainders are built of wooden beams. All arches are reinforced with steel cast-iron rods. The tower is covered with a tented octahedral roof about 10 m high. There is a brick addition to the tower 6.5 m high with covered porch. The total weight of the tower is about 30,000 kN. The foundation of the tower is a strip foundation about 2–2.5 m wide made from rubble stone and is supported by short wooden piles (up to 3 m) which cut through the fill layer and embedded in clayey soil (Fig. 2a and c). However, this has been not accurately determined. The soil of Nevyansk tower are alluvial and alluvial-deluvial clayey materials with hard and slightly plastic consistency. A sand layer is encountered. The slag was filled by the dry method. (Lushnikov et al. 1979).

The Tower’s strip foundation depth is 4.2–4.3 m, which found by charged body method and penetration test in 2000 (Slukin 2012).

2 Current Geotechnical Situation

According to the latest engineering and geological surveys (2000) which have been made nearby the Tower at the base of the foundations found following soils:

(1) IGE1 – fill soil (slag with earth mixture) of depth 4.5–7.8 m, approximately the deformation modulus E = 15 MPa;
(2) IGE2 – alluvial loam hard to high plasticity, E = 12–20 MPa;
(3) IGE3 – crushed limestone, E = 25 MPa;
(4) IGE4 – limestone bedrock, slightly weathered, medium strength, design resistance R_c = 25 MPa.

The groundwater find at the depth of 3.0–3.1 m from the surface and it linked hydraulically with a nearby pond.

Than in 2002 the Ural scientific research institute of construction conducted next observation of Tower's tilting. In the analysis of survey materials, first of all, is determined a height change of the top of bedrock in the direction of the tower tilting.

In a section of soil of 4 m in length, the difference in the bedrock level is more than 2 m (Lushnikov 2002). Accordingly, the thickness of the compressible layer increases. Thus, compressibility of ground is differential, and compressibility is increased in the direction of the tilting. This was well known before, and often shown as the main cause of the deformations of the Tower. In report of Kopeikin and Morozov (2001) shows the structure of soils in depth, according to the tracing along one of the profiles under the Tower. In the profile can be see anomalies of different densities in depth up to 5–6 m.

3 Reconstruction of the Tower Inclination and Prediction Its Deformations

In this case, to try for reproduction the stages of the construction of parts of the Tower, as well as the rate of settlement and, in particular, the difference of the settlements, and considering the hypothesis that each level was straight vertically. For absolutely reliable existing inclination of the top of the third "vosmeric" is 2.2 m from the center of the top of the basement which found trough analysis of photogrammetry survey (Alyabeva and Ivanova 2000).

Consider three hypotheses of the development of deformations, based on the mutual inclination of floors of the Tower among themselves. The spatial position of each lower floor at the time when constructed the next floor is shown in Fig. 3.

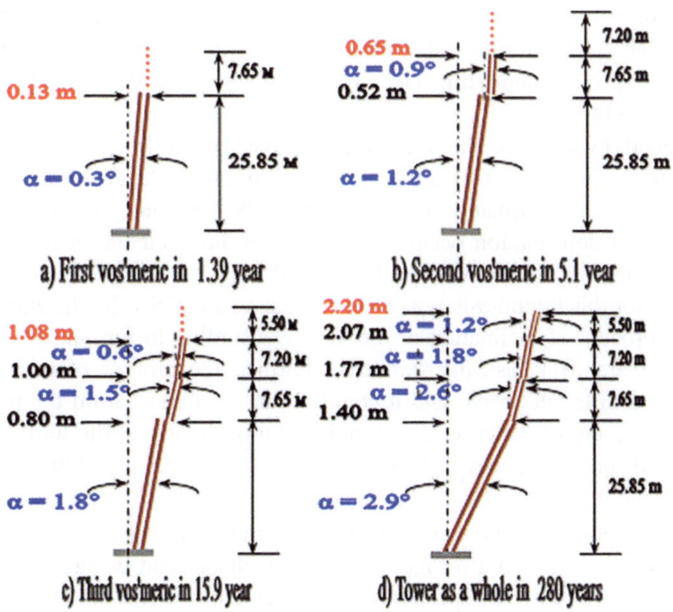

Fig. 3. Reconstruction of tower axis inclination

The first hypothesis: the inclination of the top of the upper octagon accumulates over the past 280 years.

In the development of this prediction all values of the inclinations of the Tower at various intervals were approximated by a (logarithmic) time dependence

$$\Delta = \beta \ln (t/t_o) \qquad (1)$$

where Δ is the inclination of the axis of the tower during the time t (years, days, etc.) from the beginning of the construction; t_0 - an additional parameter (1 year, 1 day, etc.). Of course, this prediction will receive the significant refinements after when reliable information of the deformation in the past is received.

This hypothesis should be considered the most possible. According to this hypothesis, the inclination of the top of the upper octagon of the Tower cannot exceed the derivative of the function (Skulsky 1965), i.e. magnitudes - $v_s = \beta/t = 1000$ $(0.3904/280) = 1.39$ mm/year, and the inclination of the roof top - $v'_s = v_s H'/H = 1.39 \times 57.5/46.2 = 1.72\,\mathrm{mm/year}$, ($H' = 57.5$ m and $H = 46.2$ m – height of Tower and brick wall height).

If to take as reliable the development of deformation under this logarithmic law, then by back substitution, sequentially declining back the top of each level by 0.13 m, 0.64 m, 1.08 m and at last 2.20 m for the third "vosmeric", one can obtain the following terms: first vosmeric – 1.39 year; second and third – 5.1 and 15.9 years, respectively (Fig. 3).

The rate of changing of the settlements of the opposite sides of the foundation, corresponding to an inclination of $\Delta = 2.2$ m, is equal to $v\Delta S = vs$ $B/H = 1.39 \times 10.0/46.2 = 0.30$ mm/year, where $B = 10$ m is the foundation side. According to this hypothesis, in 100 years (t = 380 years, i.e. approximately to 2100) the inclination of the top of the Tower can be equal to $\Delta'_{2100} = (\beta H'/H) \ln (t/t_o) = (0.3904 \times 57.5/46.32) \times \ln(380/1.0) = 2.88$ m, and after 300 years (t = 580 years, i.e. approximately to 2300) $\Delta'_{2300} = (0.3904 \times 57.5/46.32) \times \ln(580/1.0) = 3.09\,\mathrm{m}$.

The second hypothesis: the inclination of the Tower in 2 and 1/2 arshin (Δ' 1870 = 1.778 m, and not 2.43 m, as predicted by formula (1), V.I. Nemirovich-Danchenko, it is approximately 150 years from the beginning of construction; the further course of deformation occurs at such a rate that over the next 130 years the inclination of the top of the tent would $\Delta' = 2.35$ m.

The basis for this hypothesis is as follows: it is assumed that by the end of the 19th Century the process of inclination of the Tower according to formula (1) (but with a different parameter $\beta 1$) basically stabilized; Further, the process of inclination continues, but at a different speed, and it could be has an impact from the plant, power station, pumping out of water, etc. The inclination of the top of the third chetveric in this case will be $\Delta_{1870} = \Delta'_{1870} H/H' = 1.778 \times 46.2/57.5 = 1.429\,\mathrm{m}$. And $\beta_1 = \Delta'_{1870}/\ln(t/t_o) = 1.429/\ln(150/1.0) = 0.2852\,\mathrm{m}$.

From this hypothesis follows that the terms of the beginning of the construction of floors are as follows: the first octagon - 1.57 year; the second octagon - 9.4 year; the third octagon - 44.1 year. The speed of the inclination of the tent top (at T = 130 years after 1870) will be $vs' = 1000(\Delta'_{2000} - \Delta'_{1870})/T = 1000(2.35-1.778)/130 = 4.40\,mm/year$, and the rate of changing of settlement in the opposite sides of the basement will be

$$v_{\Delta s} = v_s B/H = 4.4 \times 10.0/57.5 = 0.76\ mm/year$$

This hypothesis predicts too long (and, in the authors' opinion, unrealistic) the duration of the construction of the Tower, but the highest rate of development of the precipitate vs, and especially their settlement difference vΔS. According to this hypothesis, the terms of construction of each floors must also be changed and the Tower incline would equal Δ = 2.79 m in 100 years and Δ = 3.66 m in 300 years.

The third hypothesis: as the second hypothesis, the inclination Δ'1870 = 1.778 m; but the further course of deformations occurs that (over T = 130 years), the rate of inclination of the tent top would be equal to any actually measured value of the deformation, for example, to the above rate of changing the settlement difference of opposite sides of the basement vΔs = 0.5 mm/year.

In this case, the current deformation for the last period of T = 130 years is due to the dependence $\Delta' = \Delta'_{1870} + \beta_2 T^{\alpha} = 1.778 + 0.0238 \cdot T^{0.6535}$, which will be further refined and serve as a basis for next analyses.

Inclination of the top of the tent for 100 years (T = 230 years), will be

$$\Delta'_{2100} = 1.778 + 0.0238 \cdot 230^{0.6535} = 2.61\ m,$$

For 300 years (T = 430 years), $\Delta'_{2300} = 1.778 + 0.0238 \cdot 430^{0.6535} = 3.03\,m$.

As can be seen, the inclination of the top of the tent Δ'2100 for 100 years by any of the hypotheses (Δ'2100 = 2.88, 2.79 and 2.61 m) differ insignificantly. After 300 years (Δ'2300 = 3.09, 3.67 and 3.03 m), the difference will be more significant, and, as follows from the analysis, the most dangerous is the second hypothesis. Of course, if instrumental measurements confirm a higher rate of deformation than vΔs = 0.50 mm/year (for example, vΔs = 0.90 mm /year).

As example, in Table 1 gives the results of calculations at two values VΔs,1 = 0.5 mm/year and VΔs,2 = 1.0 mm/year, based on which a conclusion can be made that at a current rate of settlements VΔs = 1.03 mm/yr. the Tower will turn out in danger approximately in 2200 when the stability coefficient can be $k_{st} = 1$.

Table 1. Tilting of the Nevyansk Tower according to different hypotheses

Hypothesis	Monitoring of inclination		Long-term prediction of inclination	
	Foundation (mm/year)	Top of tower (cm)	100 years	300 years
Historical	0.3	17.2	2.52	2.86
From observation of 1870	0.76	43.6	2.79	3.66
Settlement	0.5	28.7	2.63	3.21
Settlement	1.0	57.5	2.92	4.08

4 Conclusions

The results of the analysis (primarily on the basis of stereophotogrammetric survey) were largely unexpected: the actual tilt of the Tower turned out to be noticeably larger in comparison with earlier measurements: the inclination of the top was 2.2 m versus 1.8 m, as it was thought before. However, the main hypothesis was confirmed: different floors of the Tower have a different inclination gradually decreasing by height.

The measurements performed and their analyses do not support the hypothesis about the purposeful erection of the Nevyansk Tower as a leaning building. The "saber-like" longitudinal axis, continuously settlement and inclination of axis, different tilts of floors, door and window openings, rods, etc., indicate that the Tower was initially was built as a vertical structure. An instability of soil bed resulted in the appearance of non-uniform settlings immediately after the beginning of the construction works, and each floors were built vertically but under conditions of the continuous non uniform deformation.

At present, an important conclusion can be made on the stability and safety of the current state of the Tower; yet, it should be taken as a temporary one. This conclusion calls for a comprehensive verification, confirmation or rejection, based on the regular precise instrumental observations and careful and high-skilled analysis.

Should be organized a constant monitoring of the Nevyansk Tower, under understanding its historical and culture importance, similarly on the Leaning Tower of Pisa. Such monitoring must involve continuation of an analogous analysis (including any other hypothesizes), though allowing for a wider set of details and additional factors. A compulsory constituent of the monitoring must be the newest precise tools; in a great demand also are regular two-image photogrammetry shooting, geo-radar surveying, etc.

References

Skulsky, V.: Scientific and technical report. Observation of the foundation settlement, inclination of the inclination and deformation of the blockwork of the leaning Nevyansk Tower. Ural Promstroiniiproekt, Sverdlovsk (1965)

Slukin, V.M., Gorodetskaya, N.N.: Research of technical state of monument of architectural and historical heritage. Building science. Academic Journal of Ural Scientific Research Institute Project of RAACS, pp. 76–79 (2012)

Research report on the geotechnical investigation of soils nearby of the Nevyansk Tower. LLC "NikolaiGeo", Yekaterinburg (2000)

Lushnikov, V.V., et al.: Research report. Survey and observation of settlement of the Nevyansk Tower. Ural Polytechnic Institute Press, Sverdlovsk (1979)

Zaretskii, Y.K., Kapustin, V.K., Lushnikov, V.V., Sukhanov, B.I.: Predicting the stability of the Nevyansk Tower. Bases, Found. Soil Mech. J. **6**, 242–244 (1981)

Alyabeva, A.D., Ivanova, L.A.: Report of results of two-image photogrammetry of the Nevyansk Tower. LLC "Technology -2000", Yekaterinburg (2000)

Kopeikin, V.V., Morozov, P.A.: Research report on the results of geo-radar survey of geological structure of soils and disposition of underground spaces on the premises of the Nevyansk plant near the leaning tower. International Conversion Fund, Ekaterinburg (2001)

Numerical Analysis of Mechanically Stabilized Earth Walls in Hybrid Retaining Wall Systems

Hussein S. Abbas[1(✉)], Rami M. El-Sherbiny[2], and Abdelsalam M. Salem[2]

[1] Dar al-Handasah (Shair and Partners), Cairo, Egypt
hussein.salah@dar.com
[2] Faculty of Engineering, Cairo University, Giza, Egypt
rsherbiny@eng.cu.edu.eg

Abstract. Hybrid retaining walls comprising a lower soil-nail wall (SNW) and a mechanically stabilized earth wall (MSEW) on top are typically needed in combined cut and fill wall sections such as in road widening. However, available literature and design guidelines on MSE walls in such hybrid systems are limited. The purpose of this research is to numerically-investigate the behavior of MSE walls in hybrid retaining wall systems during and after construction. Two-dimensional finite element models of different wall configurations are analyzed using PLAXIS 2D software. The base-line model is verified against a published case history of a monitored MSE wall with good agreement. The effect of the nail length to total height (L_N/H_T), SNW facing slope (ω), reinforcements vertical spacing (V_L), reinforcements stiffness (J), and MSEW facing slope (ψ) on the tensile forces in the MSEW are studied. The results of tensile force distribution of MSE wall are presented at this paper. The calculated tensile forces from the numerical modeling are compared with calculated values from MSEW design methods using the Simplified method, Structure stiffness method, and K-stiffness method. Furthermore, the tensile forces of numerical modeling are higher than the tensile forces of the K-stiffness method (empirical method) and equal or less than the tensile forces of Simplified Method and Structure stiffness method (Limit equilibrium method).

1 Introduction

Hybrid retaining wall systems are typically used in cases where combined cut and fill are required, for example to allow for road widening using soil nail wall (SNW) as a lower wall and a mechanically stabilized earth wall (MSEW) above the soil nail wall as shown in Fig. 1. The limited use of hybrid retaining wall system in projects is largely due to lack of an established design procedure for these structures. Hybrid retaining walls are still considered "experimental" and many questions regarding its design and performance remain unanswered. Alhabshi (2006) mentioned that the basic design concept consists of transferring the tensile forces in the reinforcements into the soil through the mobilized friction at the interfaces and the factor of safety (FoS) for global stability of the reinforced soil structure highly depends on the pullout resistance or tensile strength of the reinforcements. He also studied the effect of soil nail wall

H. Shehata and B. Das (Eds.): GeoMEast 2018, SUCI, pp. 249–264, 2019.
https://doi.org/10.1007/978-3-030-01923-5_18

parameters on the behavior of hybrid retaining wall system. Jayawickrama (2009) stated that the combination of MSEW and SNW may provide a more economical design in cut/fill situations than the traditionally used full-height MSEWs or SNWs. He carried out an instrumentation and monitoring effort with the objective of improving understanding of hybrid wall design and performance with limited and questionable data points (Bathurst 2016). Wei (2013) stated that the behavior of these wall to retain a soil volume are so far similar because both techniques are based on the concept of using soil reinforcement as passive inclusions in the soil mass to create a gravity structure and hence improve soil stability.

Further development of design procedures for hybrid walls requires additional understanding of the wall behavior. This paper presents a 2D finite Element model to evaluate the performance of the MSEW in an hybrid retaining wall system. The focus of this paper is on the developed tensile forces in the MSEW and how it is affected by wall configuration. The finite element model is validated against a published case history of a monitored MSE wall. The results obtained from the validation model were compared with measurements obtained from the published field data, including wall deformations and strains in the reinforcement. The model is used to investigate the effect of nail length to total height (L_N/H_T), SNW facing slope (ω), reinforcements vertical spacing (V_L), reinforcements stiffness (J), and MSEW facing slope (ψ) on reinforcement's tensile force. The calculated tensile forces are compared with values from MSEW design methods using the Simplified method, Structure stiffness method, and K-stiffness method.

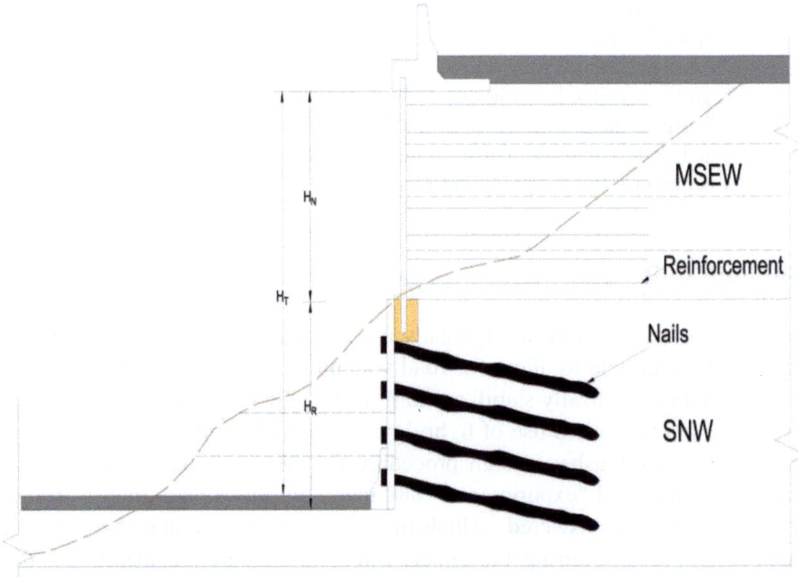

Fig. 1. Hybrid retaining wall system

2 Methodology

Two dimensional finite element software Plaxis 2D was used to model the hybrid wall system with focus on the tensile forces developed in the upper MSEW. The model was verified against a monitored case history of a full scale modular-block MSEW. The facing of the MSEW was modeled using linear-elastic non-porous solid elements, or using plate elements. Solid elements are defined using unit density, stiffness modulus, and Poisson's ratio. Plate elements are defined using axial stiffness and bending stiffness. The stiffness modulus of block facing is reduced to 1/10 of that calculated for the concrete modular block according to Yoo (2004) to take into account of the discrete nature of the modular block facing, thus is taken as 1×10^6 kPa. The soil structure interaction is modeled by interface elements with Mohr column model for the block-to-block interface and default interface with interaction coefficient (R) for the block-to-soil interface. The reinforcement layers are modeled by geogrid elements with elastic axial stiffness (EA). The soil parameters were concluded from the results of stress-strain response from plane-strain tests on the soil samples (Bathurst 2009). The hardening soil model was applied to fit measurements from laboratory tests. The wall was modeled in a staged-construction sequence in lifts considering the compaction stresses.

The modeling of SN wall was verified against a monitored case history of a full scale experimental soil nail wall was constructed as part of the French national research project on soil nail (CLOUTERRE) that was conducted between 1986 and 1991. Due to studying the behavior of MSEW, the verification for MSEW are presented here.

3 Model Verification

Bathurst (2000) reported measurements from experimental full-scale modular-block reinforced soil walls constructed as part of a research program carried out at the Royal Military College. The geogrid reinforced wall No. 1 was 3.60 m high, 3.30 m width, 6.0 m wide, with target facing batter of 8° from vertical and constrained between two lateral walls covered with plywood, Plexiglas, and lubricated polyethylene sheets over the side walls. This configuration was intended to ensure plane stain conditions and minimized the friction between the backfill soil and the sides. The base of the wall was seated on a rigid concrete floor to make construction of the wall as easy as possible and simplify the interpretation of wall performance. The wall facing consisted of a column of discrete dry-stacked, solid masonry concrete blocks with 300 mm wide, 150 mm high, 200 mm long, and weighed 20 kg. The geogrid reinforcements were made of polypropylene (PP) placed at a vertical spacing and length of 0.60 m and 2.52 m respectively. Tables 1, 2 and 3 show the properties of structural elements. Figure 2 shows a cross section of the geosynthetic reinforced wall 1. Laboratory tests were carried out on the soil backfill to measure the soil parameters. The compacted dry unit weight of the backfill was 16.8 kN/m^3, the angle of friction was determined from plan strain tests to be 44°, and the stress- strain behavior was measured by plane strain tests on sand specimens with secant stiffness E_{50} = 42.5 MPa at confining stress σ_3 = 80 kPa (Bathurst 2009). Table 4 shows the summary of the parameters that were adopted in the numerical analysis.

Table 1. The elastic parameters for facing and interface

Parameters	Unit	Modular block facing	Block to block interface
Model	–	Liner elastic	Mohr Coulomb
Drainage type	–	Non-porous	Non-porous
Unit weight, γ_{sat}	kN/m^3	21.80	21.80
Young's modulus, E	kPa	1×10^5	1×10^5
Cohesion, C′	kPa	–	46
Effective angle of shearing resistance, Ø′	Degrees	–	57
Dilatancy angle, ψ	Degrees	–	0
Poisson's ratio, ν_{ur}	–	0.15	0.15
Interface, R_{inter}	–	1	1

Table 2. The elastic parameters for plate facing

Parameters	Unit	Facing
Axial stiffness (EA)	kN/m	30000
Bending stiffness (EI)	kN/m/m	225
Poisson's ratio, ν	–	0.15

Table 3. The input parameters for geogrid element.

Parameters	Unit	Facing
Elastic axial stiffness	kN/m	97

A two dimensional finite element model was built to simulate this case study. The case study of the geosynthetic reinforced soil wall was established such that the boundary conditions do not influence the wall behavior. The length of the mesh extended from the left boundary at x = −1.0 m to the right boundary at x = 6.0 m, while the height of the mesh was falling between two boundaries at z = 0 and z = 3.60 m. Figure 3 illustrates the finite element mesh for blocks facing and plate facing. The results of the numerical mode of the wall were compared with the results of field monitoring in terms of horizontal deformation at facing (Fig. 4) and reinforcement strains at the end of construction (Fig. 5). The computed deformation shows a good match compared to the monitored values with estimated difference ranging from 7.0% to 21% for modeling using solid elements and ranging from 3.0% to 10% for modeling using plate elements as shown at Fig. 5. The computed reinforcement strains show a reasonable match compared with the monitored values with estimated difference ranged from 2% to 5% except layers 4 and 5 that show estimated difference ranged from 10% to 15%. The higher values of reinforcements strain at the facing were caused by the connection between the facing and the geogrid as shown at (Fig. 5). According to the above results, the Plaxis 2D model can be considered adequate to simulate the geosynthetic reinforced soil wall with hardening soil model and plate element for MSEW facing.

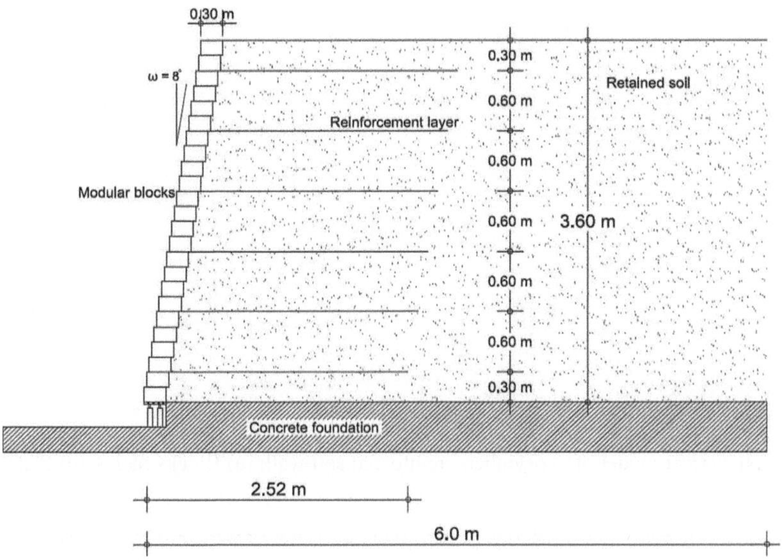

Fig. 2. Cross section of wall geometry of RMCC Wall 1.

Table 4. Summarized material properties adopted for the soil.

Parameters	Unit	Backfill
Model	–	HS
Drainage type	–	Drained
Unit weight, γ_{sat}	kN/m^3	16.80
Secant stiffness in standard drained triaxial test, E_{50ref}	MPa	42.5
Tangent stiffness for primary oedometer loading, E_{oedref}	MPa	42.5
Unloading/reloading stiffness, E_{urref}	MPa	127.5
Power for stress-level dependency of stiffness, m	–	0.5
Effective cohesion, C′	kPa	2
Effective angle of shearing resistance, Ø′	degrees	44
Dilatancy angle, ψ	degrees	11
Poisson's ratio, ν_{ur}	–	0.25
Reference pressure, P_{ref}	kPa	80
Earth pressure coefficient at rest, K_0	–	0.305
Interface, R_{inter}	–	0.67

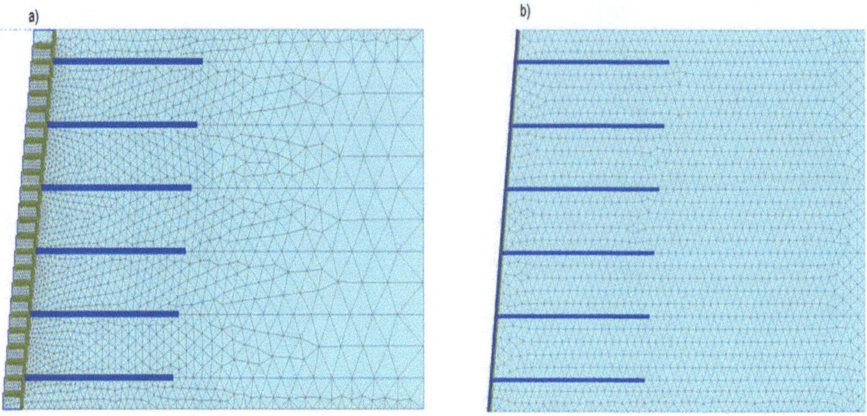

Fig. 3. Numerical model of geosynthetic reinforced soil wall; (a) Blocks facing (b) Plate facing

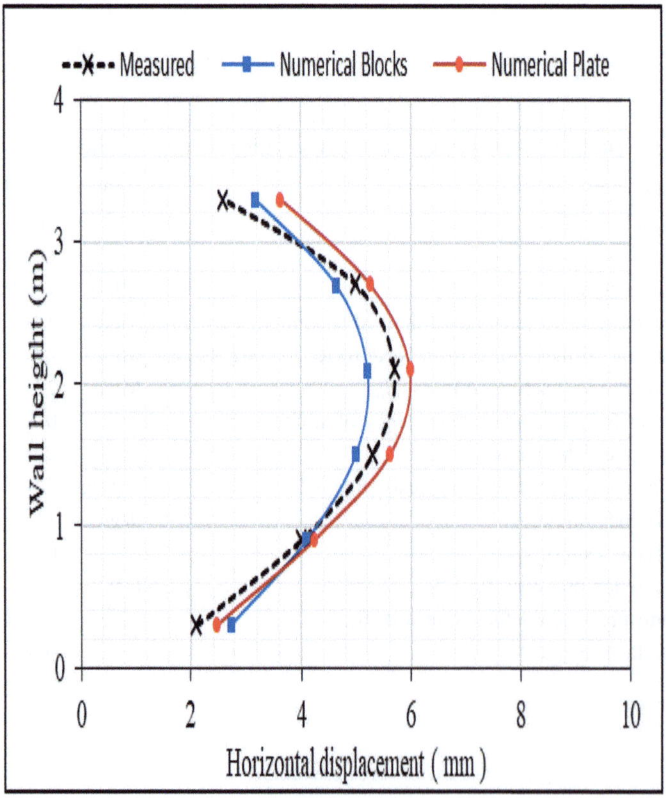

Fig. 4. Comparison between the computed lateral displacement and measured results at the wall facing

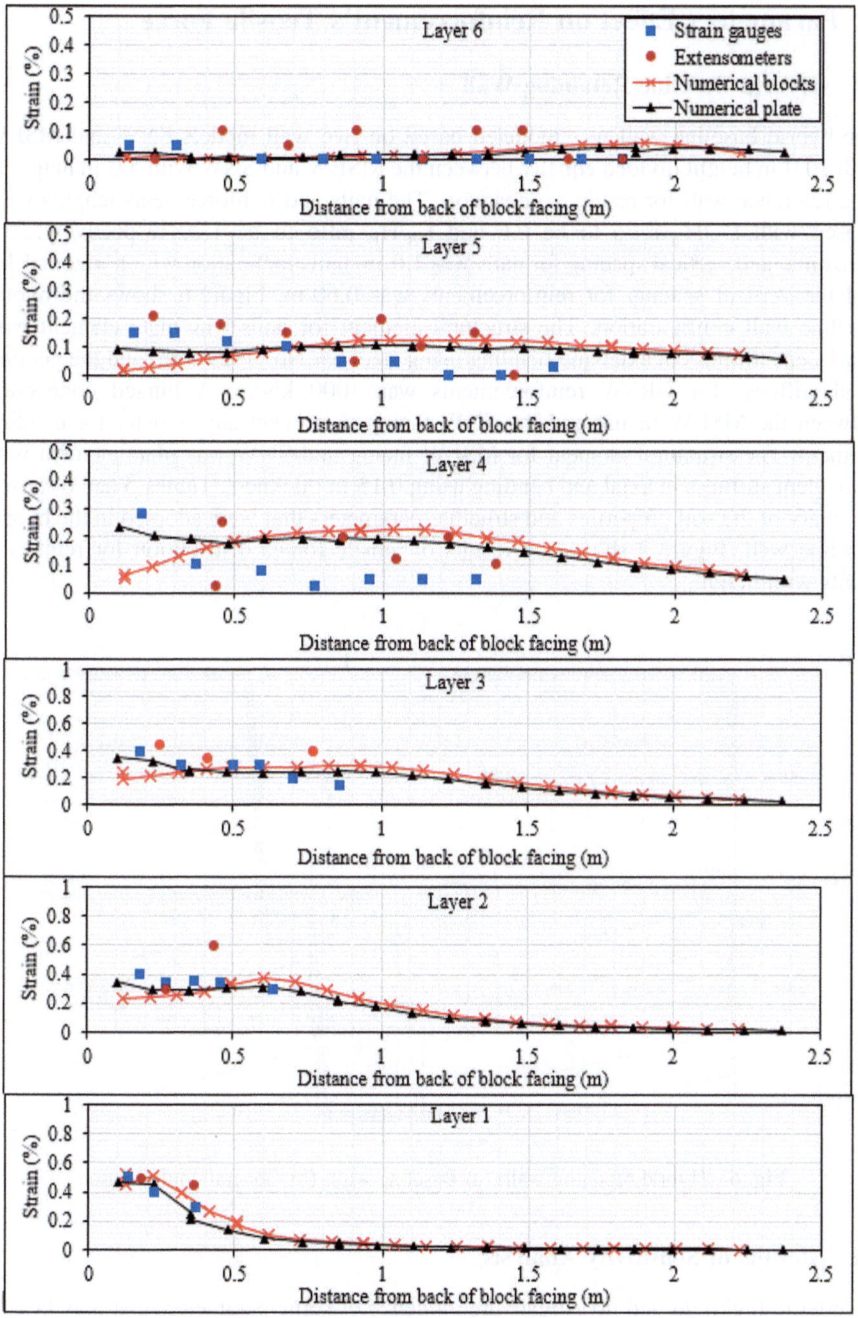

Fig. 5. Comparison between the computed reinforcements strains and measured results

4 Parameters Effect on Reinforcement's Tensile Force

4.1 Hybrid Baseline Retaining Wall

The hybrid baseline wall was modeled based on two wall models SNW and MSEW with 10.0 m height divided equally between the MSEW and SNW with 5.0 m height to be a reference wall for results comparison. The nails and reinforcements lengths were chosen with L_N/H_T ratio to be 0.6 and L_R/H_R ratio to be 1.20 respectively. The horizontal and vertical spacing for nails was 1.0 m, nails inclination with horizontal 10° and the vertical spacing for reinforcements was 0.60 m. Figure 6 shows the hybrid baseline wall configuration. The structural element for nails was plate element with equivalent stiffness in axial and bending using steel bar No. 9 (29 mm) and the geogrid axial stiffness for GRSW reinforcements was 1000 kN/m. A hinged connection between the MSEW facing and the SNW facing to prevent any transfer for bending moment. The structural element for MSEW facing and SNW was plate element with equivalent stiffness in axial and bending using 0.15 m thickness. Tables 5 and 6 show a summary of the soil properties and structure parameters that were adopted in the hybrid baseline wall. Figure 7 shows the results of tensile forces distribution for reinforcements with height.

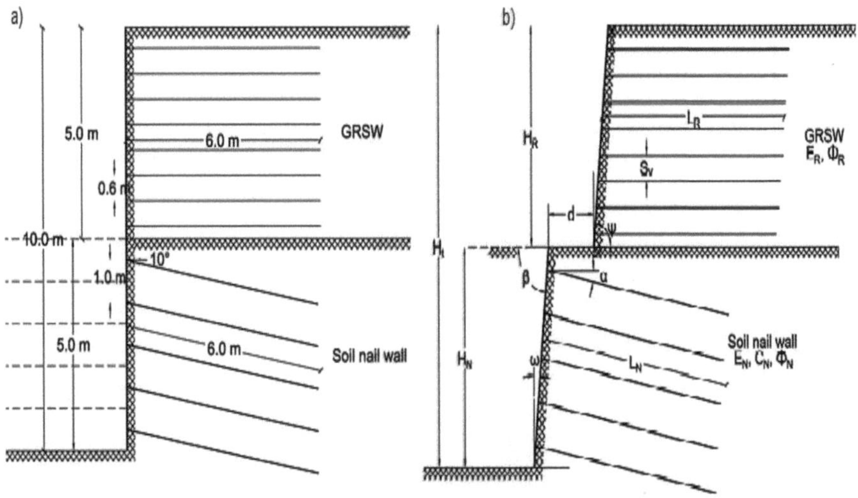

Fig. 6. Hybrid retaining wall: (a) Baseline wall; (b) General configuration

4.2 Results of Sensitivity Analysis

In order to highlight and investigate the influence of some parameters that may be of a noticeable impact on the response and results of MSEW, a numerical sensitivity study was carried out for some parameters such as; nail length to total height (L_N/H_T), SNW facing slope (ω), reinforcements vertical spacing (V_L), reinforcements stiffness (J), MSEW facing slope (ψ). The results of tensile forces are summarized and presented in

Table 5. Summarized soil properties adopted for the hybrid baseline wall

Parameters	Unit	MSEW	SNW
Model	–	HSM	HSM
Drainage type	–	Drained	Drained
Unit weight, γ_{sat}	kN/m^3	18.00	18.0
Secant stiffness in standard drained triaxial test, E_{50ref}	MPa	25	25
Tangent stiffness for primary oedometer loading, E_{oedref}	MPa	25	25
Unloading/reloading stiffness, E_{urref}	MPa	75	75
Power for stress-level dependency of stiffness, m	–	0.5	0.5
Effective cohesion, C′	kPa	2	5
Effective angle of shearing resistance, Ø′	degrees	35	35
Dilatancy angle, ψ	degrees	5	5
Poisson's ratio, ν_{ur}	–	0.30	0.30
Reference pressure, P_{ref}	kPa	100	100
Earth pressure coefficient at rest, K_0	–	0.426	0.426
Interface, R_{inter}	–	0.67	0.55

Table 6. Summarized structural properties adopted for the hybrid baseline wall

Parameters	Unit	MSEW facing	SNW facing	Nail element	Reinforcement element
Axial stiffness (EA)	kN/m	3×10^6	3×10^6	132×10^3	1000
Bending stiffness (EI)	kNm^2/m	5625	5625	6.94	–
Poisson, s, ν	–	0.15	0.2	0.2	–

the form of charts with different parameters including the baseline wall parameters with blackline as an easy reference to comparison. The variable values of effective parameters on tensile force of reinforcements are stated in Table 7, also the effect of each parameter on the tensile force distributions on MSEW are shown in Fig. 8.

The increasing of the nail length to height ratio (L_N/H_T) leads to decrease the tensile forces on MSEW till reaching nail length to height (L_N/H_T) equals 0.6 therefore any additional increase after that value will not have significant effect to tensile forces on GRSW as shown in Fig. 8a, and during the decreasing of the nail length to height ratio (L_N/H_T) from 1.0 to 0.6, the maximum tensile forces are located in the middle of wall with difference between 4.5% to 7.5% from the baseline wall forces, but at nail length to height ratio (L_N/H_T) less than 0.6, the maximum tensile forces are located in the lower part of wall with difference between 45% to 270% from the baseline wall forces. This behavior could be attributed to (L_N/H_T) ratio that acts as toe restraint for MSEW.

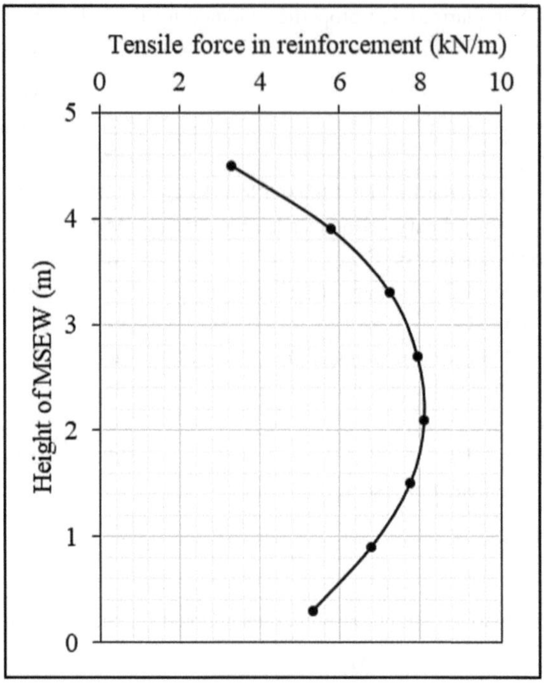

Fig. 7. The tensile force distribution of reinforcements with height

Table 7. Summarized structural properties adopted for the hybrid baseline wall

Parameter	Value					
Nail Length/Total height (L_N/H_T)	0.30	0.40	0.50	0.60	0.75	1.00
Setback distance (d, m)	0	2	4	6	10	0
SNW Facing slope (ω)	0	15	30	45		
Reinforcements vertical spacing (V_L)	0.4	0.6	0.8			
reinforcements stiffness (J)	250	500	1000	2000	3000	
MSEW facing slope (ψ)	45	60	70	80	90	

The increasing of the SNW facing slope (ω) leads to increase the tensile forces on MSEW as shown in Fig. 8b, and during increasing the facing slope, the maximum tensile forces are increasing and located in the middle of wall with difference 8% to 21% from the baseline wall forces, but the tensile forces at the lower layer are decreasing with difference 13.5% to 41% from the baseline wall forces. This behavior could be attributed to increasing the settlement below GRSW due to lateral squeeze type failure which leads to increase the tensile forces on reinforcement to resist the shear.

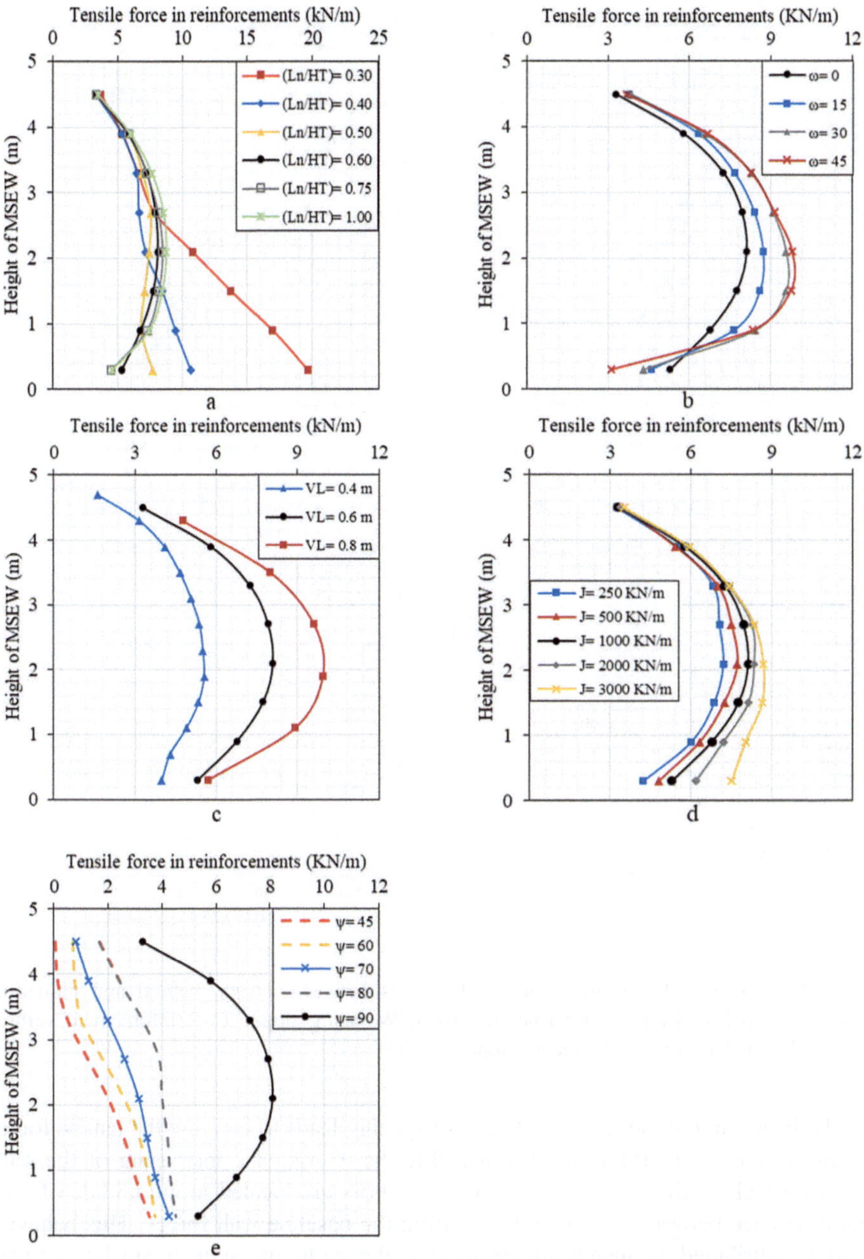

Fig. 8. The tensile force distribution of reinforcements with height at: (a) Nail length to total height; (b) SNW facing slope; (c) Reinforcement vertical spacing; (d) Reinforcement stiffness; (d) MSEW facing slope

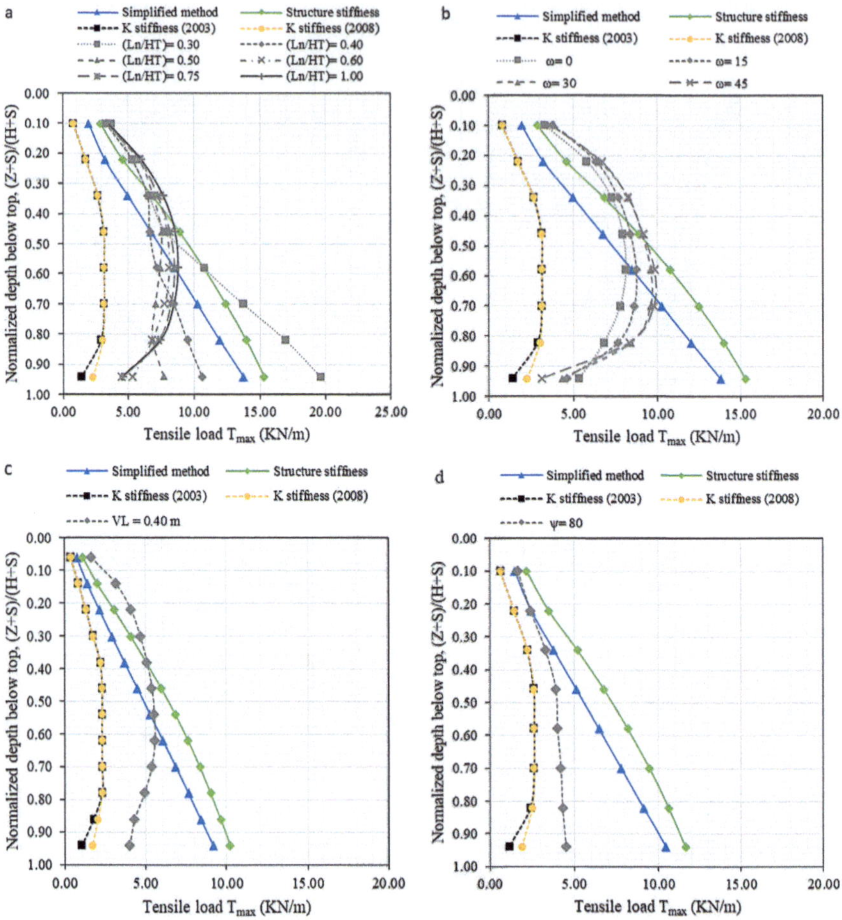

Fig. 9. Comparison of reinforcement tensile forces between design method and numerical analyses at: (a) Nail length to total height; (b) SNW facing slope; (c) Reinforcement vertical spacing $V_L = 0.4$ m; c) MSEW facing slope $\psi = 80^{\circ}$

The increasing of the reinforced vertical spacing leads to increase the tensile forces in reinforcements at MSEW as shown in Fig. 8c. During the increasing of the reinforced vertical spacing, the maximum tensile forces are located in the middle of wall with difference between 23.0% to 31.5% from the baseline wall forces. This behavior could be attributed to increasing decreasing the earth pressure at smaller vertical reinforcement spacing resulting in confining soil between reinforcements which lead to reduce the soil deformation and the mobilized tensile force in reinforcements.

The increasing of the reinforcement stiffness leads to increase the tensile forces in reinforcements at GRSW as shown in Fig. 8d. During the increasing of the reinforced stiffness, the maximum tensile forces are located in the middle of wall with difference between 7.0% to 11.0% from the baseline wall forces. This behavior could be attributed to increasing the reinforcement stiffness which leads to decrease the developed strains and increase the mobilized tensile forces on reinforcements.

Table 8. Comparison between design methods of MSEW

Method	Assumptions	Factors ignored
Simplified method	• Limit equilibrium method • Wall foundation is rigid or unyielding does not influence the magnitude and distribution of reinforcement loads • MSE structures are tolerant to deformations • Earth pressure coefficient varies with depth based on type of reinforcement stiffness till 6.0 m	• Effect of change parameters • Construction sequence/compaction effect • Different soil/facing type • Deformation • Foundation compressibility • Foundation movement (Toe restraint) • Reinforcement stiffness
Structure stiffness	• Limit equilibrium method • Wall foundation is rigid or unyielding does not influence the magnitude and distribution of reinforcement loads • MSE structures are tolerant to deformations • Earth pressure varies with depth based on reinforcement stiffness	• Effect of change parameters • Construction sequence/compaction effect • Different soil/facing type • Deformation • Foundation compressibility • Foundation movement (Toe restraint) • Reinforcement stiffness
K-Stiffness method	• Empirical method by using a database of 11 full-scale geosynthetic walls • Results of simple statistical analyses for various wall components and their influence on reinforcement loads • Wall foundation is rigid or unyielding does not influence the magnitude and distribution of reinforcement loads	• Construction sequence/compaction effect • Deformation • Foundation compressibility • Foundation movement (Toe restraint)

The increasing of the facing slope leads to increase the tensile forces on MSEW as shown in Fig. 8c. During increasing the facing slope, the maximum tensile forces are located in the lowest layer of wall till facing slope value 80° after that value the maximum tensile forces are located in the middle layer of wall. This behavior could be attributed to decreasing the value of active earth pressure coefficient which leads to decrease the acting earth pressure on reinforcements.

5 Comparison with Design Methods

A comparison was carried out between the numerical analyses results and some of design methods such as; Simplified Method, Structure stiffness method (FHWA 1990), and K-Stiffness method (Bathurst 2003; 2008) through the tensile forces on reinforcements with the major affected parameters stated earlier in the previous section as shown in Fig. 9. Table 8 shows the difference between methods assumptions and the factors ignored in the design methods which leads to tensile forces on the design methods differ from tensile forces of the numerical analyses. The numerical analyses results show that the tensile forces on reinforcements are higher than the tensile forces of the K-stiffness method (empirical method) and equal or less than the tensile forces of Simplified Method and Structure stiffness method (Limit equilibrium method).

6 Conclusions

This study is mainly concerned with numerically simulating the behavior mechanically stabilized earth wall in Hybrid retaining wall system during and after construction. A two-dimensional finite element analysis using PLAXIS 2D software was used to simulate a published case history of a monitored MSE wall No. 1 that constructed as part of geotechnical research program that carried out by the Royal Military college of Canada that was conducted between 1999 and 2000. The numerical test results were compared with the field monitoring results from horizontal deformation at facing, reinforcement strains at the end of construction which showed a good match that reflected a justified authentication for the modelling approach and software. Hybrid baseline wall was modeled based on two wall models SNW and MSEW. A numerical sensitivity study was carried out for some parameters such as; nail length to total height (L_N/H_T), SNW facing slope (ω), reinforcements vertical spacing (V_L), reinforcements stiffness (J), MSEW facing slope (ψ).

Increasing of the nail length to height ratio (L_N/H_T) leads to decrease the tensile forces on MSEW till reaching nail length to height (L_N/H_T) equals 0.6 therefore any additional increase after that value will not have significant effect to tensile forces on GRSW. Nail length to total height ratio (L_N/H_T) is acting as toe restraint/stiffness that effect on the magnitude and distribution of reinforcements. In general, at value of (L_N/H_T) ratio is greater than or equal 0.6, reinforcement load distributions at MSEW becomes trapezoidal in shape (i.e. fixed toe) and at value of (L_N/H_T) ratio is less than 0.6, reinforcement load distributions at MSEW becomes more triangular in shape. Increasing of the SNW facing slope (ω) leads to increase the tensile forces on MSEW and during increasing the facing slope, the maximum tensile forces are increasing and located in the middle of wall with difference 8% to 21% from the baseline wall forces, but the tensile forces at the lower layer are decreasing with difference 13.5% to 41% from the baseline wall forces.

Increasing of the reinforcement vertical spacing (V_L) leads to increase the tensile forces in reinforcements at MSEW. During the increasing of the reinforced vertical spacing, the maximum tensile forces are located in the middle of wall with difference between 23.0% to 31.5% from the baseline wall forces.

Increasing of the reinforcement stiffness (J) leads to increase the tensile forces in reinforcements at GRSW. During the increasing of the reinforcement stiffness, the maximum tensile forces are located in the middle of wall with difference between 7.0% to 11.0% from the baseline wall forces.

Increasing of the MSEW facing slope (ψ) leads to increase the tensile forces of reinforcements at MSEW. During increasing the facing slope, the maximum tensile forces are located in the lowest layer of wall till facing slope value 80° after that value the maximum tensile forces are located in the middle layer of wall.

Regarding to the numerical simulating of the behavior for mechanically stabilized earth wall in Hybrid retaining wall system, the numerical results are more accurate than other design methods as it considers the factors that may be neglected by these methods such as; foundation compressibility, foundation movement (Toe restraint), and MSEW deformation.

Acknowledgments. Foremost, I would like to thank my supervisor, Prof. Dr. Abdelsalam M. Salem, Professor of Geotechnical Engineering and foundations, Faculty of Engineering, Cairo University for the patient guidance, encouragement and advice he has provided throughout my time of research. Also, I would like to express my sincere gratitude to my advisor Prof. Dr. Rami M. El-Sherbiny Professor of Geotechnical Engineering and foundations, Faculty of Engineering, Cairo University, for the continuous support, his patience, motivation, enthusiasm, and immense knowledge.

References

Alhabshi, A.: Finite element based design procedures for MES/Soil-Nail hybrid retaining wall systems. Ph.D. dissertation, Department of Civil and Environmental Engineering, Texas Tech University (2006)

Allen, T.M., Bathurst, R.J., Holtz, R.D., Walters, D.L., Lee, W.F.: A new working stress method for prediction of reinforcement loads in geosynthetic walls. Can. Geotech. J. **40**(5), 976–994 (2003)

Bathurst, R. J., Peiyuan, L., Jinyuan, L.: Statistical evaluation of the FHWA simplified method and modifications for predicting soil nail loads. Int. J. Geomech. (2016). https://doi.org/10.1061/(asce)GT.1943-5606.0001614

Bathurst, R.J., Miyata, Y., Nernheim, A., Allen, T.M.: Refinement of K- stiffness method for geosynthetic reinforced soil walls. Geosynth. Int. **15**(4), 269–295 (2008)

Bathurst, R.J., Walters, D., Vlachopoulos, N., Burgess, P., Allen, T.M.: Full scale testing of geosynthetic reinforced walls. In: Proceedings Geo-Denver 2000, ASCE Special Publication No. 103, Advances in Transportation and Geoenvironmental Systems Using Geosynthetics. Invited keynote paper, pp. 201–217 (2000)

Huang, B., Bathurst, R.J., Hatami, K.: Numerical study of reinforced soil segmental walls using three different constitutive soil models. J. Geotech. Geoenviron. Eng., 1486–1498 (2009). https://doi.org/10.1061/(ASCE)GT.1943-5606.0000092

Jayawickrama, P.W., Wood, T., Wei, Y., Alhabshi, A.: Design procedures for MSE/Soil nail hybrid wall systems. In: Final Research Report (draft) Submitted to Texas Department of Transportation, Report No. FHWA/TX-08-0-5205-1 (2009)

Wei, Y.: Development of equivalent surcharge loads for the design of soil nailed segment of MSE/soil nail hybrid Retaining walls based on results from Full-scale wall instrumentation and finite element analysis. Doctor of philosophy, The Texas Tech University (2013)

Yoo. C.S.: Design of a geosynthetic reinforced segmental retaining wall in a tiered arrangement–use of numerical modeling as a design aid. In: 3rd Asian Regional Conference on Geosynthetics, GeoAsia, Seoul, Korea, pp. 173–182 (2004)

Author Index

H. Shehata and B. Das (Eds.): GeoMEast 2018, SUCI, pp. 265–266, 2019.
https://doi.org/10.1007/978-3-030-01923-5

MIX
Papier aus verantwortungsvollen Quellen
Paper from responsible sources
FSC® C105338

If you have any concerns about our products,
you can contact us on
ProductSafety@springernature.com

In case Publisher is established outside the EU,
the EU authorized representative is:
Springer Nature Customer Service Center GmbH
Europaplatz 3, 69115 Heidelberg, Germany

Printed by Libri Plureos GmbH
in Hamburg, Germany